Survival Analysis in Medicine and Genetics

Chapman & Hall/CRC Biostatistics Series

Chapman & Hall/CRC Biostatistics Series

Chapman & Hall/CRC Biostatistics Series

Survival Analysis in Medicine and Genetics

Jialiang Li
Shuangge Ma

CRC Press
Taylor & Francis Group
Boca Raton London New York

CRC Press is an imprint of the
Taylor & Francis Group, an **informa** business
A CHAPMAN & HALL BOOK

CRC Press
Taylor & Francis Group
6000 Broken Sound Parkway NW, Suite 300
Boca Raton, FL 33487-2742

First issued in paperback 2022

© 2013 by Taylor & Francis Group, LLC
CRC Press is an imprint of Taylor & Francis Group, an Informa business

No claim to original U.S. Government works

ISBN 13: 978-1-03-247748-0 (pbk)
ISBN 13: 978-1-4398-9311-1 (hbk)

DOI: 10.1201/b14978

Library of Congress Cataloging-in-Publication Data

Li, Jialiang, 1981-
  Survival analysis in medicine and genetics / Jialiang Li, Shuangge Ma.
  pages cm -- (Chapman & Hall/CRC biostatistics series)
  Includes bibliographical references and index.
  ISBN 978-1-4398-9311-1 (hardback)
  1. Survival analysis (Biometry) 2. Medical statistics. 3.
Medicine--Research--Statistical methods. I. Ma, Shuangge, 1978- II. Title.

  R853.S7.L53 2013
  610.72'7--dc23   2013010826

Visit the Taylor & Francis Web site at
http://www.taylorandfrancis.com

and the CRC Press Web site at
http://www.crcpress.com

To my parents, grandparents, uncles, aunties, and
cousins.
J. Li

To Emily and Iris.
S. Ma

Contents

Preface

In modern medicine, knowledge of human lifetime and what shortens it has been gradually accumulated over years of tremendous numerical studies. All quantitative products in those studies are well supported by a system of statistical tools, namely *survival analysis*. This book is specifically devoted to this important category of statistical methods.

A number of well-written, widely used introductory texts on this subject already exist. The first two chapters of this book may serve a similar function and provide some basic textbook-type materials for a third- or fourth-year undergraduate module or a graduate module in statistics or biostatistics. Compared to other texts, we do not repeat basic principles of regression analysis and focus mainly on the special concerns of the survival model. We cover almost all fundamental materials while confining ourselves to simple notations and plain language in these two chapters. The counting process theory is only appended in the second chapter for the sake of completeness and may entertain students who enjoy theories. According to the authors' experiences, the materials in these two chapters are usually sufficient for a one-semester module at the level indicated above. Another piece that can be easily incorporated in the syllabus of this module is the first section of Chapter 5 where elementary diagnostic medicine approaches are discussed. Currently there is, as far as we know, no formal statistical modules on diagnostic medicine in most statistics departments. Usually, such issues are addressed in epidemiology modules. We also feel this is relevant to statistical research, especially to survival analysis and thus recommend the addition. Some exercises are included at the end of the chapters. Sample lecture slides are also available on the official Web site for this book: *www.crcpress.com/product/isbn/9781439893111*.

The remainder of the book provides a nice complement to most existing texts on various important topics in survival analysis. Methodological and applied researchers may see the use of these materials in scientific practice. Specifically, the authors are both interested in interval censored data and provide a chapter-length discussion on this topic in Chapter 3. The theories and methods we include for this type of complicated incomplete data reflect the state of the art of the development of this field and may lead to new applications. Chapter 4 is about special modeling methodology including nonparametric regression, multivariate survival data, cure rate model, and Bayesian statistical methods. We wish to highlight a few features of this chapter: (i) The discussion of smoothing methods for survival data has never formally appeared in any texts or research books on survival analysis. We provide some theoretical background to link the current development to the well-established research field of nonparametric regression. (ii) For analyzing competing risks we introduce not only the traditional cause-specific hazard regression approaches but also Fine and Gray's cumulative incidence regression approach, which is absent in most textbooks on competing risks. (iii) We recognize the importance of Bayes methods for survival analysis and lay out the necessary steps to arrive at the computational results. For all the analytic problems in the section we only sketch the full conditionals, which are usually sufficient to implement a Gibbs sampler. Because we are *frequentists* by training and only occasional users of Bayes methods in our research projects, we must confess that a good deal of computational strategies are omitted and their implementation is often crucial to the success of the method.

Chapter 5 is dedicated to time-dependent diagnostic accuracy study. The materials in the first section of this chapter are quite elementary for medical students and can be used for teaching. This chapter also presents a new way of looking at how predictors are associated with censored survival time. The accuracy-based analysis differs substantially from the traditional hazard-based regression analysis and offers an alternative interpretation. Several clincal examples are included to illustrate the methods.

One primary goal of this book is to emphasize the application of survival analysis methods in genetics, which is discussed in Chapter 6. Genetics helps to answer the fundamental biological question— *What is life?*—and is believed to play an important role in all kinds of human disease research. We extract a set of statistical genetic methods particularly for censored survival time outcome from the massive research results published in recent years. When facing the sequence of 3 billion nucleotides which make up the human genome and attempting to identify individual genes within the sequence, the current state-of-the-art approach in the statistical and bioinformatic community is to translate the genomic problems to regression modeling and related variable selection problems. In the hyperactive field that genomics has become, the focus continues to evolve and the methodology updates rapidly. As we write this book, new methods are still being invented to improve or replace the *new* methods in this book which refer to activities that are two or three years old. The idea of Chapter 6 is thus to set forth the statistical framework in which the burgeoning research activity takes place and to impart a fixed portion of stable materials.

We used R as the major software to illustrate all kinds of statistical methods. Other software such as SAS and STATA are also mentioned at proper places. The computer commands and outputs referenced throughout the book are printed in a monospaced typewriter font like this. The authors may be contacted via e-mail at stalj@nus.edu.sg and shuangge.ma@yale.edu and would appreciate being informed of typos, errors, and improvements to the content of this book.

We are fortunate to have worked with many idea-motivating data sets provided by our collaborators from Australia, Bangladesh, China, Germany, India, Singapore, the United Kingdom, and the United States. Our collaborators constantly encouraged our work and their data stimulated new ideas for statistical research. We especially want to thank our colleagues and students for providing nutritious feedback to improve the presentation. Many thanks for the support of the National University of Singapore (J. Li) and Yale University (S. Ma), our professional homes, where we began this project and where we are completing it. Part of the book was written when Li took his sabbatical leave at the Department of Biostatistics in University of North Carolina, Chapel Hill. He would like to thank the department for their hospitality. We also acknowledge the grant support from the Academic Research Funds R-155-000-130-112 and the National Medical Research Council NMRC/CBRG/0014/2012 in Singapore.

MATLAB® is a registered trademark of The Math Works, Inc. For product information, please contact:

The Math Works, Inc.
3 Apple Hill Drive
Natick, MA 01760-2098
Tel: 508-647-7000
Fax: 508-647-7001
E-mail: info@mathworks.com
Web: http://www.mathworks.com

Chapter 1

Introduction: Examples and Basic Principles

Survival time is a random variable that measures how long it takes for a subject to arrive at an event of interest. The occurrence of the event is sometimes referred to as a *failure*. In the literature, survival data have also been referred to as lifetime data, failure time data, or time to event data. There is no significant distinction among terminology, and they are used interchangeably throughout this book.

In this chapter, we start with some examples to illustrate the main features of survival data in Section 1.1. These examples present data obtained from various research designs. A large amount of examples can also be found in published literature. Some practical concerns in designing a survival study are addressed in Section 1.2. Section 1.3 lays out the fundamental statistical concepts needed for modeling survival data, including distribution characterization and commonly used parameters. After that, we discuss the important issue of data incompleteness in Section 1.4.

1.1   Examples

We use a few biomedical examples to illustrate survival data in this section. These examples motivate methodological development in the downstream chapters. Survival data in these examples are obtained from various types of study designs.

The first example is taken from a typical epidemiological study. Epidemiologists are usually interested in associating a well-defined outcome variable with a set of predictor variables. The strength of association between outcome and predictors can be described by statistical measures such as regression coefficients, odds ratios or correlation coefficients. The outcome variable is usually related to a specific disease. It can be the physical measurement of the disease magnitude such as blood pressure for hypertension and MMSE (mini-mental state examination) score for dementia, the event of the occurrence of disease such as a binary indicator of diabetes status, or the time to the development of a terminal outcome such as the time to disease-specific mortality. Predictors may be broadly classified into two categories: risk factors and confounders. Risk factors have clinically meaningful indications for the disease outcome and are often on the pathway of the disease causation, such as occupational and environmental exposures (e.g., smoking, radiation) and measures of physical or biological functions (e.g., cholesterol level). Confounders are included in the analysis to control for existing heterogeneity of the population such as demographic characters (e.g., gender, race, age).

Example (Bloodstream Infection). Overuse of antibiotics is a major driver of antibiotic resistance, a growing problem in intensive care units (ICUs) worldwide. To examine the relationship between duration of carbapenem administration and subsequent nosocomial multidrug resistant (MDR) bloodstream infections (BSIs), a prospective observational study was undertaken at the National University Hospital in Singapore for all adult patients admitted to an ICU or high dependency unit (HDU) receiving more than 48 hours of a carbapenem antibiotic (Donaldson et al., 2009).

During the two-year study period, 415 patients were followed. The outcome of interest was the development of BSI. This disease may not directly lead to death. We still call the time to the occurrence of this outcome a survival time. In many medical studies, the survival time may not be a literal time to death but a time to the occurrence of a disease. The methods used to analyze these data are no different from those for analyzing a real survival time.

In this study, the research question of interest was whether the duration of carbapenem use would affect the probability of developing BSI among patients in ICU and HDU. More detailed discussions and analysis results are provided in Chapter 2. We refer to Donaldson et al. (2009) for more details on the study design.

The second example is from a genetic study. Human genetic research studies have advanced during the second half of the last century, especially after 1990 when sequencing technology greatly enhanced our understanding of gene structure. There are a large number of outstanding references on the biological aspect of such studies, see for example Snustad and Simmons (2006). In this book we focus on analyzing the numerical data collected from genetic studies and using statistical methods to generate useful information for failure time outcomes from such studies.

Example (Lung Cancer). Lung cancer is the leading cause of cancer mortality as it is associated with a low survival probability. Nonsmall cell lung cancer (NSCLC) is more common than small cell lung cancer. Gene mutations may provide prognostic value to guide proper treatment decisions. In a research study led by Tan Tock Seng Hospital in Singapore, Lim et al. (2009) used whole genome amplification (WGA) technology to investigate how specific gene mutations affected the survival probability of patients with NSCLC.

In this study, 88 advanced-stage NSCLC patients were enrolled, and their low-volume lung biopsies underwent WGA before direct sequencing for EGFR, KRAS, P53, and CMET mutations. These genes have been suggested to be associated with lung cancer in published studies. Each patient was followed from the time of diagnosis of cancer to the time of death or the end of study. The event of interest is the death outcome. The time for a patient to arrive at this outcome is usually called the *survival time*. The statistical analysis procedure to study the distribution of survival time is termed *survival analysis*. Specifically, practitioners are interested in determining the probability that a patient with NSCLS survives for a certain length of time, say one year, or five years, after the diagnosis of disease.

In addition, medical investigators aimed at comparing the survival probabilities between patients with certain genetic mutations and those without such mutations. Tumors bearing certain genetic mutations might be resistant to certain therapeutic strategies and thus should be treated with alternative procedures. Therefore, identifying such tumors that can affect the survival probability would be meaningful for cancer patients. As described earlier, this study examined a total of four types of mutations. In Chapter 2, we provide detailed statistical analysis results to address the aforementioned research questions.

Next, we consider an example from a randomized clinical trial. Clinical trials are important experimental means to establish the safety and efficacy of new therapeutic agents and treatment regimens. Acting differently from observational studies, usually investigators of clinical trials provide direct interferences to the subjects by randomly allocating them to treatment plans and compare their experimental outcomes under different treatment settings (usually a new treatment vs. standard care). Randomization plays an important role in these trials to ensure a fair comparison. We refer to publications such as Lachin (1988) for more detailed discussions on clinical trials.

Example (STAR*D Trial). Sequenced Treatment Alternatives to Relieve Depression (STAR*D) study is a multisite, prospective, randomized, multistep clinical trial of outpatients with nonpsychotic major depressive disorder (Rush et al., 2004). The study compared various treatment options for those who did not attain a satisfactory response with citalopram, a selective serotonin reuptake inhibitor (SSRI) antidepressant. The study enrolled 4000 adults (ages 18 to 75) from both primary and specialty care practices who had not had either a prior inadequate response or clear-cut intolerance to a robust trial of protocol treatments during the current major depressive episode. After receiving citalopram (Level 1), participants without sufficient symptomatic benefit were eligible for randomization to Level 2 treatments, which entailed four switch options (sertraline, bupropion, venlafaxine, cognitive therapy) and three citalopram augment options (bupropion, buspirone, cognitive therapy). Patients without sufficient improvement were further assigned into Levels 3 and 4 of the trial. In this book, we will restrict our attention to the first two levels of observations.

The primary outcome measurement is the clinician-rated, 17-item Hamilton Rating Scale for Depression (HRSD), administered at entry and exit from each treatment level through telephone interviews by assessors masked to treatment assignments. Treatment remission, for research purposes, is defined as an HRSD total score < 8 upon exit from a treatment level. The 30-item Inventory of Depressive Symptomatology (IDS), obtained by the same telephone-based interviewers, is a secondary symptom outcome. Other secondary outcomes, all obtained by an interactive voice response telephone system, include brief self-reports of depressive symptoms, function, side-effect burden, quality of life, and participant satisfaction.

Although the HRSD was the primary outcome measure used to determine response for research purposes, the decision to proceed to the next level or to follow-up depended on clinical judgment regarding the benefit of the current treatment(s) as informed by the 16-item Quick Inventory of Depressive Symptomatology–Clinical Rating (QIDS), which was obtained by the clinical staff at each treatment visit without knowledge of the primary research outcome (that is, HRSD score). The analysis we conduct in this book for this example is based on the QIDS score.

The investigators were interested in two major clinical events of the patients. The first was the response to treatment: nonresponse was defined as a < 50% reduction in symptom severity measured by the QIDS at exit. Another event of interest was the remission, that is, the absence of depressive symptoms: remission was defined as ≤ 5 on the QIDS score. Protocol medication clinic visits were scheduled at weeks 0 (baseline), 2, 4, 6, 9, and 12 at all treatment levels. The actual visits, however, could be flexible (for example, the week 2 visit might be held within ±6 days of week 2). Extra visits would be held if clinically needed. If a participant exhibited a response or remission only at week 12, two additional visits might be added to determine if that status was sustained.

In this example, we want to study the time to the first response and time to remission. The difference between this example and the previous ones is that now the events are what we "hope" to occur. The statistical methods for analyzing time to "success" is essentially the same as those for analyzing time to failure. More details on analysis results of this example are provided in the next chapter.

1.2 Design a Survival Study

All examples given in Section 1.1 are related to human subjects. In fact, the development of most survival analysis methods has been driven by the biomedical need to improve the health of patients. The application of statistical methods in human medicine is an emphasis of this book. We briefly comment on a few practical issues for the design of a scientific study, which may generate data for survival analysis.

A study that generates survival time data is necessarily longitudinal in nature, in contrast to a cross-sectional study where data is collected at a fixed point of the time line. The time measurements and other data for sampled subjects in a survival study can be collected retrospectively, as in a case-control study, or prospectively (Lilienfeld and Lilienfeld, 1980). A prospective survival study can be an observational cohort study, as in the bloodstream infection and lung cancer examples, or a randomized clinical trial, as in the STAR*D example. An observational follow-up study allows us to characterize the natural history of the disease in quantitative terms. After we describe the severity of a disease, priorities may then be established for clinical services and public health programs. On the other hand, we usually conduct a trial to modify the natural history of a disease so as to prevent or delay an adverse outcome such as death or disability and to improve the health of an individual patient or the general population. A carefully designed randomized trial can evaluate the effectiveness and side effects of new forms of interventions. Currently, both observational studies and randomized clinical trials are being widely practiced in medicine, and the choice of design depends on the goal of the study, specific disease, and practical limitations.

Ethical issues frequently arise in survival studies involving human subjects, especially in clinical trials. Ethics may be thought of as a discipline for dealing with what is good or bad, a set or system of moral values, or a guiding philosophy. The earliest ethical regulation in modern scientific history was the Nuremberg Code after atrocities committed by Nazi physicians in World War II alerted the world to the dreadful potential for abuse in medical research. The Nuremberg Code requires informed and voluntary consent and stresses the need to minimize risk to the subjects and to utilize appropriate methodologies. Since then many other ethical principles have been developed such as the Declaration of Helsinki, issued by the World Medical Assembly in 1964 and revised in 1975 and 1981. In April 1990 at a meeting in Brussels, regulatory authorities from Europe, the United States, and Japan formed the International Conference on Harmonization (ICH, *http://www.ich.org*) and subsequently established the Good Clinical Practices (GCP) guidelines. The GCP includes guidelines on safety, quality, efficacies, and multidisciplines and represents an international uniform standard for regulatory agencies to accept the results from clinical trials.

Statisticians can play an important role to protect patient safety and rights in the course of a research study which usually consists of multiple phases. Even though clinical trial may sound like a less frightening term than medical experiment, it essentially places human subjects under a testing environment. The hazards may or may not be known, for example, in a Phase I trial that examines the toxicity of an agent. Randomization, blinding, and treatment termination procedures should all be carefully devised to respect the well-being of subjects. It has been a difficult dilemma as to whether a patient, especially one with a high-morbidity disease, should receive a random treatment allocation or follow the physician's judgment about what is best for the patient (Taylor et al., 1984; Zelen, 1979; Weinstein, 1974). Providing a short-term solution to fully respect individual ethics may undermine the generalizability of the scientific study in a long run and eventually violate collective ethics (Bartroff and Lai, 2011). Further discussion on this topic is beyond the scope of this book, and interested readers may refer to Levine (1996), Schafer (1982), and Schaffner (1993).

Another design aspect relevant to statisticians is the calculation of a proper sample size. The calculation approaches may be based on a required precision of an interval estimate of the population parameter, or a required significance and power for establishing the alternative hypothesis. One additional numeric specification for sample size calculation is how large an effect size we expect to detect. Further computational details for sample size analysis are given in Section 2.2.2. Sometimes the limited resources allow only a moderate sample size for the design, and accordingly we may evaluate the power of the study under such a design. Either a sample size calculation or power analysis is necessary at the planning stage.

It is now increasingly recognized that a clinical trial with a fixed sample size is not the most prevalent case because of the high demand for time and expense and also because of ethical considerations. Sometimes a trial may continue too long, resulting either in excessive harm if the new therapy is of no benefit, or in a delay of new therapy to later patients should it prove effective. Such concerns have driven the development of sequential designs and adaptive designs in the literature. Simply speaking, sequential designs examine and compare treatment groups at predefined steps and may terminate the study earlier than originally planned if interim analysis declares sufficient significance. Adaptive designs are a similar flexible procedure that starts out with a small up-front commitment of sample size and then extends by assigning new subjects to treatment groups based on the information generated by the study to that point. These designs are well supported by theories (Zelen, 1969; Bauer and Kohne, 1994; Muller and Schafer, 2001; Jennison and Turnbull, 2003; Tsiatis and Mehta, 2003). In particular, sequential designs have been incorporated in ICH-E9 with detailed procedural guidelines.

1.3 Description of Survival Distribution

Suppose that we are interested in investigating a population of subjects who are at risk of developing a failure outcome. For a subject randomly selected from the population, we may denote his or her time to the failure event as T. We need to determine the distribution of the random variable T in order to answer all kinds of practical questions, for example, "How long can a patient with lung cancer survive with a 90% probability?" "Will treatment A be significantly more beneficial to the patients with a chronic mental illness than treatment B?" "What are the key factors that can lead to a longer survival after a heart surgery and can we provide necessary health care to patients to reduce the risk factors?" A statistical distribution of a random variable includes the specification of the range of possible values the variable may take (which is also referred to as *support*) and the corresponding frequencies at all such possible values.

From the examples given in the previous section, we may notice that T is usually a positive continuous variable. The support of this distribution is naturally taken to be $\mathbb{R}^+ = [0, \infty)$. Following a common practice for continuous variables, we assume that the distribution function for T is $F(t) = P(T \leq t)$ with a density function $f(t) = d\{F(t)\}(dt)^{-1}$. Using calculus we can easily compute $F(t)$ from a given $f(t)$ by noting that $F(t) = \int_0^t f(u)du$.

According to the definition, we may interpret $F(t)$ as the probability of a randomly selected subject dying before time t. We are usually more interested in the complement

$$S(t) = 1 - F(t) = P(T > t),$$

which states the probability of observing a survival time longer than a fixed value t. This is primarily because in most biomedical studies, knowing how quickly subjects die may not be as informative as knowing how long they can still be alive. Efficacy and efficiency of health care procedures are frequently assessed by how much they improve the survival probability, which is directly reflected in the function $S(t)$. $S(t)$ is called the *survival function*.

Another important related function for describing the survival distribution is the hazard function. We define the hazard function as

$$h(t) = \lim_{\Delta t \to 0} \frac{P(T \in [t, t + \Delta t)|T \geq t)}{\Delta t}, \qquad (1.1)$$

where Δt is an infinitesimal increment of time. The hazard may be regarded as the changing rate of the *conditional* probability of dying at time t given the survival time is no less than t. Because of this interpretation, $h(t)$ sometimes is also called *instantaneous failure rate*.

Using the definition of $f(t)$ and $S(t)$, we can easily show that $h(t) = f(t)/S(t)$. Furthermore, we may denote $H(t) = \int_0^t h(u)du$ to be the cumulative hazard function and obtain the following equation

$$S(t) = \exp\{-H(t)\}$$

by noticing that $f(t)/S(t) = -d\{\log S(t)\}(dt)^{-1}$.

In summary, knowing one of the four functions, $f(t)$, $S(t)$, $H(t)$, and $h(t)$, allows us to derive the other three. These functions can serve an identical purpose of describing the distribution of survival time. Conventionally, $f(t)$ and $S(t)$ are used to form the likelihood functions for estimation and test, while $h(t)$ is usually reserved to present a regression model.

In classic statistical analysis, a distribution is usually specified with a known functional form and some unknown parameters. The parameters may determine the distribution completely. Some specific parameterized distribution families are introduced in Sections 2.1.2 and 2.3.4, and Table 2.1 contains a list of commonly assumed density functions for survival data.

The moments of the probability distributions are usually important summary measures of distribution characteristics. Mean and variance are familiar examples, and are the first-order and second-order moments, respectively. These two moments can completely specify the normal and many other two-parameter distributions. However, for most survival distributions, it is inappropriate to assume a symmetric distribution such as normal. In practice, it is more relevant to summarize the survival distribution by the moments for the truncated distribution. One such useful moment parameter is the mean residual life, which is defined as

$$r(t) = E(T - t | T \geq t). \tag{1.2}$$

By definition, $r(t)$ can be interpreted as the expected lifespan after a subject has survived up to time t. The mean of the overall survival time is simply $r(0)$. Since we care about survival expectancy progressively, only reporting $r(0)$ is not so conclusive as $r(t)$ for a series of time points of interest. Therefore mean survival time $E(T)$ is not a popular summary measure in survival data analysis. Simple calculation leads to the equation

$$r(t) = \int_t^\infty S(u)du/S(t).$$

Higher-order moments $E(T^k | T \geq t)$ are employed sometimes to describe the distribution features (for example, variability and skewness) of the remaining life.

Another type of distribution parameters is the quantile. The $100p$th percentile of the survival distribution $F(t)$ is given by

$$F^{-1}(p) = \inf\{t : F(t) > p\}. \tag{1.3}$$

The 50th percentile is the median. The median survival time is the time at which half of the subjects in the study population can survive. It offers two advantages over the mean survival time. First, its sample estimate is less affected by extremely large or small values, while the mean estimate may be sensitive to a small number of outliers. Even one or two subjects with very long survival times can drastically change the sample mean but not the sample median. Second, we must observe all the deaths in the study population in order to evaluate the sample mean. On the other hand, to calculate the median survival of a sample, we only need to observe the deaths of half of the group.

In many clinical studies, particularly cancer studies, investigators frequently express the prognosis of the disease with the term *five-year survival*. This parameter is simply $S(t)$ evaluated at $t = 5$ years. Mathematically and biologically there is no reason to only focus on a particular time point to examine the natural history of the disease. Another problem with five-year survival is that its definition excludes those who survive less than five years from consideration. Therefore, if we want to assess the effectiveness of a new therapy introduced less than five years ago, we cannot use this criterion as an appropriate measure. Other similar measures include *two-year survival*, *ten-year survival*, and others.

1.4 Censoring Mechanisms

One remarkable feature of survival data is that they may be subject to *censoring* and hence provide incomplete information. In the lung cancer example introduced in Section 1.1, some patients were followed until the end of the study without developing the mortality outcome. This was fortunate for the patients but placed a practical obstacle for data analysts and statisticians. Basically we ended up collecting incomplete information about the survival experiences for these patients, which in turn reduced the effective sample size of the study. Completely ignoring these patients may lead to a biased description of the survival experience. It would also be inefficient for us to throw away data which may contain valuable information for survival, in which case we may lose a substantial amount of efficiency. Statistically, we may still be able to make use of data on those censored subjects as we do observe lower bounds of their survival times.

We say that a datum point for survival time T is *right censored* if the exact T is known only to exceed an observed value. There are many practical scenarios that may generate right censored data. In most situations, right censoring arises simply because the individual subject is still alive when the study is terminated. The BSI example is a clear demonstration. In other instances, some subjects may move away from the study areas for reasons unrelated to the failure time, and so contact is lost. In yet other instances, individuals may be withdrawn or decide to withdraw from the study because of a worsening or improving prognosis.

Traditionally, we classify right censoring patterns into the following three categories:

1. Type I censoring: Many studies only have limited funds, and investigators cannot wait until all the subjects develop the event of interest. It is thus agreeable to observe for a fixed length of time, for example, nine months or five years. Survival times for the subjects that have developed the outcome of interest during the fixed study period are the exact uncensored observations. The survival times for the subjects who are still alive at the end of the study are not known exactly but are recorded in the data set as the length of the study period. Under type I censoring, the censoring time is always equal to the total length of the study time and thus makes the follow-up computation such as parameter estimation fairly simple.

2. Type II censoring: Type II censoring usually appears in laboratory studies with nonhuman subjects such as mice. The event of interest is often the fatal outcome. The study proceeds until a fixed proportion or number of subjects have died, for example, 45% of the sample or 120 subjects. In this case, the censoring time is always equal to the largest uncensored survival time. Because the exact number of events can be achieved, the power of the follow-up hypothesis test can be readily satisfied at a pre-specified level.

3. Type III censoring: In most clinical and epidemiologic studies, the period of follow-up study is fixed in calendar, say, year 2006 to 2012, and patients may be recruited into the study at different times during the study period. Some may develop the outcome of interest before the study end point and thus provide exact survival times. Some may withdraw during the study period and are lost to follow-up thereafter. Their survival times are at least from their entry to the last contact. And others may never develop the outcome of interest by the end of the study, and so their survival times are at least from entry to the end of the study. Such incomplete follow-up observations are censored. Under type III censoring, the censoring times are not identical for all censored subjects and behave like a random variable.

We will assume type III censoring for methodological development in survival analysis throughout the book since the other two types of censorings are much easier to deal with. For example, type I censoring can actually be incorporated in type III censoring as a special case. The event of censoring can thus be regarded as random.

In a study with n subjects, the ith subject ($i = 1, \cdots, n$) has a time T_i to failure and a time C_i to censoring. Usually, it is assumed that the failure times are independent and identically distributed (i.i.d.) with distribution F and density f, and the censoring times are i.i.d. with distribution G and density g. Under right censoring, for the ith subject, we only observe $Y_i = \min(T_i, C_i)$

in practice. Usually, an event indicator $\Delta_i = I(T_i \leq C_i)$ is also included in the data file. Assuming that T_i and C_i are independent, we may then obtain the likelihood function based on the sample $\{(Y_i = y_i, \Delta_i = \delta_i) : i = 1, \ldots, n\}$ as

$$\mathcal{L} = \prod_{i=1}^{n} f(y_i)^{\delta_i} (1 - F(y_i))^{1-\delta_i} g(y_i)^{1-\delta_i} (1 - G(y_i))^{\delta_i}. \qquad (1.4)$$

The *independent censoring* assumption is crucial in the above construction, since it allows us to use the marginal distributions F and G instead of considering more complicated joint distributions. Many procedures for survival analysis introduced in the following chapters are based on this assumption (or a conditional independence assumption when covariates are present).

Usually, we are not interested in G if it does not contribute any information to the knowledge of survival distribution. This could be true if the parameters that specify the distribution function F are completely unrelated to the parameters that specify the distribution function G. Such a scenario is referred to as *noninformative censoring*, under which we can simplify the above likelihood function further as

$$\mathcal{L} \propto \prod_{i=1}^{n} f(y_i)^{\delta_i} (1 - F(y_i))^{1-\delta_i}. \qquad (1.5)$$

In addition to right censoring, there are also situations where we have left or interval censored observations. Left censoring occurs when it is known that the event of interest happens prior to a certain time t, but the exact time of event is unknown. In this case, we only know that the true time to event is shorter than t. Loosely speaking, the same techniques used to analyze right censored data can be applied to left censored data in a parallel manner. It is not common to observe left censored data in most medical studies, and thus we omit further discussion on this type of incomplete data. Interval censoring occurs when the event of interest is known to happen between two time points. This kind of censoring mechanism has become prevalent in, for example, cancer studies, and is now receiving extensive attention in the literature. We deliver more attention to interval censoring in Chapter 3.

Censoring may be considered as a special case of missing data (Little and Rubin, 2002). A hierarchy of three different types of missingness is usually distinguished: (1) Missing completely at random (MCAR); (2) Missing at random (MAR); (3) Not missing at random (NMAR). MCAR is a missing data mechanism completely independent of the failure time process. For example, type I right censoring, type III random right censoring and interval censoring induced by preplanned intermittent observations all satisfy MCAR.

It is sometimes unrealistic to assume MCAR for many practical survival data. MAR is introduced as a relatively weaker condition which assumes that the probability of missing may depend on observed data but not on unobserved data. For example, all Type II and Type III right censoring and case I interval censoring belong to this type. Under either MCAR or MAR, we do not need to model the data missing process and thus simplify the likelihood-based analysis.

The third type of missing data is NMAR where the probability that failure times are missing is related to the specific values that should have been observed. An NMAR mechanism is often referred to as *nonignorable* missingness, and we have to incorporate the missing distribution in the likelihood to correct estimation bias. When the independent censoring assumption is violated, we face NMAR and can use almost none of the traditional statistical tools to be introduced in the next chapter.

Another important concept of data incompleteness in survival data analysis is *truncation*. *Left truncation* occurs when the event time of interest in the study sample is greater than a left truncation value. Consider, for example, a study of cancer survival in a retirement community, where only senior people over age 65 are allowed in the community. Here the survival time is left truncated by the entering age of subjects (which is 65 with this specific example). Ignoring the truncation mechanism may lead to biased estimation and inference.

In a parallel manner, right truncation occurs when the event time of interest is less than a right truncation value. A special example is where all subjects in the study sample have already experienced the event of interest. Consider an AIDS study where the event time of interest is defined as the period between infection with AIDS virus and the clinical onset of AIDS. In a prospective study, subjects infected with AIDS virus enter the study and are followed until some or all of them develop AIDS. However, such a study can be prohibitively long and expensive. As an alternative, we may study patients who have already developed AIDS and try to retrospectively collect information on AIDS infection (for example, from blood transfusion). Note that such a study strategy has a certain similarity with case-control design.

Left truncation is also called *delayed entry* and may appear in prospective studies. There is also a right truncation issue in some retrospective studies because of a similar selection limitation. Truncation affects the sample acquisition plan, and ignoring truncation makes the subsequent inference biased. Usually when truncation is presented, only *conditional* analysis can be conducted. Extrapolating to the whole cohort may demand additional information which may or may not be available. Since we are not providing any further details on truncation in this book, see other studies for more discussions (Cnaan and Ryan, 1989; Kalbfleisch and Lawless, 1989; Keiding and Gill, 1990; Gijbels and Wang, 1993; Gurler, 1993, 1996; van der Laan, 1996b; Gijbels and Gurler, 1998; He and Yang, 1998; Huang et al., 2001; Tse, 2003; Shen, 2005; Tse, 2006; Shen and Yan, 2008; Dai and Fu, 2012).

1.5 Exercises

1. TRUE or FALSE?

   (a) Survival function is bounded between zero and one.

   (b) Hazard function is bounded between zero and one.

   (c) The hazard of an exponential distribution is constant.

2. Give short answers to the following questions.

   (a) What is the probabilistic interpretation of the hazard function?

   (b) Why is normal distribution not a good candidate to model survival data?

   (c) What are the practical advantages of using median survival time over mean survival time?

   (d) What is ten-year survival?

3. For a positive random variable T, show that $E(T) = \int_0^\infty S(t)\, dt$.

4. For a positive random variable T and $p \geq 1$, show that $E(T^p) = \int_0^\infty pt^{p-1}S(t)\, dt$.

5. Suppose that we have a survival data set taken from a Weibull distribution. The Weibull density function is given in Table 2.1 in Chapter 2. Compute its survival function.

6. Suppose that we have a sample $\{t_1, t_2, \cdots, t_n\}$ from the Weibull distribution. Assume $a = 0$ and $\gamma = 2$. Find the maximum likelihood estimator for σ.

7. Suppose that we have a sample $\{t_1, t_2, \cdots, t_n\}$ from the Weibull distribution. Assume $a = 0$ and $\sigma = 1$. Find the maximum likelihood estimator for γ.

8. Suppose that for a sample $\{t_1, t_2, \cdots, t_n\}$ from the Weibull distribution, we only observe a censored sample $\{(y_i, \delta_i) : i = 1, \cdots, n\}$, where $y_i = \min(t_i, c_i)$, $\delta_i = I(t_i \leq c_i)$, and c_i is the random censoring time. Assume $a = 0$ and $\gamma = 2$. Find the maximum likelihood estimator for σ.

9. Suppose that for a sample $\{t_1, t_2, \cdots, t_n\}$ from the Weibull distribution, we only observe a censored sample $\{(y_i, \delta_i) : i = 1, \cdots, n\}$, where $y_i = \min(t_i, c_i)$, $\delta_i = I(t_i \leq c_i)$, and c_i is the random censoring time. Assume $a = 0$ and $\sigma = 1$. Find the maximum likelihood estimator for γ.

10. For a standard normal random variable Z, denote its distribution function and density function by $\Phi(z)$ and $\phi(z)$, respectively. Show that $E(Z|Z \geq z) = \phi(z)/(1 - \Phi(z))$ for $z > 0$. The density function of a standard normal distribution is given in Table 2.1 in Chapter 2 with $\mu = 0$ and $\sigma = 1$.

11. For a standard normal random variable Z, show that $E(Z^2|Z \geq z) = 1 + z\phi(z)/(1 - \Phi(z))$ for $z > 0$.

12. For a standard logistic random variable Z, show that $E(Z|Z \geq z) = \frac{1}{1+e^z}\{\frac{z}{1+e^z} - \log(\frac{e^z}{1+e^z})\}$. The density function of a standard logistic distribution is given in Table 2.1 in Chapter 2 with $\mu = 0$ and $\sigma = 1$.

13. Show that, if T is continuous with distribution F and cumulative hazard H and $Y = \min\{T, c\}$ for a fixed c, then $E\{H(Y)\} = F(c)$.

14. Suppose that T is an exponential variable whose density function is given in Table 2.1 with $a = 0$ and $\sigma = 1/\lambda$. Consider a fixed censoring time c. Show that the probability of censoring is $\pi = e^{-\lambda c}$. Furthermore, if $Y = \min\{T, c\}$, prove $E(Y) = (1 - \pi)/\lambda$.

Chapter 2

Analysis Trilogy: Estimation, Test, and Regression

Survival analysis involves a large collection of statistical methods to deal with survival data introduced in the previous chapter. Statisticians are usually interested in studying the unknown numeric characteristics of a population, or the so-called parameters. To achieve a close and meaningful "guess" of the unknown parameters, we may have to consider some computational methods whose theoretical properties are justifiable. The process of acquiring such a meaningful "guess" is called an estimation. In survival analysis, it is often impossible to simplify the estimation task to a couple of distribution parameters, and we are interested in determining the complete distribution of survival time. The objective is thus not a few parameters, and we have to provide the *functional* estimate of the whole survival distribution curve at all time points. The estimation method for this kind of problem is called nonparametric estimation, since the estimand is not a traditional finite-dimensional "parameter." The nonparametric estimator will be introduced in Section 2.1.

Once an estimation is achieved, further efforts are required for statistical inference. In Section 2.2, the comparison procedure of two sample survival distributions is discussed where we present the log-rank test. The hypothesis is posed regarding two distribution functions without any specific parameterization. The log-rank test is constructed in a nonparametric manner and provides more flexibility than parametric testing procedures such as the two-sample t-test. Censored observations can also be incorporated easily in the log-rank test. Such a test has become a standard numerical approach in medical research for right censored survival data.

A great emphasis in the current survival analysis research community is to study the effects of multiple predictive variables on the survival time with a mathematical model. We introduce standard regression techniques in Section 2.3, including semiparametric models and parametric models. The semiparametric model proposed by David Cox (1972) is especially important and often chosen as the default model in routine data analysis. Other than the proportional hazards assumption, the model does not place any other restrictive distribution assumptions for data. Fitting the semiparametric model yields useful inferences and leads to sensible interpretation for survival data. Model selection and model diagnosis are necessary steps involved in order to build an appropriate regression model. When the proportional hazard assumption fails, one usually has to choose alternative modeling tools such as the parametric models. We select a few popular parametric models and use them to illustrate the applications of specific cases in survival analysis.

2.1 Estimation of Survival Distribution

2.1.1 Nonparametric Estimation

In Chapter 1, we introduced the survival function $S(t) = P(T > t)$, which measures the probability of observing a survival time longer than a fixed value t. In many medical studies, it is not appropriate to assume any parametric form such as normal for the distribution of T. Therefore, the estimation of $S(t)$ cannot be simplified to an estimation problem with a finite number of unknown parameters. A nonparametric or distribution-free method has to be used to determine $S(t)$ for all possible t. We consider estimation for this unknown function in this section.

Suppose that we obtain a sample $\{T_i;\ i = 1, \cdots, n\}$ with exact survival times for a total of n independent subjects. Then an empirical estimator for $S(t)$ can be constructed as

$$\hat{S}^*(t) = n^{-1} \sum_{i=1}^{n} I(T_i > t), \tag{2.1}$$

where $I(A)$ is an indicator function for event A. The summation given in (2.1) leads to the total number of subjects whose survival times are longer than t. Such an empirical estimator possesses many nice statistical properties such as unbiasedness, $\sqrt{n}$ consistency, and asymptotic normality. Checking the unbiasedness is elementary. Since $EI(T_i > t) = S(t)$, $E\hat{S}^*(t) = S(t)$. The proof of consistency and asymptotic normality of (2.1) can be found in many introductory probability textbooks where the authors illustrate asymptotic convergence theories for random functions (see, for example, Durrett, 2005). Statisticians have already used (2.1) as basic building elements to construct the modern empirical process theories.

The difficulty encountered in a real medical study is that we usually cannot observe the exact survival times for all the n subjects. Instead, a realistic sample includes a portion of censored survival times. As introduced in Section 1.3 of Chapter 1, under right censoring, the data set we have contains $\{(Y_i = \min(T_i, C_i), \Delta_i = I(T_i \le C_i)); \ i = 1, \cdots, n\}$ as opposed to $\{T_i; \ i = 1, \cdots, n\}$. Formula (2.1) can only be used for a group of subjects with complete survival times and is no longer applicable for the whole sample. In order to use information from censored subjects as well, we have to choose an alternative procedure.

The Kaplan-Meier (KM) nonparametric estimator (Kaplan and Meier, 1958) for $S(t)$ is especially developed for right censored data and constructed as follows. We denote observed survival times in the sample as $t_1 < t_2 < \cdots < t_k$ ordered from the smallest to the largest. The set of these k time values is usually a proper subset of $\{Y_i = y_i; \ i = 1, \cdots, n\}$. Let n_i be the number of subjects at risk of dying at t_i (being alive any time before t_i) and d_i be the number of observed deaths at t_i ($i = 1, \cdots, k$). Suppose that t_m ($m \le k$) is the largest time point that is no greater than a fixed value t. The estimated survival function at time t is then

$$
\begin{aligned}
\hat{S}(t) &= \left(\frac{n_m - d_m}{n_m}\right) \times \left(\frac{n_{m-1} - d_{m-1}}{n_{m-1}}\right) \times \cdots \times \left(\frac{n_1 - d_1}{n_1}\right) \times 1 \\
&= \prod_{i=1}^{m} \frac{n_i - d_i}{n_i}.
\end{aligned}
\tag{2.2}
$$

We may use an inductive idea to obtain the estimate in (2.2). First, we notice that at the baseline $t = 0$, the KM estimator gives $\hat{S}(0) = 1$ and remains the same until the first failure moment t_1. Then at t_i ($i \ge 1$), the conditional probability for the n_i subjects who have survived until this moment (and also $t_{i-1}, t_{i-2}, \cdots, t_0$) to be able to survive t_i is naturally estimated by $\frac{n_i - d_i}{n_i}$. The overall survival probability at t_i is then the survival probability at t_{i-1} multiplied by this conditional probability. The recursive formula may also be written as

$$
\hat{S}(t_i) = \frac{n_i - d_i}{n_i} \hat{S}(t_{i-1}).
$$

We thus have an estimate for each t_i. The survival curve between t_{i-1} and t_i

may not be well defined. Usually we take it as a constant. If at the largest observed survival time t_k all n_k subjects die, then the KM estimator reaches zero; on the other hand, if at t_k some subjects are still alive, then the KM estimator $\hat{S}(t_k) = \prod_{i=1}^{k} \frac{n_i - d_i}{n_i}$ and is undefined for all $t > t_k$. In the literature, setting $\hat{S}(t)$ (for $t > t_k$) equal to $\hat{S}(t_k)$ or 0 has been suggested. We note that $\hat{S}(t = \infty) > 0$ is closely related to the so-called cure rate model, which will be discussed in Chapter 4, Section 4.3. Because of its specific format, KM estimator is also referred to as a product-limit estimator.

The aforementioned derivation is intuitive. A more rigorous derivation of the KM estimator can be based on maximizing the nonparametric likelihood function (see Section 2.5). Alternatively, a redistribution method suggested by Efron (1967) can be used to derive the same estimator.

We can easily see that the KM estimator reduces to the empirical function (2.1) when there is no censored observation between t_{i-1} and t_i and hence $n_{i-1} - d_{i-1} = n_i$. The formula of KM estimator (2.2) incorporates right censored observation for estimating the survival function. The calculation of the KM estimator has a long history, but its statistical properties were only justified after Kaplan and Meier's paper and other contributors' works. In general, we cannot provide any finite sample properties such as unbiasedness for the KM estimator. The immaculate performance of the KM estimator in real applications is guaranteed by its large sample properties.

Assume that the event and censoring processes are independent. Under some other mild regularity conditions which are usually satisfied in practice, we can show that $\sqrt{n}(\hat{S}(t) - S(t))$ for any fixed t converges in distribution to a normal distribution with mean zero. Such an asymptotic distribution result can be retained by following counting process theories (Fleming and Harrington, 1991). In fact a stronger result, which states that the stochastic process $\{\sqrt{n}(\hat{S}(t) - S(t))\}$ indexed by t converges to a mean zero Gaussian process, is achievable. The asymptotic variance of the KM estimator may be estimated by the following Greenwood formula

$$\widehat{var}\{\hat{S}(t)\} = \hat{S}^2(t) \sum_{i=1}^{m} \frac{d_i}{n_i(n_i - d_i)}. \tag{2.3}$$

The derivation of this variance estimator can also rely on the counting process theories. With these results, one may then form a $(1 - \alpha)100\%$ confidence interval for $S(t)$ at any t. An interval estimate may provide more information about the survival distribution since sampling variability is reflected by the width of the interval.

Sometimes the asymptotic confidence intervals may include impossible values outside the range $[0, 1]$. It is recommended to first apply the asymptotic normal distribution to a transformation of $S(t)$, such as $\log\{-\log S(t)\}$ or $\log \frac{S(t)}{1 - S(t)}$, with unrestricted range and then back-transform the interval for

$S(t)$. For example, the variance for $\log\{-\log S(t)\}$ may be estimated as

$$\widehat{var}[\log\{-\log\hat{S}(t)\}] \quad = \quad [\log\{\hat{S}(t)\}]^{-2}\sum_{i=1}^{m}\frac{d_i}{n_i(n_i-d_i)}, \qquad (2.4)$$

and asymptotic $100(1-\alpha)\%$ confidence interval is given by

$$[s_L, s_U] = \log\{-\log\hat{S}(t)\} \pm z_{\alpha/2}\sqrt{\widehat{var}[\log\{-\log\hat{S}(t)\}]}. \qquad (2.5)$$

This results in an interval estimation for $S(t)$ as

$$\left[\exp\{-\exp(s_U)\}, \exp\{-\exp(s_L)\}\right] \qquad (2.6)$$

which is always retained in $[0, 1]$.

The above point-wise confidence interval is valid for $S(t)$ at a particular point t. For a finite number of time points, the joint confidence interval can be constructed based on the joint normality distribution result. Simultaneous confidence band for the whole survival curve $S(t)$ may be needed, and its construction can be based on the bootstrap or the Hall and Wellner's method (Hall and Wellner, 1980; Borgan and Leistol, 1990).

Example (Lung Cancer). We now return to the lung cancer example described in Section 1.1. Among the 88 patients with NSCLC, 32 eventually developed the death outcome, while 56 stayed alive until the end of the study. The KM estimator for this data set is displayed in Figure 2.1.

The graph was generated in R using the `survfit` function in the `survival` package. We refer to Robert and Casella (2010) for an introduction of the R software and *http://cran.r-project.org/web/packages/survival/index.html* for the `survival` package. The key component of this function is the specification of a correct `formula`. For the current example, if the survival time Y and censoring status Δ are recorded as `time` and `status` in the data set, then the correct R program to produce Figure 2.1 is to first use

```
survfit(Surv(time, status) ~ 1, conf.type='plain' )
```

to generate an R object and then use the function `plot` in R. The results from this function involve all the n_i, d_i, and $\hat{S}(t_i)$, and the function `plot` can make use of the results to generate the graphical display as in Figure 2.1.

The $\sim$ sign in R is usually regarded as an equality sign that bridges the response to its left and the covariates to its right. In the above computer code, one can replace the "1" on the right-hand side of $\sim$ with other categorical variables to produce separate KM curves for different categories. The 95% confidence intervals based on the Greenwood formula (2.3) for standard errors are also produced in the graph by specifying the `conf.type` option in `survfit`. To provide more stable confidence intervals for probabilities, it is sometimes recommended to transform the probability to a log scale or complementary log-log scale. One can achieve this by setting `conf.type` equal to `'log'` (the default setting) or 'log-log'. One can also choose not to present the interval by choosing conf.type equal to 'none'. There are other options in function survfit which allow users to produce desirable results for their analyses. Some of them will be mentioned in the following discussion.

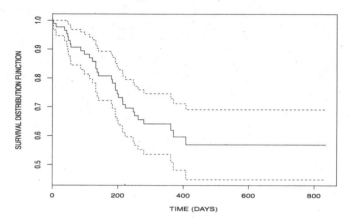

FIGURE 2.1: The Kaplan-Meier estimator of the survival function and its point-wise 95% confidence interval for the lung cancer data.

As we observe from Figure 2.1, the KM estimator decreases gradually as time increases. Since the estimated survival curve is based on a random sample of observations and reflects the overall survival probability in the general population, the point estimation may not yield an exact prediction for the survival probability of an individual patient. An interval estimation may be more meaningful for personalized health care in practice. For example, based on Figure 2.1 we can confidently claim that the probability for a randomly selected individual to survive 200 days is between 60% and 80%. We also notice that the confidence intervals are not of equal width over time. As time increases, the number of subjects at risk n_i decreases, leading to less information and a greater variability for the estimation of survival function.

Example (STAR*D Trial). At Level 2 of STAR*D introduced in Section 1.1, 565 adult outpatients who had nonpsychotic major depressive disorder without remission despite a mean of 11.9 weeks of citalopram therapy (Level 1), received augmentation of citalopram with sustained-release bupropion (279 patients) or buspirone (286 patients). The aim of the augmentation treatment was set *a priori* as a remission of symptoms—defined as a score of 5 or less on the 16-item Quick Inventory of Depressive Symptomatology (QIDS), which assesses the core symptoms of major depression and on which scores range from 0 to 27, with higher scores indicating greater symptom severity. Clinical visits were recommended at baseline and weeks 2, 4, 6, 9, and 12.

Survival analysis for this data set was first conducted by Trivedi et al. (2006). We plot the estimated probability functions of remission for the two treatment arms separately. We note that, in this case, we are more interested in the occurrence of the remission outcome, and therefore show the plot of $F(t) = 1 - S(t)$ (the distribution function) which is now a gradually increasing curve (Figure 2.2). The two treatment arms have similar cumulative chances of remission as the study time goes on. By the end of the study, the remission probability is almost 50% for either treatment group. The exact proportions of remission for the two arms are 39.0% for sustained-release bupropion group (solid line) and 32.9% for buspirone group (dotted line), respectively. These two proportions are not significantly different with a p-value 0.13 (computed using a chi-squared test). Having said that, we notice that a simple comparison between these two remission rates is not conclusive and may lose important temporal information that is available in this data set. To fully compare the two treatment arms for their remission probabilities across the whole time range, we have to employ a rigorous test procedure introduced in the next section.

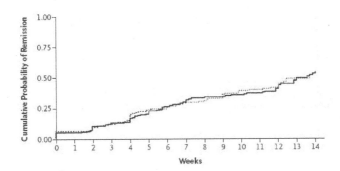

FIGURE 2.2: The Kaplan-Meier estimator of the remission probabilities for the two treatment groups in STAR*D. The solid line is for the sustained-release bupropion group, while the dotted line is for the buspirone group.

The estimated $S(t)$ may lead us to find the estimates of other functions (for example, $h(t)$, $H(t)$) by using the mathematical relationships introduced in Section 1.3. Alternatively, we may first estimate $H(t)$ and then use its relation to $S(t)$ to deduce a sensible estimator for $S(t)$. Adopting the same notations as before, a natural estimator for $H(t)$ is given by

$$\hat{H}(t) = \frac{d_m}{n_m} + \frac{d_{m-1}}{n_{m-1}} + \cdots + \frac{d_1}{n_1}, \tag{2.7}$$

since the cumulative hazard only increases at each observed failure moment, and d_i/n_i provides an empirical estimate for the instantaneous failure probability at t_i. The estimator (2.7) has been referred to as the Nelson-Aalen

(NA) estimator for the cumulative hazard (Nelson, 1969, 1972; Aalen, 1975, 1976) and leads to an estimator of the survival function $\tilde{S}(t) = \exp\{-\hat{H}(t)\}$. The resulting survival function estimator has been referred to as the Fleming-Harrington (FH) estimator by many authors (Fleming and Harrington, 1991), and we adopt this name in this book. Inference can be based on the following asymptotic normality result. Under independent censoring and some other mild regularity conditions, it can be shown that the NA estimate is $\sqrt{n}$ consistent and point-wise asymptotically normally distributed. At a fixed time point t, the variance of the NA estimate can be consistently estimated by

$$\widehat{var}\{\hat{H}(t)\} \;=\; \sum_{i=1}^{m} \frac{d_i}{n_i^2}, \tag{2.8}$$

or another equally valid formula

$$\widehat{var}\{\hat{H}(t)\} \;=\; \sum_{i=1}^{m} \frac{d_i(n_i - d_i)}{n_i^3}. \tag{2.9}$$

These theoretical results have been established in early publications such as Breslow and Crowley (1974) and Aalen (1976), and have also been implemented in most statistical packages. For R, we may change the default setting (KM estimator) in `survfit` to `type='fleming-harrington'`. Figure (2.3) gives a comparison between the KM and FH estimators for the lung cancer example. The difference between the two estimates is extremely small. Such similarity can be expected for most practical data.

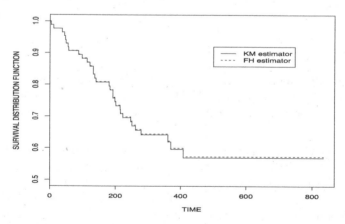

FIGURE 2.3: KM and FH estimates of the survival probability for the lung cancer data.

One may notice that FH estimates always lie above KM estimates. This can be verified straightforwardly by applying the inequality $\exp(-x) \geq 1 - x$. In practice, we usually prefer using the KM estimator for survival function and the NA estimator for cumulative hazard function since they are both non-parametric maximum likelihood estimators for the corresponding functions.

Remark. A critical condition for the validity of the KM estimate is that the censoring must be independent of survival time. In other words, the reason that an observation is censored must be unrelated to the cause of event. This assumption is true if the patient is still alive at the end of the study period, or if the patient has died of a cause other than the one under study (for example, death due to ischaemic heart disease in a cancer survival study). However, the assumption may be violated if the patient develops severe adverse symptoms from the treatment and is forced to leave the study before death. The assumption is also violated when the patient is removed from the study after achieving sufficient improvement. In the presence of such *dependent censoring*, the KM method is no longer appropriate. Under such a scenario, joint modeling of the event and censoring processes is needed to obtain an appropriate estimate of the survival function.

In addition to the above statistical methods, we note that there is another common estimation method for survival distribution used in demography and life insurance. Such a so-called "actuarial" method requires the construction of a table of survival time distribution for (usually) equally spaced time intervals. Investigators may identify the failure risks for individuals falling in different time intervals from the life table. Most statistical software also provide functions for this method. The actuarial method was also used in medical studies in old days when computing technology was less advanced and KM estimators were expensive to calculate (especially for a sample with a large number of failure times). Because the grouping is somewhat arbitrary and the estimator assumes homogeneity in survival probability within each interval, this method is usually deemed crude and has been replaced by the KM estimator in the recent biostatistical literature. We thus omit further discussion on this estimator in this book. Interested readers may refer to Chapter 3 of Lawless (2003) for more details.

The median of the survival distribution can be more informative a measure of location than the mean. Specifically, the median survival time can be estimated based on a KM estimator as

$$\hat{t}_{0.5} = \inf\{t : \hat{S}(t) \leq 0.5\}. \tag{2.10}$$

The interval estimation for median survival is discussed in Brookmeyer and Crowley (1982) by inverting the confidence interval of $S(t)$.

Furthermore, a $100p$th percentile can be estimated by

$$\hat{t}_p = \inf\{t : \hat{S}(t) \leq 1 - p\}, \tag{2.11}$$

and its asymptotic variance is given by

$$\text{var}(\hat{t}_p) = \frac{\text{var}(\hat{S}(t_p))}{f^2(t_p)}. \tag{2.12}$$

These quantities can all be found in the output of standard software such as STATA (StataCorp, 2011).

2.1.2   Parametric Approach

In more recent studies of survival distribution, semiparametric and non-parametric methods have been popular because of their flexibility in modeling data. Historically when KM estimators were still difficult to evaluate, parametric methods had been popular, and many distributions found them good applications in practice. With a given parameterized density, inference is obtained after estimation of a relatively small number of unknown parameters. Theoretical results are usually transparently easy to understand, and implementation is much simpler than calculating KM curves in earlier days.

Parametric estimators arise naturally when certain assumptions for the distributions are plausible. For example, the survival time for a population with constant hazard rate over time can be suitably modeled by the exponential distribution. Such an ageless property may not hold for the general population but may be appropriate for a short follow up and when the risk is homogeneous for the whole period. Populations with strictly increasing or decreasing hazard values over time can be modeled by the Weibull or gamma distributions. Increasing hazards may be common among elder subjects while decreasing hazards may be common among younger subjects. When a study is clearly focused on an old or young population, it might be convenient to adopt Weibull or gamma distributions. Populations with a unimodal hazard over time can be modeled by the inverse Gaussian, log-normal or log-logistic distributions. This type of hazard shape can be observed in clinical trials, where subjects admitted in the study with initially worsening disease conditions receive proper health care and then improve their conditions with gradually decreasing failure rates. There are many other methods to describe specific patterns of the hazard function and arrive at new parametric distribution families. Examples may include piecewise constant hazards, hazards as a linear function of time, bathtub-shaped hazards, and others. Alternatively, sometimes one can take a mixture of two or more parametric distributions to accurately describe a survival data set and/or reflect the heterogeneity nature of some populations. All these parametric methods may be associated with certain settings in real life. We provide some remarks for a few popular distributions next.

Exponential Distribution. This may be the simplest parametric distribution, and quite often we assume $a = 0$ in its density given in Table 2.1 to allow all positive values for T. That leaves a single parameter σ which is the mean as well as the standard deviation of the distribution. We note that $\sigma = 1$ corresponds to the unit exponential distribution. Most interesting, the hazard of this distribution does not depend on time. The distribution of the residual life, after truncation, is identical to the original distribution. That is, the distribution of $T - t$ conditional on $T \geq t$ is identical to the original distribution of T. Further, the distribution of $\lambda \times T$, where λ is a positive constant, is still exponential with mean $\lambda\sigma$.

We may illustrate the parametric approach with this distribution. Given the density function in Table 2.1, we can easily derive the survival function and hazard function to be $S(t) = \exp(-(t-a)/\sigma)$ and $h(t) = \sigma^{-1}$. Now suppose we know $a = 0$ and are only interested in estimating the unknown σ. According to (1.5) in Section 1.4, we construct the likelihood function to be

$$\mathcal{L}(\sigma) = \prod_{i=1}^{n} \left\{ \sigma^{-1} \exp(-t_i/\sigma) \right\}^{\delta_i} \left\{ \exp(-t_i/\sigma) \right\}^{1-\delta_i}. \tag{2.13}$$

We may easily derive the maximum likelihood estimator for σ to be $\hat{\sigma} = \sum_{i=1}^{n} t_i / \sum_{i=1}^{n} \delta_i$. The parametric estimate for the survival function is then $\hat{S}(t) = \exp(-t/\hat{\sigma})$.

Weibull Distribution. This is an extension of the exponential distribution, which allows for a power dependence of the hazard on time. Using notations in Table 2.1, we may write the survival function, hazard function, and cumulative hazard function as $e^{-(t/\sigma)^\gamma}$, $\sigma^{-\gamma}\gamma t^{\gamma-1}$, and $\sigma^{-1}t^\gamma$, respectively. It is quite clear that $\gamma = 1$ corresponds to the exponential distribution. The parameter γ is called a *shape parameter* as its value determines the shape of the hazard. For example, $\gamma < 1$ leads to a decreasing hazard, while $\gamma > 1$ leads to an increasing hazard. The log-transformed variable $\log(T)$ has a standard Gumbel distribution or the extreme-value distribution. The distribution of T^γ is exponential with mean σ^γ, and this fact enables us to evaluate any order moment easily with the formula $E(T^s) = \sigma^s \Gamma(s/\gamma + 1)$ for $s > -\gamma$ (the moment is infinite if $s \leq -\gamma$). The $100p\%$th percentile of the Weibull distribution is $\{\sigma \log[1/(1-p)]\}^{1/\gamma}$. As Weibull distribution is skewed, it is more appropriate to consider percentiles than moments. An interesting property of this distribution is that the minimum of n i.i.d. Weibull variables is still Weibull with parameters σ/n and γ. The distribution is named after Waloddi Weibull (1951).

Gompertz Distribution. The density function of this distribution is not provided in Table 2.1 as it is quite complicated. It is relatively easy to define this distribution by a hazard as

$$h(t) = \exp(\alpha + \beta t). \tag{2.14}$$

When $\beta > 0$, the hazard is increasing (positive aging); when $\beta < 0$, the hazard is decreasing (negative aging); when $\beta = 0$, it reduces to the exponential distribution. This distribution may be more appropriate to model lifetime than a linear hazard

$$h(t) = \alpha + \beta t, \tag{2.15}$$

TABLE 2.1: Some Common Distributions for Continuous Random Variables (In all density functions, t is the generic argument. $\Gamma(t) = \int_0^\infty e^{-x} x^{t-1}\, dx$ is the gamma function.)

Distribution	Density	Mean	Variance
Normal	$\frac{1}{\sqrt{2\pi}\sigma}\exp\{-\frac{(t-\mu)^2}{2\sigma^2}\}$	μ	σ^2
Logistic	$\frac{\sigma^{-1}\exp\{-(t-\mu)/\sigma\}}{(1+\exp\{-(t-\mu)/\sigma\})^2}$	μ	$\frac{\sigma^2\pi^2}{3}$
Gumbel	$\exp\{\frac{t-\mu}{\sigma}-\exp\{-\frac{t-\mu}{\sigma}\}\}/\sigma$	μ	$\frac{\sigma^2\pi^2}{6}$
Exponential	$\sigma^{-1}\exp\{-(t-a)/\sigma\}I(t>a)$	$\sigma+a$	σ^2
Weibull	$\frac{\gamma t^{\gamma-1}}{\sigma^\gamma}\exp\{-(\frac{t}{\sigma})^\gamma\}I(t>0)$	$\sigma\Gamma(\gamma^{-1}+1)$	$\sigma^2\{\Gamma(2\gamma^{-1}+1)-\Gamma(\gamma^{-1}+1)^2\}$
Pareto	$\sigma a^\sigma t^{-(\sigma+1)}I(t>a)$	$\frac{a\sigma}{\sigma-1}$	$\frac{a^2\sigma}{(\sigma-1)^2(\sigma-2)}$
Log-normal	$\frac{1}{\sqrt{2\pi}\sigma t}\exp\{-\frac{(\log t-\mu)^2}{2\sigma^2}\}I(t>0)$	$\exp(\mu+\sigma^2/2)$	$e^{2\mu+\sigma^2}(e^{\sigma^2}-1)$
Log-logistic	$\frac{\beta t^{\beta-1}\alpha^{-\beta}}{(1+(t/\alpha)^\beta)^2}I(t>0)$	$\frac{\alpha\pi}{\beta\sin(\pi/\beta)}$	$\alpha^2\{\frac{2\pi}{\beta\sin(2\pi/\beta)}-\frac{\pi^2}{\beta^2\sin^2(\pi/\beta)}\}$
Cauchy	$\frac{1}{\pi\sigma}\{1+(\frac{t-\mu}{\sigma})^2\}^{-1}$	∞	∞
F	$\frac{n^{\frac{n}{2}}m^{\frac{m}{2}}\Gamma(\frac{n+m}{2})t^{\frac{n}{2}-1}}{\Gamma(\frac{n}{2})\Gamma(\frac{m}{2})(m+nt)^{\frac{n+m}{2}}}I(t>0)$	$\frac{m}{m-2}$	$\frac{2m^2(n+m-2)}{n(m-2)^2(m-4)}$
Gamma	$\frac{\lambda(\lambda t)^{k-1}e^{-\lambda t}}{\Gamma(k)}I(t>0)$	$\frac{k}{\lambda}$	$\frac{k}{\lambda^2}$
Inverse Gaussian	$(2\pi t^3\nu)^{-1/2}\exp\{-(1+\mu t)^2/(2\nu t)\}I(t>0)$	μ^{-1}	ν/μ^3

which may also be used to model positive or negative aging population. The reason is that the hazard of Gompertz decreases ($\beta < 0$) in the short term and then becomes constant, whereas a linear hazard can be unbounded creating constrained estimation in practice. The Gompertz survival function can be obtained as

$$S(t) = \exp\left\{ -\frac{e^{\alpha}}{\beta}(e^{\beta t} - 1) \right\}. \tag{2.16}$$

The density may be obtained from $h(t)S(t)$ or differentiation of $-S(t)$. This distribution is named after Benjamin Gompertz (1825).

Bathtub-Hazard Distribution. The hazard is defined as

$$h(t) = \alpha t + \frac{\beta}{1 + \gamma t}. \tag{2.17}$$

There is no simple form for the survival function or the density function. It turns out to be rather helpful in human health study as it closely mimics the aging process and can be used as a tool for the study of long-term survival. It was proposed in Hjorth (1980). Occasionally, a quadratic hazard function

$$h(t) = \alpha + \beta t + \gamma t^2 \tag{2.18}$$

may be used to approximate the decreasing-and-then-increasing pattern (Gaver and Acar, 1979). However, the use of this hazard may have the same problem as the linear hazard, and sometimes the hazard value can be negative.

Inverse Gaussian Distribution. Consideration of this distribution is not based on hazard but on the time scale. Consider a first-hitting-time (FHT) model which has two basic components: (i) a parent stochastic process and (ii) a boundary set. When the parent process first encounters the boundary set is considered the occurrence of the threshold event (Lee and Whitmore, 2006). Such a probabilistic framework has been widely applied in survival analysis since it has a close resemblance to the lifetime process in human health studies and other practical failure time processes (Lee and Whitmore, 2010). A common parent process considered in practice is the Wiener process equipped with a positive initial value and certain mean and variance parameters. This process is appropriate since researchers have found that daily or hourly improvements and decrements in patient health can be accommodated by the bi-directional movements of the Wiener process (Lee and Whitmore, 2006). The boundary is the zero level of the process. It is shown that the time required for the process to reach the zero level for the first time has an inverse Gaussian distribution if the mean parameter is negative so that the process tends to drift toward zero. This distribution can accommodate a wide variety of shapes and is a member of the exponential family. Tweedie (1945) investigated its intriguing mathematical properties and applied it in a clinical trial study. He found great success in applying the inverse Gaussian to model the survival distribution for a series of patients treated for cancer while log-normal and Weibull distributions fit such data poorly. Since then many pioneering studies have been followed to extend the application of this distribution in scientific studies. See Chapter 1 of Seshadri (1993) for a detailed historical survey.

Although there are a large number of distributions suggested in the literature, particular parametric distributions are usually only appropriate for specific examples and have very narrow generalizability. To prevent model misspecification, we advise to always check a chosen parametric distribution against the empirical KM estimator. If the estimated survival curve resulted from an assumed parametric distribution is too different from the KM curve, the assumption for such a parametric approach may be questionable. On the other hand, it is also worth noting that with practical data sets in various fields, plenty of parametric modes have led to estimated survival curves reasonably close to the KM estimates.

Even though parametric methods are no longer the main analysis tool to estimate survival distribution, they are still widely used in simulation experiments to study theoretical properties of survival data. Commonly assumed distributions for survival data include Weibull distribution, log-normal distribution, log-logistic distribution, and others. See Table 2.1 for their parametric distribution forms.

2.2 Two-Sample Comparison

2.2.1 Log-Rank Test

In the lung cancer example described in Section 1.1, investigators needed to examine the survival probabilities and make a comparison between patients with certain gene mutation and those without such a mutation. If the distributions for the two samples are similar, the mutation will be considered as not related to the disease progress, and subjects with or without the mutation should be treated in the same way. On the other hand, if the distribution for the mutation group shows a significantly different survival curve from the non-mutation group, investigators may then choose to treat the two groups with different strategies, according to the estimated survival probabilities of the individual groups. This essentially leads to a *two-sample comparison problem*. In practice, we may need to determine whether two samples of failure times could have arisen from the same underlying distribution. The comparison is required to be made between two continuous functions across the whole observed time range. This task demands functional comparison and is substantially more sophisticated than two-sample comparison of location parameters such as means and medians.

Denote the underlying survival functions for the two sample groups to be $S_1(t)$ and $S_2(t)$. We introduce a nonparametric test for the following hypothesis:

$$H_0 : S_1(t) = S_2(t) \qquad v.s. \qquad H_1 : S_1(t) \neq S_2(t). \qquad (2.19)$$

Note that under the alternative, we do not specify the exact time point of deviation. Otherwise, a simple parametric test would be sufficient.

Still using the notations in the preceding section, we may construct a two by two contingency table, as shown in Table 2.2, at each observed failure time t_i. Let n_{1i} and n_{2i} be the numbers at risk for the two groups and d_{1i} and d_{2i} be the numbers of death for the two groups. Given n_{1i}, n_{2i}, and d_i, the observed death outcome for sample group 1, d_{1i}, follows a hypergeometric distribution with mean $m_{1i} = n_{1i}d_i/n_i$ and variance $v_{1i} = n_{1i}n_{2i}d_i(n_i - d_i)/\{n_i^2(n_i - 1)\}$ under the null (for more detailed discussions, see Shao, 1999). Using such an exact distribution result, one can easily perform a Fisher exact test for the equivalence of $S_1(t_i)$ and $S_2(t_i)$ at a single time point t_i. This is a familiar nonparametric procedure introduced in elementary statistics modules (see for example, Agresti, 1990 and Everitt, 1992).

TABLE 2.2: Two-Sample Comparison of Survival Functions at Observed Time t_i

Outcome	Sample 1	Sample 2	Total
Death	d_{1i}	d_{2i}	d_i
Survival	$n_{1i} - d_{1i}$	$n_{2i} - d_{2i}$	$n_i - d_i$
At risk	n_{1i}	n_{2i}	n_i

We intend to combine information from tables like Table 2.2 for all observed failure times t_i. If the null hypothesis (2.19) is true, we would expect the difference between d_{1i} and its expectation m_{1i} to be small at all t_i, relative to the magnitude of the variance v_{1i}. A sensible test statistic may then be constructed as

$$L = \frac{\left\{\sum_{i=1}^{k}(d_{1i} - m_{1i})\right\}^2}{\sum_{i=1}^{k} v_{1i}}. \qquad (2.20)$$

Here, k is the total number of distinct event times. Though the means and variances of all d_{1i} are exact from the corresponding hypergeometric distributions under the null, the distribution of L can only be derived in an asymptotic manner. As $n \to \infty$, the test statistic L asymptotically follows a χ^2 distribution with one degree of freedom under the null hypothesis, and such a distribution result can then be applied to calculate the test P-value as $P(\chi_1^2 \geq L)$. The summation in (2.20) indicates that a large discrepancy between the observed and expected death numbers at any time point can contribute to a large L and thus tends to declare a significant result.

The test statistic L has been referred to as a *log-rank test*, due to Peto and Peto (1972), but also closely related to Savage's earlier work (1956). Inspecting its construction in (2.20), we realize that it does not involve a naive calculation of the logarithm of the ranked observations, as might be suggested from its name. It has become a convention to report such a test result for comparing survival distributions between two distinct groups. We also take a note that in the statistics literature, this test has sometimes been called Mantel-Haenszel test, generalized Savage test, or generalized Wilcoxon test.

There are crucial assumptions to support the validity of this test. First, under a common requirement for any asymptotic test, the log-rank test requires that the sample sizes $\sum_{i=1}^{k} d_i$ and $\sum_{i=1}^{k} m_i$ go to infinity. This is usually not a practical concern when we study diseases with high mortality rates or when we study subjects with high risks of developing the failure outcomes. Second, to arrive at the χ^2 distribution under the null, we have to assume that the censoring is independent for the two sample groups. That is, the reason a subject's survival time is censored should not be due to the fact of being in one group instead of the other. When this condition is not met, a nonnegligible bias may occur in the analysis and lead to an incorrect test result.

Since its invention, the log-rank test statistic L has been extended to various versions. Most of the extensions bear the following form

$$L_w = \frac{\left\{ \sum_{i=1}^{k} w_i (d_{1i} - m_{1i}) \right\}^2}{\sum_{i=1}^{k} w_i^2 v_{1i}}, \tag{2.21}$$

where w_i is a possibly data-dependent positive weight, and L_w has been referred to as a weighted log-rank test. L is a special case of L_w when all w_is are equal to one. Different ways of constructing w_is can be used to emphasize the comparison in different time regions. For example, Gehan (1965) and Breslow (1970) proposed the weight $w_i = n_i$ where more weights are given to earlier death time points. This kind of weight is applicable when short-term survival is of more interest. On the other hand, L puts equal weights to all death time points and thus may favor long-term survival more than Gehan-Breslow's L_w. As a compromise, Tarone and Ware (1977) proposed an intermediate weight $w_i = \sqrt{n_i}$ which penalizes long-term survival times less heavily than Gehan-Brewslow's weight. Another commonly adopted weight is $w_i = \prod_{j=1}^{i} (1 - d_j/(n_j + 1))$, which was proposed by Peto and Peto (1972) and Prentice (1978). This weight function has been conveniently substituted with the KM estimate of $S(t_i)$ by many packages for survival analysis. Asymptotically, it can be proved that if the weights are generated from a predictable process, all these extended log-rank tests follow the same χ^2 distribution under the null hypothesis.

For a real application, we may need to choose among the different versions of tests. According to their construction, it is not hard to realize that these tests may be sensitive to different alternative hypotheses when $S_1(t)$ and $S_2(t)$ differ in different ways. For example, Gehen-Breslow's test is more likely to detect early differences in the two survival curves, while the unweighted log-rank test is more capable of finding differences in the tails. If heavy censoring exists, Gehan-Breslow's test is dominated by a small number of early failures and thus may have very low power to check the complete range. Therefore, we might prefer to use the log-rank test with no weight if we only have a vague idea of the shape of the distributions.

The log-rank test is most powerful when the hazard ratio between the two groups is a constant over time. Also, if the true survival distribution is exponential or Weibull for each of the two groups, the log-rank test has greater testing power than all other weighted versions. Other weighted tests are favored, when the hazard ratio is not a constant or the underlying distribution is log-normal. Finally, we notice that all the tests will fail if the two survival curves cross each other at certain point, that is, the true survival probability for one group is above that for the other group before a time point and goes below that for the other group after the time point. Empirically, we may detect such a problem when the estimated KM curves for the two groups cross. Consequently in the numerical calculation, one may have some terms $d_{1i} - m_{1i}$ being positive and others negative in the numerators of (2.20) and (2.21). Thus, even though the two distributions would be quite different, the test L or L_w could only produce a small value and fail to detect the difference. It is therefore difficult to conclude whether the two curves are significantly different from each other.

Example (Lung Cancer). We now use the lung cancer example described in Section 1.1 to illustrate the aforementioned tests. There are four genes in this study: EGFR, P53, KRAS, and CMET. For each gene, we carry out the log-rank test to see if the survival distributions for individuals with and without mutation at this gene are different. We report the test results from L and L_w with Peto-Prentice's weights in Table 2.3. For genes EGFR and P53, neither the log-rank test nor the Peto-Prentice weighted log-rank test suggests any significant difference, indicating that mutations of these two genes do not significantly modify the survival distribution. On the other hand, two versions of log-rank tests declare significant results for genes KRAS and CMET. Mutations at these two genes thus deserve further attention as they may lead to a significantly different survival distribution.

The above results have been generated in R by using the function `survdiff` contained in the package `survival`. The choice between log-rank test and Peto-Prentice test can be made by specifying the option `rho` where a value of 0 corresponds to the log-rank test and a value of 1 corresponds to the Peto-Prentice test. If we use `EGFR` to denote the binary indicator of presence of the EGFR gene mutation in the data file, then the proper R code to produce an unweighted log-rank test is

```
survdiff(Surv(time, status) ~ EGFR, rho = 0).
```

TABLE 2.3: Test Statistics and P-Values for the Four Gene Mutations in the Lung Cancer Data

Gene	Test Version	Test Statistic	P-Value
EGFR	Log-rank	1.23	0.268
	Peto-Prentice	1.50	0.219
P53	Log-rank	0.40	0.506
	Peto-Prentice	0.51	0.485
KRAS	Log-rank	7.79	0.005
	Peto-Prentice	9.30	0.002
CMET	Log-rank	4.82	0.029
	Peto-Prentice	4.04	0.045

Here, time and status represent the observed time and event status, respectively. The variable specified on the right-hand side of $\sim$ can be binary or categorical variables with more than two classes, such as multiple age groups or different ethnicity groups.

Implementations of other weights are currently not available in R but available in Stata (with options logrank, w, tw and p etc. in sts test) and SAS (via the available options for test in PROC LIFETEST). One can easily find useful examples from their help manuals.

When comparing two samples for their survival distributions, it is advised to plot the KM estimates separately for the two samples. In Figure 2.4 we present the KM curves for the two samples (KRAS mutation positive and negative). Eyeballing this figure we notice that the survival curve for patients with KRAS mutation descends rapidly from the baseline while that for patients without KRAS mutation follows closely to the overall survival curve. Such graphical displays can help to rule out the crossover of survival functions and confirm the test results.

Example (STAR*D). Consider the STAR*D example described in Section 1.1. We stated in the previous section that a comparison between the two treatment arms should be done using a log-rank test. In this case, the log-rank test gives a test value 0.0024 with a nonsignificant P-value 0.96, indicating that the distributions for the two arms are not significantly different. This result agrees with the observation from Figure 2.2, where the two remission probability curves are close to each other for the entire study time range.

Thus far, we have focused on two sample comparisons. In practice, there may be multiple groups under comparison. For example, many clinical trials have multiple treatment arms, representing different treatment strategies or different doses of the same treatment strategy. It is possible to extend the log-rank test described above to a multisample comparison problem. For an M-sample comparison problem, we may modify the two-by-two contingency table into a two-by-M table and obtain test statistic in a similar manner. The resulting test is still conventionally called a log-rank test (Kalbfleish and Prentice, 1980). Under the null, the test statistic is still χ^2 distributed, with degree of freedom $M - 1$.

2.2.2 Sample Size Calculation

Sample size calculation is an important aspect of study design. In survival analysis, an unrealistically large sample size involving unnecessarily many human subjects may be unethical, while a small sample size may be insufficient to gather information on failure time distributions. Therefore now it is well accepted that sample size determination has both an ethical as well as an efficacy imperative. Specifically, to compare survival probabilities for two groups (say, the treatment and control groups in a clinical trial), we have to determine the minimum sample size requirement for a study under prespecified type I and type II errors (Schoenfeld, 1981, 1983).

We first calculate the number of dead subjects d needed to declare the hazards ratio between two groups to be significantly different from one. Usually the effect size under the alternative is specified in terms of hazard functions instead of survival functions. We conventionally denote the hazard ratio for the two groups under the alternative hypothesis to be $\psi = h_1(t)/h_2(t)$. Note that it is not hard to translate this value into an expression for survival functions, that is, $\psi = \frac{\log S_2(t)}{\log S_1(t)}$. We aim at testing the hypothesis (2.19) with a log-rank test under a significance level α and power $1 - \beta$. In practice, we may adopt $\alpha = 0.05$ and $\beta = 0.1$ or 0.2. We note that the log-rank test statistic L can be written as $L = U^2/V$, where $U = \sum_{i=1}^{k}(d_{1i} - m_{1i})$ and $V = \sum_{i=1}^{k} v_{1i}$. The rejection region defined by the event $\{L > \chi^2(\alpha)\}$ may also be equivalently

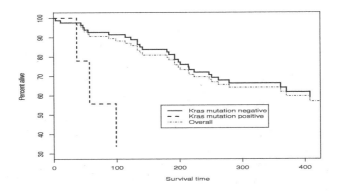

FIGURE 2.4: Kaplan-Meier estimates of the survival probability for the lung cancer data.

defined by using the event $\{|U| > u\}$ for a cut-off u such that the two events have the same probability. Therefore, we may describe the probabilities of this rejection region under the null and alternative as

$$P(|U| > u; H_0) = \alpha, \tag{2.22}$$
$$P(|U| > u; H_1) = 1 - \beta. \tag{2.23}$$

It can be shown that asymptotically under the null, U follows a normal distribution $N(0, V)$, while under the alternative U follows a normal distribution $N(V \log \psi, V)$. The probabilities in the above two equations can thus be approximated by two normal distributions, respectively,

$$\Phi(u/\sqrt{V}) = 1 - \alpha/2, \tag{2.24}$$
$$\Phi((-u - \log \psi V)/\sqrt{V}) = \beta. \tag{2.25}$$

Furthermore, V can be approximated as

$$V \approx \sum_{i=1}^{k} \frac{n_{1i} n_{2i} d_i}{n_i^2} \approx \pi_1 \pi_2 \sum_{i=1}^{k} d_i = \pi_1 \pi_2 d. \tag{2.26}$$

The two proportions π_1 and π_2 are the allocation proportions for the two samples. The two equations (2.24) and (2.25) thus only involve two unknown quantities: the total number of deaths d and the rejection cut-off u. We may easily solve the equations and obtain the required total number of deaths as

$$d = \frac{(z_{\alpha/2} + z_\beta)^2}{\pi_1 \pi_2 (log \psi)^2}. \tag{2.27}$$

In some clinical trials, we may have the simple setting with $\pi_1 = \pi_2 = 0.5$, under which

$$d = \frac{4(z_{\alpha/2} + z_\beta)^2}{(log \psi)^2}. \tag{2.28}$$

It can be proved that the sample size achieved in such a balanced design is the smallest, provided other conditions to be the same.

As a hypothetical example, suppose that we are going to examine the survival difference between two groups of subjects in a clinical trial. The trial has aimed to establish that a new treatment can provide more benefits than the standard treatment regime. The three-year survival probability for the group with the standard treatment regimen is documented to be $S_1(t = 3) = 0.20$, while that for the group with the new treatment is expected to be $S_2(t = 3) = 0.40$. The hypothesized hazards ratio is thus $\psi = \log 0.40/ \log 0.20 = 0.569$. When carrying out this study, to ensure a significance level $\alpha = 0.05$ and power $1 - \beta = 90\%$ with equal allocation for the two treatment groups, we need a minimum of $d = 132$ deaths according to expression (2.27). Roughly 66 deaths are required for each group.

The final sample size for recruiting subjects is d divided by the overall probability of death in the entire follow-up. In the above example, if the overall mortality rate in the population is 10%, we will need 660 for each group. This prevalence inflation method is simple to use but sometimes considered as a crude practice. Some conservative approaches can also be considered (Li and Fine, 2004).

We note that in practice, sample size calculation can be considerably more complicated. When designing a clinical trial, for example, we may need to take into consideration staggering recruitment, informative and noninformative dropouts, and other factors. In addition, some modern clinical trials take a sequential design, spending type I and type II errors over multiple interim analyses, making the sample size calculation much more complicated. The survival analysis techniques described above can be the building blocks for sample size calculation under complicated setup. When there is a lack of simple sample size formula, one may have to resort to simulation-based approaches.

2.3 Regression Analysis

The goal of regression analysis is to study the relationship between a response variable and a group of covariates. In survival analysis, the response variable is the survival time, while the covariates may include many factors such as demographic variables, prognostic characteristics, treatment types, and others. In this section we first consider a semiparametric regression model to study how covariates affect the distribution of survival time. Model selection and diagnostics are followed for this model. We then introduce a parametric regression model which may serve as an alternative to the semiparametric model in some cases.

2.3.1 Proportional Hazards (PH) Model

2.3.1.1 PH Model and Partial Likelihood

Although distribution specification can be made equivalently among the survival function, hazard function, cumulative hazard function, and density function, the convention has been to choose the hazard function to describe the covariates dependence in survival analysis. The Cox proportional hazards model (Cox, 1972) assumes that the hazard function at time t for a given p-vector of covariates $\mathbf{X} = (X_1, \cdots, X_p)^T$ is

$$h(t; \mathbf{X}) = h_0(t) \exp(\mathbf{X}^T \boldsymbol{\beta}), \qquad (2.29)$$

where $h_0(t)$ is the unknown baseline hazard function and $\boldsymbol{\beta} = (\beta_1, \cdots, \beta_p)^T$ is a vector of unknown parameters. This model effectively partitions the contribution of time and covariates to the hazard function into two multiplicative components:

(i) The function $h_0(t)$, commonly referred to as the baseline hazard, specifies how the hazard function changes as a function of time when there is no covariate effect.

(ii) The exponential term $\exp(\mathbf{X}^T\boldsymbol{\beta})$ involves a simple linear combination of the covariates through coefficients $\boldsymbol{\beta}$ and does not depend on time. The coefficients $\boldsymbol{\beta}$ characterize the relative effects of different covariates on the hazard.

Using this model, we may provide practical interpretations on how covariates affect the survival distribution. For example, a positive β_k for a continuous covariate X_k would indicate that increasing the value of X_k leads to a higher level of hazard.

The nomenclature of model (2.29) follows from the fact that the ratio of hazards $h(t; \mathbf{X})/h(t; \mathbf{X}^*)$ is a constant over time t when $\mathbf{X}$ and $\mathbf{X}^*$ are fixed. In fact, suppose that $\mathbf{X} = X_1$ is one dimensional and refers to a binary indicator of a subject being in the treatment group ($X_1 = 1$) or control group ($X_1 = 0$). Then model (2.29) gives $h(t; X_1 = 1)/h(t; X_1 = 0) = \exp(\beta_1)$ which is independent of time. The parameter β_1 is then interpreted as a constant of log hazards ratio for comparing a subject from the treatment group to a subject from the control group. This ratio is a fixed proportion and does not vary over time. Because hazard is sometimes approximately regarded as risk in practice, model (2.29) has also been called a relative risk model by many statisticians.

Fitting model (2.29) involves the complete determination of unknown function $h_0(t)$ and unknown parameters $\boldsymbol{\beta}$. In practice knowing the parameter $\boldsymbol{\beta}$ can lead to a clear answer on the relationship between $\mathbf{X}$ and survival outcome. The functional h_0 is thus treated as a nuisance and is only estimated if we want to draw the survival curve. To this end, we may employ a maximum likelihood estimation (MLE) approach to estimate $\boldsymbol{\beta}$. Cox (1972) has shown that the estimation of $\boldsymbol{\beta}$ can be based on the following so-called partial likelihood function:

$$L(\boldsymbol{\beta}) = \prod_{i=1}^{m} \frac{\exp(\mathbf{x}_i^T\boldsymbol{\beta})}{\sum_{j \in R(t_i)} \exp(\mathbf{x}_j^T\boldsymbol{\beta})}, \tag{2.30}$$

where $\mathbf{x}_i$ is the observed vector of covariates for the individual who dies at the ith ordered death time t_i, $R(t_i)$ is the index set for all individuals who are alive and uncensored at a time prior to t_i, and the product is over all m distinct event times. A simple sketch to derive (2.30) is appended in Section 2.5. The partial likelihood function (2.30) incorporates all the essential information for $\boldsymbol{\beta}$ from data and thus a value $\hat{\boldsymbol{\beta}}$ that maximizes (2.30) is believed to be the most possible value of the parameter.

The maximization can be carried out through numerical methods such as the Newton-Raphson procedure. The procedure follows a recursive algorithm so that for an estimate at current iteration $\hat{\beta}_c$ we update to the next estimate by calculating

$$\hat{\beta}_{c+1} = \hat{\beta}_c + \mathbf{I}^{-1}(\hat{\beta}_c)\mathbf{U}(\hat{\beta}_c), \tag{2.31}$$

where

$$\mathbf{U}(\hat{\beta}_c) = \frac{\partial \log L}{\partial \beta}\Big|_{\beta=\hat{\beta}_c}$$

is the score function, and

$$\mathbf{I}(\hat{\beta}_c) = -\frac{\partial^2 \log L}{\partial \beta \partial \beta^T}\Big|_{\beta=\hat{\beta}_c}$$

is the observed information matrix evaluated at the current estimate. With this approach, an initial value is needed. As the objective function is concave, choosing the initial value may affect the number of iterations needed for convergence but not convergence itself. Based on such gradient and Hessian quantities, the algorithm converges to the maximizer of (2.30) at a quadratic speed.

The score function may be written as

$$
\begin{aligned}
\mathbf{U}(\beta) &= \sum_{i=1}^{m}\left[\mathbf{x}_i - \sum_{j\in R(t_i)} \mathbf{x}_j \frac{\exp(\mathbf{x}_j^T\beta)}{\sum_{j\in R(t_i)}\exp(\mathbf{x}_j^T\beta)}\right] \\
&= \sum_{i=1}^{m}\left[\mathbf{x}_i - \sum_{j\in R(t_i)} \mathbf{x}_j w_{ij}\right] \\
&= \sum_{i=1}^{m}[\mathbf{x}_i - \bar{\mathbf{x}}(t_i)]
\end{aligned}
$$

where

$$w_{ij} = \frac{\exp(\mathbf{x}_j^T\beta)}{\sum_{j\in R(t_i)}\exp(\mathbf{x}_j^T\beta)},$$

and

$$\bar{\mathbf{x}}(t_i) = \sum_{j\in R(t_i)} \mathbf{x}_j w_{ij},$$

and interpreted as the distance between the sum of covariates among the deaths and the weighted sum of covariates for the survivors in the risk sets. We note that the score function does not involve the actual value of the failure time t_i in its calculation. The consistency of the estimates $\hat{\beta}$ follows since the score function evaluated at the true value of β converges to zero as sample size goes to infinity. Also under the large sample assumption, the resulting MLE of β can be shown to follow a multivariate normal distribution with mean equal to the true β and covariance matrix $E\{\mathbf{I}^{-1}(\beta)\}$, which may be suitably estimated by $\mathbf{I}^{-1}(\hat{\beta})$ (Royall, 1986). The detailed proof is based on the counting process theory, which we briefly sketch in Section 2.5 of this chapter. Most statistical packages for survival analysis can perform the numerical calculations. Inferences may then follow from the outputs provided therein.

Example (Bloodstream Infection). In the BSI study introduced in Chapter 1, in addition to the duration of carbapenem use, there are other factors that may affect the survival probability. It is important to control for the effects of all factors in a regression model. We include the following variables in **X**: days of carbapenem use, sex, age of the subject, indicator of co-morbidity, indicator of ventilation >96 hours, length of stay, and malignancy. Assume the PH model. The corresponding coefficients are estimated and summarized in Table 2.4.

TABLE 2.4: Fitted Results from the Proportional Hazards Model for the BSI Data

Variable	$\hat{\beta}$	Standard Error	$\exp(\hat{\beta})$	P-Value
Days of carbapenem	-0.067	0.037	0.935	0.070
Sex (Male)	-0.020	0.372	0.980	0.960
Age	-0.021	0.011	0.979	0.065
Any co-morbidity	0.509	0.445	1.664	0.250
Ventilation > 96 hours	0.994	0.416	2.703	0.017
Length of stay	-0.009	0.007	0.991	0.230
Malignancy	0.768	0.392	2.157	0.050

The coefficient for days of carbapenem is estimated to be -0.067 with a standard error 0.037. The P-value is for testing whether the coefficient is equal to zero and based on the fact that the estimated coefficient has an asymptotically normal distribution. In this case, holding other factors constant, increasing days of carbapenem by one unit would result in multiplying the hazard by 0.935, suggesting a slight beneficial effect on the development of BSI. However, since the P-value for this coefficient is quite large (greater than the nominal level 0.05), we fail to declare there is a significant relationship between the duration of carbapenem and the failure outcome. Coefficients for other factors can be similarly interpreted.

The results in Table 2.4 have been obtained in R using the `coxph` function in the `survival` package. In general, for a survival time variable `time`, a censoring indicator `status`, and a set of covariates `x1`, `x2`, `x3`, we may call `coxph` using the following line:

```
coxph(Surv(time, status) ~ x1 + x2 + x3, data = set1)
```

where `set1` stores all the variables mentioned in this statement. One can also use `stcox` in `STATA` or `PROC PHREG` in `SAS` to fit the Cox PH model and arrive at the same results.

The P-values for the preceding example are obtained from a Wald test where the test statistic for the jth coefficient is computed by $\hat{\beta}_j$ divided by its standard error. The test statistic has a standard normal distribution for a large sample. For example, to test the significance of Sex in the above BSI example, we may compute the test statistic to be $-0.020/0.372 = -0.054$, and the two-sided P-value is $2\Phi(-0.054) = 0.960$. The test is useful to check the significance of an individual variable given the presence of other variables in the model.

In addition to such a conditional test, one may also test the significance of the predictor variable by fitting a Cox PH model with a single predictor variable. Such a marginal test may be important for the purpose of screening out useless predictors and will be emphasized in Chapter 6, where we discuss the analysis of extremely high dimensional data. The marginal test agrees with the conditional test only when the sample is homogeneous or the variables have some orthogonal properties. This may be achievable in some randomized clinical trials. Otherwise the two tests should be interpreted differently. When the predictor is a categorical variable such as sex or treatment group, the marginal test based on a Cox model is essentially the same as the log-rank test introduced in Section 2.2.1.

We now consider another example and study the interaction effects in the Cox regression model.

Example (Accident). A multicenter cross-sectional health survey was conducted in 12 universities and 4 universities of applied sciences in North Rhine-Westphalia, Germany in 2006 and 2007 (Faller et al., 2010). The aim of this study was to assess the general prevalence of accidents among university students, to describe the specific kinds of accidents, and to analyze associated factors. A total of 2855 students participated and completed the self-administered questionnaire, where 252 (8.8%) had experienced an accident in the context of their studies. Since the cumulative risk of having experienced a study-related accident increases with the number of years at a university, a survival analysis was applied to the survey data.

We first plot the survival function for the time to the first accident. The stratified survival curves for different gender, university type (university or applied university), and faculty (sports-related or not) are shown in Figure 2.5. The two curves for male and female students in sports-related faculty drop much faster than other curves. This plot suggests that students enrolled in a faculty with more sports activities might have a higher risk for accidents than other students.

A Cox PH model is fitted. We first consider a model with interaction terms and find that the interaction terms are not significant (Table 2.5). We thus exclude them from further analysis and subsequently fit a model with only the main effects. In this fitted model, Age is significantly associated with the time to accident: the negative sign indicates that given all other covariates equal, younger students tend to run into accidents faster than older students; the hazard ratio is 0.944, indicating that the hazard will be multiplied by 0.944 for one year increase in age. Another significant variable is Sport, which has been expected after a visual inspection of Figure 2.5. The hazard ratio is 6.685, indicating the hazard for students in sports-related faculty is 6.685 times of that for students in nonsports-related faculty with other variables held constant. The gender and whether the university is applied or not are not significantly associated with the time to accident.

In a Cox PH model, the baseline hazard function $h_0(t)$ is a nuisance parameter and usually does not draw our interest directly. For the sake of completeness, we still present its estimation method here. The key quantity that needs to be estimated is the transitional or conditional survival probability

$$\eta_i = P(T > t_i | T > t_{i-1})$$

for ordered observed survival times $t_1 < t_2 < \cdots < t_m$. The estimate of η_i given the estimate for the parametric regression coefficient $\hat{\boldsymbol{\beta}}$ is

$$\hat{\eta}_i = \left(1 - \frac{\exp(\mathbf{x}_i^T \hat{\boldsymbol{\beta}})}{\sum_{j \in R(t_i)} \exp(\mathbf{x}_j^T \hat{\boldsymbol{\beta}})}\right)^{\exp(-\mathbf{x}_i^T \hat{\boldsymbol{\beta}})}, \tag{2.32}$$

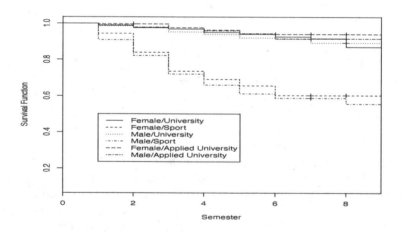

FIGURE 2.5: Kaplan-Meier estimates of the survival functions for the accident study.

TABLE 2.5: Fitted Results from the Cox Proportional Hazards Model for the Accident Data

Variable	$\hat{\beta}$	Standard Error	$\exp(\hat{\beta})$	P-Value
With interaction				
Age	−0.0570	0.1938	0.945	0.014
Male	0.0386	0.2197	1.039	0.860
Applied University (FH)	−0.3818	0.3622	0.683	0.290
Sport	1.8814	0.1938	6.563	< .001
Male:FH	0.4293	0.4604	1.536	0.350
Male:Sport	0.0593	0.2758	1.061	0.830
Without interactions				
Age	−0.0578	0.0232	0.944	0.013
Male	0.1167	0.1273	1.124	0.360
Applied University	−0.1257	0.2209	0.882	0.570
Sport	1.8999	0.1392	6.685	< .001

where $\mathbf{x}_i$ and $R(t_i)$ are defined in the same way as in (2.30). The estimated baseline hazard is simply

$$\hat{h}_0(t_i) = 1 - \hat{\eta}_i, \tag{2.33}$$

for the m distinct observed survival time points t_i $(i = 1, 2, \cdots, m)$. For all other time points, we take the convention that $\hat{h}_0(t) = 0$.

In practice, $\hat{H}_0$ is more useful than $\hat{h}_0$ because the estimated cumulative hazard can be less noisy. Using the result (2.33), we may immediately obtain the estimated cumulative baseline hazard and survival functions as

$$\hat{H}_0(t) = -\sum_{i=1}^{k} \log \hat{\eta}_i, \tag{2.34}$$

$$\hat{S}_0(t) = \prod_{i=1}^{k} \hat{\eta}_i, \tag{2.35}$$

for $t_k \leq t < t_{k+1}$, and here we define $t_{m+1} = \infty$. We note that this model-based estimator $\hat{S}_0(t)$ is constructed in a similar way as the KM estimator. Furthermore, one can obtain a model-based estimator of the survival function for the ith subject as

$$\hat{S}_i(t) = \hat{S}_0(t)^{\exp(\mathbf{x}_i^T \hat{\beta})}. \tag{2.36}$$

This estimator can also be calculated for censored subjects and will be used for model checking and diagnostics.

2.3.1.2 Adjustment for Ties

The partial likelihood function (2.30) needs an adjustment when there are ties among the uncensored failure times. Three adjustments methods are popular in practice. The first approach is an exact adjustment. The basis for this approach is to assume that the d_i tied observations at a particular time t_i are due to a lack of measurement precision, and the ties could have been observed in any of the $d_i!$ possible arrangements. The denominator in (2.30) can thus be modified by averaging across all possible permutations. This approach is the most accurate but also the most time-consuming. The other two adjustments are Breslow and Efron approaches (Peto, 1972; Breslow and Crowley, 1974; Efron, 1977). The expressions in these two approaches can be more easily evaluated than the exact approach. Specifically, the Breslow method modifies the partial likelihood function (2.30) to be

$$L(\boldsymbol{\beta}) = \prod_{i=1}^{m} \frac{\exp(\mathbf{x}_i^{*T}\boldsymbol{\beta})}{\{\sum_{j \in R(t_i)} \exp(\mathbf{x}_j^T \boldsymbol{\beta})\}^{d_i}}, \tag{2.37}$$

where $\mathbf{x}_i^*$ is the sum of the covariates of d_i individuals observed to fail at time t_i. The Efron method modifies the partial likelihood function (2.30) to be

$$L(\boldsymbol{\beta}) = \prod_{i=1}^{m} \frac{\exp(\mathbf{x}_i^{*T}\boldsymbol{\beta})}{\prod_{k=1}^{d_i}\{\sum_{j \in R(t_i)} \exp(\mathbf{x}_j^T \boldsymbol{\beta}) - \frac{k-1}{d_i} \sum_{j \in D(t_i)} \exp(\mathbf{x}_j^T \boldsymbol{\beta})\}}, \tag{2.38}$$

where $D(t_i)$ is the set of labels of d_i individuals failing at time t_i. Both (2.37) and (2.38) give asymptotically biased estimates for $\boldsymbol{\beta}$ since the score equations resulted from differentiating the likelihood functions do not converge to zero as sample size tends to infinity. Empirically, we may observe a substantial bias from the Breslow method when the number of ties are large. The performance of Efron method is intermediate among the three, slightly less biased than the Breslow method while less computationally intensive than the exact method. All three methods are available in R by specifying the `method` option in `coxph`. With the (time, status) dual and three covariates, a sample R code is as follows

```
coxph(Surv(time, status) ~ x1+x2+x3, method = efron)
coxph(Surv(time, status) ~ x1+x2+x3, method = exact)
coxph(Surv(time, status) ~ x1+x2+x3, method = breslow)
```

The default setting is the Breslow method due to its low computational burden. Most practitioners prefer to take this approach when the number of ties is small.

When there are tied observations, the conditional probability η_i cannot be estimated by (2.32) and must be obtained via a numerical method by solving the following equations:

$$\sum_{j \in R(t_i)} \exp(\mathbf{x}_j^{*T}\hat{\boldsymbol{\beta}}) = \sum_{j \in D(t_i)} \frac{\exp(\mathbf{x}_j^{*T}\hat{\boldsymbol{\beta}})}{1 - \eta_i^{\exp(\mathbf{x}_j^{*T}\hat{\boldsymbol{\beta}})}}, \qquad i = 1, 2, \cdots, m. \tag{2.39}$$

The estimated baseline hazard and survival functions can be modified accordingly.

2.3.1.3 Time-Dependent Covariates

We consider another factor that may lead to more complicated likelihood functions than (2.30). The covariates $\mathbf{X}$ in model (2.29) are assumed to the same at all observation times and so usually measured at the start of the follow-up. This assumption may hold for factors such as sex, race, body height of an adult, and other time-invariant characteristics observed in the study. However, there are often some explanatory variables which are also varying with the study time, such as a patient's blood pressure in a hypertension study, the serum creatinine level in a renal study, or simply the current age (not the age at the entry of the study, since age at the failure time may be more relevant) for a long follow-up study. All such information can be collected and used for better personalized treatment depending on a dynamic assessment of the prognosis of the patient. When such time-dependent covariates are present, the model and procedure used to fit the model need to be modified.

The time-dependence could be internal or external. An internal time-dependent covariate is usually subject-specific, and its measurement requires the presence of the subject in the study. For example, the blood pressure and serum creatinine level for a subject are internal time-dependent. In contrast, an external time-dependent covariate could arise from several practical sources. One type of such covariates is a time-dependent therapeutic or treatment agenda, which is usually under the control of the investigators. In particular, switching treatments, as in the STAR*D trial, may be necessary for both ethical consideration and experimental efficacy. Another type might be exemplified by an occupational health study, where the environmental hazards such as air or water pollution degrees might vary from time to time. All these external time-dependent covariates could be measured without actually observing individual subjects. Of note, age is usually considered as an external time-dependent variable. In fact, once we know the birth date of the subject, the age at any follow-up time can be easily calculated.

The Cox model (2.29) incorporating time-dependent covariates $\mathbf{X}(t)$ may be written as

$$h(t) = h_0(t) \exp(\mathbf{X}^T(t)\boldsymbol{\beta}). \qquad (2.40)$$

We note that now the relative hazard is $h(t)/h_0(t) = \exp(\mathbf{X}^T(t)\boldsymbol{\beta})$ which is no longer a constant independent of time. The model is thus not a PH model. For example, the first element β_1 in the vector $\boldsymbol{\beta}$ should be interpreted as the log-hazard ratio for two individuals whose value of $X_1(t)$ at time t differs by one unit, holding all other covariates to be the same at that time.

To fit model (2.40), we modify the partial likelihood function as

$$L(\boldsymbol{\beta}) = \prod_{i=1}^{n} \frac{\exp(\mathbf{x}_i^T(t_i)\boldsymbol{\beta})}{\sum_{j \in R(t_i)} \exp(\mathbf{x}_j^T(t_i)\boldsymbol{\beta})}, \qquad (2.41)$$

and the same computing procedure for fitting the time-independent PH model can be applied to maximize (2.41). We note that in practice it is more important for data analysts to prepare the data in a proper form to reflect the time-dependent nature of the covariates and then apply the software to perform the numeric calculation. One has to work out his or her own data-manipulation program since the R code for fitting model (2.40) is not simply a statement with an interaction between time and covariates. More efforts should also be placed upon the interpretation of the attained coefficients. Time-dependent covariates may introduce complications in practice. Consider the study of smoking on lung cancer survival. Smoking status, which can be measured with the number of cigarettes smoked per day or per week, is time-dependent. When incorporating this covariate in a Cox model, it may be more sensible to use the cumulated number of smoked cigarettes as opposed to the number measured at a specific time point.

We hereby point out a couple of medical examples dealing with time-dependent covariates. First, in the classical Stanford Heart Transplant study (Crowley and Hu, 1977), the time to death T for each patient was the outcome of interest. From the time of admission to study until the time of death, a patient was eligible for a heart transplant. Whether a patient actually had a transplant in this study depended on whether his waiting time W was shorter than the observed event time. A time-dependent covariate $X(t) = I(W < t)$ was thus created in the analysis to study the effect of transplant. The data is publicly available in R as heart in the package survival. A second example is a clinical trial in Chronic Myeloid Leukemia (CML) that was reported in the Benelux CML Study Group (de Bruijne et al., 2001). One important biomarker to predict the survival of the patient was white blood cell (WBC) counts measured every one to two months. The current value of WBC was more meaningful than the baseline value. Since the exact value of WBC at the failure event might not be available, the last observation was taken as a proxy in the analysis. The data is available in R as wbc2 in the package dynpred.

The time-dependent analysis sometimes invites criticism for their interpretation. For example, a time-dependent measure $X(t)$ may be taken at or close to the event time because of the foreseeable progress of the disease status in lieu of the prescheduled measurement occasions. One would then question whether the covariate causes the event or the event changes the covariate and the routine analysis simply muddies the causal relation. A possible remedy is to replace the value taken at a crisis $X(t)$ by a time-lagged version $X(t - \Delta t)$. More discussion can be found in Andersen and Liestol (2003).

Another important contribution of incorporating time-dependent covariates is for model diagnostics introduced in Section 2.3.3. In addition, in industrial life-examination, we can often measure the wear or damage sustained by components while they are functioning, and such time-dependent measurements are very helpful covariates to predict the failure risk. Some practical experiences on including time-dependent covariates in a PH model can be found in Altman and de Stavola (1994).

2.3.1.4 PH Model versus Logistic Model

To study the dependence of mortality or other time-related dichotomous outcome on covariates in a follow-up period, another widely used model is the logistic regression model. Logistic regression does not examine individual event time points and only uses the binary outcome in a fixed period as the response variable. There are many excellent books devoted to logistic regression such as Collet (2002) and Hosmer and Lemeshow (2000), among others. Specifically, using the same notations as in survival analysis, we have the following logistic regression model:

$$\log \frac{1 - S(C)}{S(C)} = \alpha + \mathbf{X}^T \beta_L, \tag{2.42}$$

where C is the total follow-up time for the study or a prefixed time point. The parameter estimate $\hat{\beta}_L$ in this model can usually be obtained from maximizing the likelihood function built upon the underlying independent Bernoulli distribution.

On the other hand, according to (2.29), we have

$$S(C) = \{S_0(C)\}^{\exp(\mathbf{X}^T \beta)}. \tag{2.43}$$

Substituting this expression for $S(C)$ into the left-hand side of (2.42), we have

$$\log \frac{1 - S(C)}{S(C)} = \log \left[\{S_0(C)\}^{-\exp(\mathbf{X}^T \beta)} - 1 \right]. \tag{2.44}$$

Taking a Taylor expansion and keeping only the first order term, we have

$$\log \frac{1 - S(C)}{S(C)} \approx \log \frac{1 - S_0(C)}{S_0(C)} - \mathbf{X}^T \beta \frac{\log S_0(C)}{1 - S_0(C)}. \tag{2.45}$$

Thus, we can find the equivalence between the PH model and logistic regression model if we let

$$\alpha = \log \frac{1 - S_0(C)}{S_0(C)}, \qquad \beta_L = -\beta \frac{\log S_0(C)}{1 - S_0(C)}.$$

Furthermore, it can be shown that the asymptotic covariances for the MLEs $\hat{\beta}$ and $\hat{\beta}_L$ are

$$\text{cov}(\hat{\beta}) = \frac{\mathbf{V}}{n(1 - S_0(C))}, \qquad \text{cov}(\hat{\beta}_L) = \frac{\mathbf{V}}{n S_0(C)(1 - S_0(C))},$$

where $\mathbf{V}$ depends on the design matrix and is identical in the two formula.

Therefore, the asymptotic relative efficiency (ARE) of the logistic regression model to the PH model is given by

$$\text{ARE} = \frac{S_0(C)}{(1 - S_0(C))^2} \{\log S_0(C)\}^2.$$

In general, the ARE is smaller than one, indicating that the PH model is more efficient than the logistic regression model, that is, the estimated effect has a smaller sampling variability. Such a result is intuitively reasonable, considering that the Cox model uses information across the whole time interval, whereas the logistic regression model only uses information at one time point. On the other hand, we note that when C is small (a short follow-up period) and $S_0(C)$ is high (low incidence of event occurrence), the ARE would be very close to one. Also, empirical studies usually find that the estimated coefficients from the two models are similar and thus provide comparable interpretations.

2.3.2　Model Selection

Sometimes it is necessary for investigators to agree on a proper model form. If a model contains too many covariates, many problems may arise, including the following. (i) Too many variables may cause the computation of estimated parameters infeasible. When facing a large number of independent variables, the sample correlation may be rather substantial, even though the variables in nature may be uncorrelated. Such a multicollinearity issue can directly lead to the divergence of the computation algorithm and yield unreasonable estimation results. (ii) A model with numerous covariates may be more difficult to interpret than a parsimonious model. Each regression slope for an independent variable is usually interpreted by holding other independent variables constant. Simply including all variables we collect in the data may introduce a counterfactual model with no explicit conditional interpretation. For example, in a hypertension study, we may collect both the diastolic blood pressure and systolic blood pressure for a patient. Putting both variables in a model can be confusing, since we can almost never modify one while keeping the other fixed. Another similar example is to involve both the body weight and weight of a certain part of the body. (iii) The requirement of estimating too many parameters may decrease the efficiency of estimates and inflate the sampling variability. This is an obvious statistical reason since the influence on variance can immediately be transferred to confidence intervals and hypothesis tests.

On the other hand, we are reluctant to lose available information for all the covariates without any justification. There are major risk variables that we want to examine their associations with the survival outcome; there are treatment variables that we want to investigate its efficacy in a clinical trial; there are important confounding variables (patient characteristics, environmental factors, controlling variables, etc.) that we must include in the model to completely specify the disease prognosis. All these practical considerations and the belief of a *true* model, which is the most comprehensive and saturated, drive us to proceed in an opposite direction and favor a model with a large number of variables.

A balance must be achieved by using statistical principles. To select among various candidate models, we may restrict our attention to nested models. In a hypothetical setting, we may compare two competing models M_1 and M_2 for describing the data: M_1 contains the full set of covariates $\mathbf{X}$, while M_2 contains only part of $\mathbf{X}$, say, without loss of generality $\mathbf{X}_1 = (X_1, \cdots, X_{p_1})$ where $p_1 < p$. We may then carry out a statistical test for the following hypothesis:

$$H_0 : M_2 \text{ is sufficient.} \quad v.s. \quad H_1 : M_1 \text{ must be chosen.} \qquad (2.46)$$

We notice that model M_2 can be regarded as a special case of M_1 where all the coefficients for $(X_{p_1+1}, \cdots, X_p)$ are equal to zero. Therefore, model M_2 is said to be *nested* within model M_1. A common procedure for testing (2.46) is a likelihood ratio test given by

$$2\mathrm{logLik}(M_1) - 2\mathrm{logLik}(M_2), \qquad (2.47)$$

where $\mathrm{logLik}(M_i)$ is the maximized log-likelihood function for model M_i ($i = 1, 2$), i.e., the likelihood function (2.30) where we replace the unknown parameters with their MLEs. When the sample size $n \to \infty$, the likelihood ratio test has a χ^2 distribution with degree of freedom $p - p_1$ under the null hypothesis. Test P-value may then be evaluated under the asymptotic null distribution. However, we notice that in many applications the exact null distribution is more complicated than a χ^2 distribution. As was noted in empirical studies (Stram and Lee, 1994; Satorra and Bentler, 2001; Zhang, 2005) for similar settings, more accurate approximation can be made with a mixture of χ^2 distributions.

If the test result is not statistically significant, the two models will be judged to be equally suitable for describing data. Other things being equal, the model with fewer parameters is usually preferred. On the other hand, a rejection of the null hypothesis indicates that the fitted likelihoods from the two models are significantly different. The test thus encourages the inclusion of additional parameters and the adoption of a more complex model.

Example (Bloodstream Infection). We turn to the BSI example for which we have provided the Cox PH model fit in the previous section. We note that Sex, Any co-morbidity, and Length of stay are not significant at the significance level 0.10. We thus may want to compare the full model with a reduced model containing all other variables except these three.

The log-likelihood for the full model (M_1) is -158.29 while that for the reduced model (M_2) is -160.09. The likelihood ratio test is thus equal to 3.593 which follows a χ^2 distribution with 3 degrees of freedom. The P-value is obtained to be 0.3089, indicating that the two models are not significantly different from each other. Therefore, we can use the simplified model without the three variables and will not lose a significant amount of information. The fitted results for model M_2 are summarized in Table 2.6. The estimates are close to those under the full model.

TABLE 2.6: Fitted Results from the Proportional Hazards Model with a Reduced Set of Covariates for the BSI Data

Variable	$\hat{\beta}$	Standard Error	$\exp(\hat{\beta})$	P-Value
Days of carbapenem	−0.072	0.035	0.930	0.0386
Age	−0.018	0.011	0.981	0.0823
Ventilation > 96 hours	0.932	0.413	2.539	0.0242
Malignancy	0.901	0.383	2.464	0.018

Practitioners may find the above test to be limited because it can be used to compare only a pair of nested models. An alternative method is to use data to select an appropriate model based on some optimality criterion. Under this setup, all plausible combinations of predictor variables are competing against each other, and the one that optimizes the criterion is selected. This idea follows a standard practice in optimal design literature where the design is chosen based on some predefined optimality criterion or criteria.

We focus on two criteria for model selection: Akaike's information criterion (AIC) and Schwartz's Bayesian information criterion (BIC). These model selection tools have been extensively used in the literature for a large number of applications including simple regression analysis, analysis of longitudinal data, generalized linear regression, and others. Each criterion selects the optimal model with the smallest criterion value. The two criteria are defined by

$$\text{AIC:} \qquad -2\log\text{Lik} + 2\#, \qquad\qquad (2.48)$$

$$\text{BIC:} \qquad -2\log\text{Lik} + \#\log n. \qquad\qquad (2.49)$$

Here, "#" denotes the number of unknown parameters of the model, and logLik is the maximized log-likelihood function.

AIC and BIC were proposed based on entirely different objectives. AIC uses available information to identify the most suitable model for approximating the observed data with a fixed sample size, while BIC tends to select the model that can fully adapt to the distribution of data as sample size approaches infinity. Philosophical debates on the two criteria were provided at length in Burham and Anderson (2004).

There are additional variants of these criteria in the literature. For example, Atkinson (1988) proposes using "3#" instead of "2#" in the AIC definition. There are still quite a few other ways to define BIC as mentioned in Li and Wong (2010) and literature cited therein. Specifically, with the Cox PH model, Kass and Wasserman (1995) note that the "sample size" n, which appears in the penalty term of BIC definition, must be carefully chosen. In censored-data models, subjects contribute a different amount of information to the likelihood function, depending on whether or not they are censored. Volinsky and Raftery (2000) find that in the penalty term of BIC, substituting d, the number of uncensored events, for n, the total number of individuals, results in an improved criterion without sacrificing the asymptotic properties shown by Kass and Wasserman (1995).

Many of these criteria values are available in commonly used statistical packages, such as R, STATA, and SAS. These various definitions do not usually matter much in practice because it is the difference in the criterion values between models that are of interest. We may adopt empirical guidelines to interpret the severity of the difference: absolute difference in the BIC values between two models of 2 units or less is weak, between 2 to 6 is positive, 6 to 10 units is strong, and higher than 10 is very strong.

Different versions of AIC and BIC criteria do not always result in the same best model. In general, BIC places more penalty on models with a larger number of parameters than AIC. Consequently, we can achieve a relatively more concise description of the covariates dependence using BIC. There are other theoretical and numerical results that suggest BIC performs better than AIC in finding the true model (Shao, 1997). For this reason, we may favor the model selected by BIC over that selected by AIC when the two criteria do not pick the same model. Our preference of BIC over AIC is not because of the difference between frequentist and Bayesian's opinions (see Section 4.4). In fact, even AIC can be associated with a savvy prior and retain a Bayesian interpretation.

The choice between AIC and BIC is mainly conceived as an art by most practitioners. Still there are some theoretical discussions on an objective preference. It is now agreed among some statisticians that AIC has an asymptotic optimal property when the true model is infinite dimensional, while BIC is consistent when the true model is finite dimensional (Shao, 1997; Yang, 2007). Because in practice it is unknown whether a true model is nonparametric or parametric, efforts can be spent on detecting which state is more likely by using some mathematical index (Liu and Yang, 2011).

When the likelihood ratio test and model comparison criteria are appropriate for choosing between nested covariate structures, it is desirable that conclusions be based on both the test and the model selection results. Wolfinger (1996) discusses these two procedures for linear regression models, and one can extend the reasoning therein to the Cox PH regression models as well. However, we note that the null distribution for the likelihood ratio test may differ from the nominal χ^2-distribution and consequently produce p-values from a misspecified reference distribution. Therefore, when there are conflicting results from the two procedures, we base our conclusions on the model selection criteria.

In addition to BIC, there are many other Bayesian methods for model selection in the literature, including the evaluation of posterior model probabilities

(Ibrahim and Chen, 1998), the conditional predictive ordinate statistics (Dey et al., 1997), and the model averaging approach (Taplin and Raftery, 1994). The implementation of these special selectors may be relatively more difficult than the simple information criterion while some of them have appeared in standard packages. These Bayesian methods are proved to offer theoretical and practical advantages in a wide range of problems. As they are less popular, further details are omitted.

In Chapter 6, with extremely high-dimensional covariates, we discuss various variable selection approaches based on the notion of penalization. With low-dimensional data, both approaches discussed in this section and those in Chapter 6 are applicable. Of note, although they all enjoy a certain asymptotic consistency property, with practical data, their results do not necessarily match. With high-dimensional data, the approaches described above may not be applicable. For example, when the number of covariates is larger than the sample size, the likelihood ratio test cannot be computed. Thus we may have to resort to the approaches described in Chapter 6.

2.3.3 Model Performance

In linear regression analysis, model diagnostics is usually conducted using residuals. With the Cox model, we can assess the overall goodness-of-fit of the model using the Cox-Snell residuals (Cox and Snell, 1968; Crowley and Hu, 1977; Collet, 1994) defined by

$$c_i = \delta_i + e^{\mathbf{x}_i^T \hat{\beta}} \log\{\hat{S}_0(y_i)\}, \tag{2.50}$$

for the ith subject, where δ_i is the censoring indicator, y_i is the observed survival time, $\hat{\beta}$ is the estimated regression coefficient, and $\hat{S}_0(t)$ is the estimated baseline survival function given in (2.35). We note that c_i can alternatively be defined as $\delta_i + \log \hat{S}_i(y_i)$ where $\hat{S}_i(t)$ is the model-based estimation of the survival function given in (2.36). The construction of (2.50) actually has a support from probability theories and may be motivated from the martingale introduced later in Section 2.5. The Cox-Snell residual is sometimes also called the martingale residual in the literature.

When the model fitted to the observed data is overall satisfactory, the Cox-Snell residuals follow a unit exponential distribution with density function $f(t) = e^{-t}$. This is due to the fact that if T follows a survival distribution $S(t)$, then the random variable $-\log S(T)$ follows a unit exponential distribution. If the survival time for an individual is right censored, then the corresponding value of the residual c_i is also right censored. The set of residuals $\{c_i\}$ can thus be regarded as a censored sample from the unit exponential distribution. A common diagnostic procedure is thus to form a KM estimator for c_i to be $\hat{S}_c$ and plot $-\log \hat{S}_c(c_i)$ against c_i. One should observe a straight line in this plot if the original Cox model is appropriate.

We next focus on checking the assumption of the model. Proportional hazards assumption nicely allows us to separate the effects of time and covariates in the estimation procedure. When this assumption fails for the jth covariate X_j, the effects of such a covariate may be different for survival probabilities at different time points. To check this assumption, we may construct a time-dependent coefficient as

$$\beta_j^*(t) = \beta_j + \gamma_j g(t), \tag{2.51}$$

where β_j and γ_j are unknown parameters and $g(t)$ is a known function of time t. As the slope for X_j, now β_j^* is dependent on time unless $\gamma_j = 0$. Therefore, we may create a new covariate $X_j g(t)$, add it to the original covariate vector $\mathbf{X}$, and refit the Cox PH model. From the fitted results, we may obtain an estimated value $\hat{\gamma}_j$ and its associated significance test P-value. A significant P-value for $\hat{\gamma}_j$ indicates a violation of the PH assumption for the jth covariate. On the other hand, we accept the PH assumption when $\hat{\gamma}_j$ is not significantly different from zero.

Implementation of the above approach demands properly specifying $g(t)$. Common choices for $g(t)$ may include t and $\log t$. Under both choices, the fitting of the Cox PH model involving $X_j g(t)$ is more complicated than what we describe before. The main issue is that the at-risk sets in the denominator of the partial likelihood (2.30) are different for different covariate values now. Fortunately, most software (particularly including R and STATA) allow us to directly create an interaction term in the model statement as if it were just the product of X_j and failure time (while in fact it is not). Despite the relative simplicity of this approach, it can only detect deviation in a certain "direction." Performance of this approach under "misspecification" of $g(t)$ is very complicated.

Example (Bloodstream Infection). We consider checking the PH assumption for the reduced model M_2, which contains only four predictors. We thus create new variables by using the four variables multiplied by t. The fitted results are given in Table 2.7. Examining the interaction terms between time and covariates, we note that the coefficients for Days, Age, and Ventilation are all highly significant although the estimated coefficients are rather small. The validity of PH assumption for this data set is thus debatable.

In addition to the numerical approach, we may also resort to a graphical diagnostic tool by creating a residual plot from the fitted regression model. When the PH assumption does not hold, we may expect to detect it from the residual plot.

There are many suggested residuals in the literature (Collet, 1994; Hosmer and Lemeshow, 1999). The residuals function in the survival package in R allows eight different residuals to be evaluated. We introduce the following scaled Schoenfeld residual (Schoenfeld, 1982; Hosmer and Lemeshow, 1999), which has proved to be useful to check the PH assumption in empirical studies. The Schoenfeld residual for the ith covariate of the jth subject whose observed

TABLE 2.7: Fitted Results from the Proportional Hazards Model with Interaction between Time and Covariates for the BSI Data

Variable	$\hat{\beta}$	Standard Error	$\exp(\hat{\beta})$	P-Value
Days of carbapenem	0.1322	0.0621	1.1414	0.0334
Age	0.0521	0.0197	1.0534	0.0084
Ventilation > 96 hours	2.38	0.76	10.86	0.0017
Malignancy	0.1877	0.6668	1.2065	0.7782
t:Days of carbapenem	−0.0067	0.0025	0.9934	0.0105
t:Age	−0.0036	0.0008	0.9964	< .0001
t:Ventilation > 96 hours	−0.0663	0.0294	0.9358	0.0243
t:Malignancy	0.0201	0.0322	1.0203	0.5329

survival time is t_j is given by

$$\hat{s}_{ij} = x_{ij} - \frac{\sum_{l \in R(t_j)} x_{il} \exp(\mathbf{x}_l^T \hat{\boldsymbol{\beta}})}{\sum_{l \in R(t_j)} \exp(\mathbf{x}_l^T \hat{\boldsymbol{\beta}})}.$$

For subjects with censored survival times, we do not define their Schoenfeld residuals. It can be easily verified that $\sum_j \hat{s}_{ij} = 0$ for each i, and thus each $\hat{s}_{ij}$ indeed acts like a residual. Denote the covariance matrix of $\hat{\mathbf{s}}_j = (\hat{s}_{1j}, \cdots, \hat{s}_{pj})$ by Σ and its estimator by $\hat{\Sigma}$. Then each component of the transformed residual $\hat{\mathbf{s}}_j^* = \hat{\Sigma}^{-1} \hat{\mathbf{s}}_j$ is called the scaled Schoenfeld residual.

When the scaled Schoenfeld residual $\hat{s}_{ij}^*$ is obtained from the PH model with estimated coefficient $\hat{\beta}_i$ while in fact the true coefficient should be $\beta_i^*(t)$ depending on t, it can be shown that

$$E(\hat{s}_{ij}^* + \hat{\beta}_i) \approx \beta_i^*(t).$$

Thus, we may plot $\hat{s}_{ij}^*$ against t_j, the observed survival time for all subjects whose survival times are completely known. Under the PH assumption, the residuals should scatter smoothly around the horizontal line at 0 for the ith covariate. Any noticeable pattern or trend would indicate a time-varying effect for the covariate.

The scaled Schonfeld residuals for the four predictors in the reduced model M_2 for the BSI example are displayed in Figure 2.6. To help visualize any systematic patterns in the residual plots, a nonparametric lowess smoother (solid line) has been added in each panel along with the 95% confidence band (dashed lines). In all these plots, residuals scatter about zero with some noticeable departures from uniformity. In particular, the residual plot for Days shows that the residual increases with time when time is small and then levels off around zero. Such a pattern appears in other plots as well, suggesting that the proportional hazards assumptions may not hold for these variables. Such graphical diagnostic results agree with the numerical results reported in Table 2.7.

In R, the function cox.zph can test the proportional hazards assumption with the scaled Schonfeld residuals. This program is essentially a test for nonzero slope in a generalized linear regression of the scaled Schonfeld residual on chosen functions of time (Grambsch and Therneau, 1994). Such a test is similar to a score test resulted from maximizing the partial likelihood and therefore is different from the Wald test in Table 2.7. The program also offers various choices for the function of time $g(t)$. The same test has been implemented in STATA in the stphtest command. The algorithms that STATA uses are slightly different from the algorithms used by R and therefore the results from the two programs may differ slightly.

Example (Accident). We return to the accident study for university students in Germany. The survival analysis conducted in Section 2.3.1 is based on the proportional hazards assumption. We consider the model without interactions. To check the PH assumption, we use cox.zph to obtain the following output.

```
> day.coxph = coxph(Surv(time,cen)~factor(Gender) + factor(uni-
type) + factor(faculty) + Age , data = acci, method = 'breslow')
```

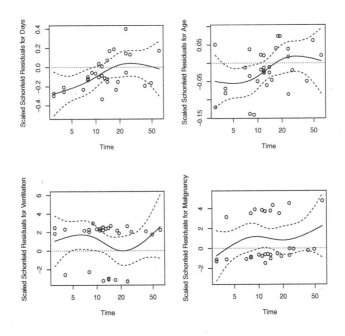

FIGURE 2.6: Graphs of the scaled Schonfeld residuals and their lowess smoothers for the BSI data set.

```
> cox.zph(day.coxph, transform = 'log')
                      rho chisq       p
factor(Gender)2   -0.1139  3.63 0.05665
factor(unitype)2   0.0757  1.56 0.21124
factor(faculty)2  -0.0888  2.06 0.15165
Age               -0.0814  2.13 0.14448
GLOBAL                 NA 13.71 0.00829
```

The $g(t)$ function is set to be $\log(t)$ in this case. Tests for all four individual predictors have P-values greater than 0.05, though the GLOBAL test is rather significant. The GLOBAL test gives an overall assessment on the PH assumption of the model, indicating that at least one of the predictors does not satisfy the proportional hazards assumption. We then plot the scaled Schonfeld residuals for each of the four covariates in Figure 2.7. All residuals except those for Faculty type seem to be very close to zero and show no obvious time-dependent pattern. Judging from these graphs, we may conclude that the PH assumption is reasonably preserved for age, gender, and university type in the accident study. PH assumption may not be appropriate for the Faculty variable.

Remark. We note that in addition to the Schoenfeld residuals, there are also other residuals given in the literature (Barlow and Prentice, 1988; Therneau et al., 1990; Collet, 1994; Hosmer and Lemeshow, 1999). They may be used for various diagnostic purposes. The use of these residuals in the diagnostic graphs helps investigators identify modeling problems as well as outliers or influential observations. For example, martingale residuals (Cox-Snell residuals) can be used to assess model adequacy (available by specifying type='margingale' in function residual in R), while score residuals can be used to detect overly influential observations (available by specifying type='score' in function residual). A remarkable feature of all these residuals is that they are defined in complicated functional forms and no longer a simple difference as the observed response minus the predicted response, as prescribed in simple regression analysis with continuous variables.

In addition to residual analysis, we have another set of tools to assess the overall performance of the Cox model, which is to examine how accurately or how closely the fitted model predicts the actual outcome.

To fix ideas, we note that the linear predictor $\mathbf{X}^T\boldsymbol{\beta}$ in the Cox model is often called a *prognostic index*. One subject's failure tendency due to his/her particular covariate profile is completely incorporated in this index. Two types of prediction accuracy measures can be constructed based on the prognostic index. The first type is the time-dependent receiver operating characteristic (ROC) analysis that will be introduced with great details in Chapter 5 of this book. Simply speaking, we may consider calculating a concordance measure to estimate $P(\mathbf{X}_1^T\boldsymbol{\beta} > \mathbf{X}_2^T\boldsymbol{\beta}|T_1 < T_2)$, where the subscripts indicate two randomly selected subjects from the population and we assume a larger prognostic index indicates a higher risk of failure. If a regression model is helpful explaining the relationship between survival and covariates, we would expect the above conditional probability to be large. See Chapter 5 for more discussions.

The second way to evaluate prediction accuracy is to draw an analogue from simple linear regression and compute a distance measure

$$d\{\mathsf{S}(t_0), \hat{S}(t_0|\mathbf{x})\},$$

where $d(\cdot, \cdot)$ is a well-defined distance function, $\mathsf{S}(t_0) = I(T > t_0)$ is the survival status at time t_0, and $\hat{S}(t_0|\mathbf{x})$ is the estimated survival probability at t_0 for a subject with covariate $\mathbf{X}$ based on the fitted Cox model. The formula of $\hat{S}(t_0|\mathbf{x})$, as is given in (2.36), is also a monotone function of the estimated prognostic index $\mathbf{X}^T\hat{\boldsymbol{\beta}}$. The most popular distance is based on the Euclidean norm, and the so-constructed performance measure is usually referred to as a Brier score (Graf et al., 1999). The sample estimates of such a score is

$$\frac{1}{n}\sum_{i=1}^{n} I(\delta_i = 1 \text{ or } y_i > t_0)\frac{\{\mathsf{S}(t_0) - \hat{S}(t_0|\mathbf{x}_i)\}^2}{\hat{C}(\min(y_i-, t_0)|\mathbf{x}_i)},$$

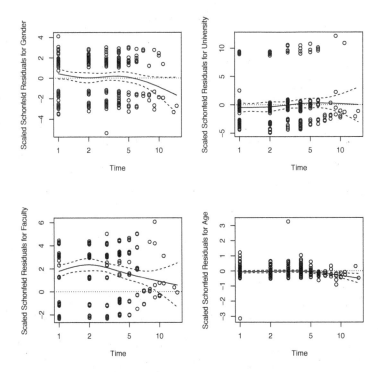

FIGURE 2.7: Graphs of the scaled Schonfeld residuals and their lowess smoothers for the German accident study.

where $\hat{C}(t|\mathbf{x})$ is an estimator of the survival function of the censoring time given $\mathbf{x}$. This kind of inverse probability weighting method is a well-known approach to adjust for censoring (or more generally missingness) in data. A good performance of the regression model can then be reflected in a small Brier score. To avoid overfitting, it is recommended to use a cross-validation version of the Brier score, where $\hat{S}$ is replaced by $\hat{S}_{-i}$ obtained from fitting the model without the ith observation (van Houwelingen and Thorogood, 1995). In addition to quadratic distance, some authors have also considered the absolute value distance and Kullback-Leibler distance (Graf et al., 1999; van Houwelingen and Putter, 2012).

When the PH assumption does not hold, it will be dangerous to still use the Cox PH model, and practitioners usually turn to alternative modeling methods such as the logistic regression model. If we still want to make use of the information about survival time, we may pose some distribution shape conditions and consider parametric model choices introduced in the following section. Sometimes these models appear to fit the analysis goal very well.

2.3.4 Parametric Regression Model

One may be attempted to borrow familiar tools from linear regression models to analyze the association between survival time and covariates. In fact survival time, like all other time variables we encounter in the physical world, is a continuous response so we do not even need to worry about any discreteness issue as involved in generalized linear models. A linear model seems to be a legitimate solution if we ignore three important features of survival time: (i) the time to the development of an event of interest is always positive and cannot take all the numbers in the real line; (ii) the survival time is usually not symmetrically distributed as a normal curve and thus places a constraint on the choice of error distribution; (iii) the survival time cannot be completely observed for the whole sample. To address these three issues, we may resolve by the following three approaches: (i) take a logarithm transformation of the survival time, so that the transformed variable can take any real number; (ii) assume a proper error distribution which comes from a larger class than the normal family; (iii) use a likelihood approach to accommodate censored observations.

Following these ideas, for the ith subject, we can now construct a model similar to the linear regression model as follows:

$$\log(T_i) = \mathbf{X}_i^T \boldsymbol{\beta} + \epsilon_i, \tag{2.52}$$

where ϵ_is are i.i.d. from a known parametric distribution f with mean zero. In this section, $\mathbf{X}_i$ is a $(p+1)$-dimensional vector whose first element is always one, and $\boldsymbol{\beta} = (\beta_0, \beta_1, \cdots, \beta_p)^T$ is the corresponding coefficient vector. In this model, each coefficient β_j $(1 \leq j \leq p)$ indicates directly the change in log survival time for each unit change of X_{ij}. That is, the effect of the covariate on the response is to either accelerate ($\beta_j < 0$) or decelerate ($\beta_j > 0$) the failure time. Model (2.52) is thus commonly referred as an *accelerated failure time* (AFT) model. The model is sometimes also called a log-linear model since it assumes that a log-transformed response variable is linearly dependent on covariates. However, log-linear model is such a general name and has been used by many authors in other places such as referring to the Poisson regression in generalized linear models. To avoid any confusion, we do not use such a term in this book.

To completely specify the AFT model, we need to select a distribution density f for the model error ϵ. Various distributions have been suggested in the literature, including normal distribution, logistic distribution, and Gumbel distribution, among others. The density functions of these three distributions are provided in Table 2.1. We notice that all these three distributions belong to the location scale family of distributions, with a location parameter μ and a scale parameter σ. Usually, the mean of the distribution of ϵ is fixed at zero ($\mu = 0$) for the sake of identifiability when there is an intercept β_0 in the model. Therefore, the parametric AFT model (2.52) involves only $p + 2$ parameters $\boldsymbol{\beta}$ and σ. This shows a parsimonious structure comparing to the semiparametric model where one has to estimate the unknown function $h_0(t)$ at all possible t.

Given a density function f, we may then construct a likelihood function for observed and censored survival times according to model (2.52). For the ith subject, let C_i be the censoring time, $Y_i = \min(T_i, C_i)$, and $\Delta_i = I(T_i \leq C_i)$. Assume that T_i and C_i are independent. For an observed sample $\{Y_i = y_i, \Delta_i = \delta_i, \mathbf{X}_i = \mathbf{x}_i : i = 1, 2, \cdots, n\}$ where δ_i is an indicator of censoring $\delta_i = 1$ or not $\delta_i = 0$, we may write the following likelihood function:

$$L(\boldsymbol{\beta}, \sigma) = \prod_{i=1}^{n} f\left(\frac{\log y_i - \mathbf{x}_i^T \boldsymbol{\beta}}{\sigma}\right)^{\delta_i} \left[1 - F\left(\frac{\log y_i - \mathbf{x}_i^T \boldsymbol{\beta}}{\sigma}\right)\right]^{1-\delta_i}, \quad (2.53)$$

where $F = \int f$ is the cumulative distribution function. Analytic solutions for MLE may not be available but numeric solutions can be obtained without any problem using most statistical packages, such as function `survreg` in R, procedures `streg` in STATA, and `PROC LIFEREG` in SAS.

It is straightforward to use gradient-based numerical procedures such as the Newton-Raphson algorithms to maximize this likelihood function to obtain the MLE $\hat{\boldsymbol{\beta}}$ and $\hat{\sigma}$. Under regularity conditions that are generally satisfied in practice, we can show that these MLEs are asymptotically normally distributed with mean equal to the true parameters and covariance equal to the inverse of the information matrix. Inference for model parameters can then be easily established.

Some AFT models share an interpretation similar to that of the PH model. For example, when the error term in (2.52) follows the one-parameter Gumbel distribution

$$f(t) = \exp\left\{\frac{t}{\sigma} - \exp\left\{-\frac{t}{\sigma}\right\}\right\} / \sigma,$$

we can derive the hazard function of T to be

$$h(t; \mathbf{X}) = \sigma^{-1} t^{\sigma^{-1}-1} \exp\{-(\beta_0 + X_1\beta_1 + \cdots + X_p\beta_p)/\sigma\}. \qquad (2.54)$$

We may define

$$h_0(t) = \sigma^{-1} t^{\sigma^{-1}-1} \exp\{-\beta_0/\sigma\}$$

to be the baseline hazard function which does not involve any covariate. The model based on the hazard function is now expressed in exactly the same form as a PH model

$$h(t; \mathbf{X}) = h_0(t) \exp\{X_1\theta_1 + \cdots + X_p\theta_p\}, \qquad (2.55)$$

where each $\theta_j = -\beta_j/\sigma$. The regression parameter θ_j can thus be interpreted as the log hazard ratio.

This model is sometimes called a Weibull model since the distribution of T given $\mathbf{X}$ is in fact the Weibull distribution. It can be fitted in R by specifying the option `dist="weibull"` when using `survreg`, in STATA by specifying the option `dist(weib)` when using `streg`, or in SAS by specifying the option `distribution=weibull` when using `PROC LIFEREG`.

For parametric models, we can determine the distribution percentile with an analytic form. For example for the Weibull model, we can find the $100P$th percentile as

$$t_P = \{-\log(1-P)\}^\sigma \exp\{\beta_0 + X_1\beta_1 + \cdots + X_p\beta_p\}. \qquad (2.56)$$

Under this expression, each β_j may be interpreted as exactly how much survival time (in log scale) would be increased or decreased for one unit change of the corresponding covariate X_j with a fixed probability P, holding other covariates constant. The sample estimate of t_P can be obtained by replacing all the unknown parameters with their MLEs.

Other choices of parametric distributions may not lead to a PH model but other interesting expressions for interpretation. When the error term in (2.52) follows the one-parameter logistic distribution

$$f(t) = \frac{\sigma^{-1} \exp\{-t/\sigma\}}{(1 + \exp\{-t/\sigma\})^2},$$

we can derive the survival function of T to be

$$S(t; \mathbf{X}) = \left[1 + \exp\left\{\frac{\log t - (\beta_0 + X_1\beta_1 + \cdots + X_p\beta_p)}{\sigma}\right\}\right]^{-1}. \qquad (2.57)$$

We then notice that the odds of a survival time of at least t is

$$\frac{S(t; \mathbf{X})}{1 - S(t; \mathbf{X})} = \exp\left\{\frac{\log t - (\beta_0 + X_1\beta_1 + \cdots + X_p\beta_p)}{\sigma}\right\}. \qquad (2.58)$$

Let $w_0(t) = \exp\{\frac{\log t - \beta_0}{\sigma}\}$ be the baseline odds function. We may rewrite the above equation as

$$\frac{S(t; \mathbf{X})}{1 - S(t; \mathbf{X})} = w_0(t) \exp\{X_1 \theta_1 + \cdots + X_p \theta_p\}. \tag{2.59}$$

The regression coefficient $\theta_j = -\beta_j/\sigma$ may be interpreted as the change in log odds for one unit change of the corresponding covariate. The regression model under this formulation is often called a *proportional odds* model due to the interpretation of its coefficients. We note that in statistics the proportional odds (PO) model also finds its application for modeling ordinal multicategory response variables (Stokes et al., 2000; Lu and Li, 2008; Liu et al., 2009).

In practice, the PO model can be fitted in R by specifying the option dist="loglogistic" when using survreg, in STATA by specifying the option dist(llog) when using streg, or in SAS by specifying the option distribution=llogistic in PROC LIFEREG.

For this model, we can find the $100P$th percentile as

$$t_P = \left\{ \frac{P}{1 - P} \right\}^\sigma \exp\{\beta_0 + X_1 \beta_1 + \cdots + X_p \beta_p\}. \tag{2.60}$$

Similar to the Weibull model, each β_j may be interpreted as exactly how much the survival time (in log scale) would be increased or decreased for one unit change of the corresponding covariate X_j with a fixed probability P, holding other covariates constant. The sample estimate can be obtained by replacing all the unknown parameters with their corresponding MLEs.

It is not hard to see that the distribution of T would be log-normal, log-logistic, or Weibull when the distribution of ϵ is chosen as normal, logistic, or Gumbel, respectively. These three distributions are widely used to model survival data but are by no means the only possible choices. Model (2.52) associates the mean of the log survival time with the regression component $\mathbf{X}^T \boldsymbol{\beta}$. One may alternatively model the distribution of T directly with known distributions and let regression components enter the distribution of T in another sensible manner. This approach turns out to be even more versatile because (i) more distributions can be considered directly for T, including gamma, F, Pareto, Gormpertz, inverse Gaussian and others, which are all excluded in the log-linear type regression model; (ii) the distribution characteristic used to link the regression component can be the survival function, hazard function, mean of the survival time, quantiles of the survival time, or functions of the above quantities. This leads to an enormous number of parametric models for survival time, and it is impossible for us to enumerate all of them in this chapter.

We may exemplify the above idea with one specific approach of using the inverse Gaussian distribution for regression analysis. This model can be motivated by a first-hitting-time (FHT) model with a Wiener parent process equipped with a positive initial value and certain mean and variance parameters (say, μ and σ^2). It is known that the time required for the process to reach the zero level for the first time has an inverse Gaussian distribution if the mean parameter μ is negative so that the process tends to drift toward zero (for this reason μ is also called a drifting parameter). The density of the inverse Gaussian distribution is given by

$$f(t; \mu, \sigma) = (2\pi t^3 \sigma^2)^{-1/2} \exp\{-(1 + \mu t)^2/(2\sigma^2 t)\}, \qquad y > 0. \qquad (2.61)$$

In this case, μ turns out to be the reciprocal of the distribution mean, and σ^2 is the so-called volatility parameter. This distribution can accommodate a wide variety of shapes and is a member of the exponential family.

Previous studies have suggested several approaches to incorporate covariate effects for this distribution. Specifically, we may model the parameter μ as a regression function of the covariates, that is, $\mu_i = \mathbf{X}_i^T \boldsymbol{\beta}$ for the ith subject (Whitemore, 1983; Lee et al., 2010; Li and Lee, 2011) and then construct the parametric likelihood function according to (2.61). Though μ does not have a direct interpretation in terms of the instantaneous failure rate or mean survival time, it does have a direct interpretation in terms of what is happening to the underlying health process. More specifically, the parameter μ for the latent health process suggests how rapidly or slowly the sample path approaches the event threshold. Therefore, modeling how covariates affect μ may be more informative or fundamental in this case.

Another attractive property of the inverse Gaussian distribution is that it can incorporate cure rate models easily. In fact, in the above discussion if the drifting parameter μ is positive, then the probability $P(Y = \infty) = 1 - \exp(-2\mu/\sigma^2)$, which is nonzero. It is thus possible that some individuals may never develop the failure outcome. Such a phenomenon may be observed in practice when some subjects are completely cured or immune from the disease. The inverse gamma distribution thus offers a flexible description for such observations. Detailed discussions of the cure rate model will be provided in Section 4.3.

We hope to offer the following remarks which are applicable to all parametric regression models.

1. All the parametric models are based on explicit assumptions on the probability distributions, and thus likelihood functions can be written similar to (2.53). Depending on the selected distribution, the parameters involved may be more than $p+2$ as in (2.53). Nonetheless, the numerical maximization program can be carried out in the same way, and the theoretical justification for asymptotic consistency and normality follows the same argument lines. Therefore, the inference results can be similarly obtained and interpreted.

2. For all parametric models, it is important to conduct diagnostic procedures to check the distribution assumptions. Graphical tools such as quantile-quantile (QQ) plots are usually helpful for this purpose. Formal statistical tests such as the Kolmogorov-Smirnov or Cramer-von Mises

test (D'Agostino and Stephens, 1986) may also be employed to check if the data distributions are significantly different from the assumed distributions.

3. We note that one specific parametric model may be successful for analyzing one particular data set but completely fail for other data sets. Choosing the proper distribution in a parametric survival model for a given data set is thus an art. Subject matter expert opinions are usually sought after, and historical evidences may also be used to support the model choice. The appropriate model selected should ideally also be easy to interpret and provide an insight into the scientific study.

Finally, the most common normal distribution, though maybe not appropriate for failure time, is widely used in econometrics and social science and censored (or truncated) normal distribution has been thoroughly studied when data are observed with upper or lower limits. Applications of these methods appear occasionally in medicine as well (Li and Zhang, 2011).

2.3.5   Other Nonproportional Hazards Models

There are a large number of alternative models to choose from, when the PH assumption does not hold and the Cox model (2.29) cannot apply. In addition to parametric models given in the previous section, there are also other semiparametric models we may consider.

Extending the log-linear model, we may have a *transformation model* given by

$$g(T_i) = \mathbf{X}_i^T \boldsymbol{\beta} + \epsilon_i, \qquad i = 1, 2, \cdots, n, \tag{2.62}$$

where ϵ_is are the i.i.d. random errors. One possibility is to make a parametric distribution assumption on f, the density function of ϵ_i, but make no parametric assumption on the transformation function g. The function g is only assumed to be increasing and otherwise unspecified, and needs to be estimated. The log-linear AFT model (2.52) may be regarded as a special case of (2.62) when g is taken to be the log transformation. The estimation of (2.62) is usually based on the concordance of ranks of the observed times and $\mathbf{X}_i^T \boldsymbol{\beta}$. See Kalbfleisch (1978b), Dabrowska and Doksum (1988), Cuzick (1988), Fine and Bosch (2000), Zeng and Lin (2006), and Zeng et al. (2009) for more details.

With the model form (2.62), we may make an alternative assumption: g is known with a parametric form, but f is not fully specified. The transformation function $g(.)$ may take a parametric form, such as the well known Box-Cox power transformation

$$g(t; \lambda) = \begin{cases} (t^\lambda - 1)/\lambda & \lambda \neq 0; \\ \log(t) & \lambda = 0. \end{cases}$$

Under such a specification for $g(.)$, when the distribution of ϵ is unknown, estimation methods of model (2.62) have been studied by Han (1987), Newey (1990), Wang and Ruppert (1995), and Foster et al. (2001). Of note, usually some mild assumptions, for example assumptions on moments or a sub-Gaussian distribution, are still needed for the error distribution to ensure satisfactory statistical properties.

The Cox model may be viewed as a multiplicative model in the sense that the effects of different predictors on the hazard function are multiplied on the baseline hazard. In general, we may consider a model

$$h(t; \mathbf{X}) = h_0(t) r(\mathbf{X}^T \boldsymbol{\beta}), \qquad (2.63)$$

where $r(.)$ is a known function that specifies the relative hazards between $h(t; \mathbf{X})$ and the baseline $h_0(t)$. Model (2.29) is a special case of (2.63) where $r(.)$ is chosen to be the exponential function. Other choices have been proposed in the literature. For example, choosing $r(x) = 1 + x$ in the specification of (2.63) leads to the additive relative hazard model. Another suggested function is $r(x) = 1 + \exp\{x\}$. However, due to computation difficulty and practical reasons, these models have not been popular among clinical investigators.

An *additive risk model* or *additive hazard model* takes the form:

$$h(t; \mathbf{X}) = h_0(t) + \mathbf{X}^T \boldsymbol{\beta}. \qquad (2.64)$$

This model has been studied by Breslow and Day (1987) and Lin and Ying (1994a). The regression coefficient in this model is thus interpreted in terms of the change in the hazard function with a unit change of the corresponding covariate. Furthermore, Aalen (1989, 1993) considers more general forms of (2.64) where components of $\boldsymbol{\beta}$ are also time-dependent. Lin and Ying (1994a) show that a closed form solution of regression coefficients can be derived, and their estimates resemble a weighted least squares estimator. One can fit this model by using the function `aalen` in the R package `timereg`.

Unlike many models described above, the additive risk model does not belong to the family of transformation models and does not provide a relative risk interpretation for regression coefficients. The additive risk model is not as popular as the Cox model, partly because with a finite sample the estimate of $\boldsymbol{\beta}$ does not guarantee that the hazard is positive at all time points for all $\mathbf{X}$. This is most relevant when the goal of the analysis is to make predictions for individuals. To ensure a positive hazard, constrained estimation may have to be conducted. In Chapters 3 and 6, we offer a brief review on the estimation strategy under the additive risk model with right and interval censored data.

To reflect its structural difference from the additive model, sometimes the Cox PH model is called a multiplicative model. Though the Cox model is dominant in the current medical research community, other models studied in this section have shown successes here and there, and should not be completely ignored. Which type of model is considered is usually a decision made by the investigators at the design stage. We strongly suggest practitioners examine as many candidates as possible when analyzing a survival data set and consider changing their initial plan when an alternative model form seems more promising. Having said that, we cannot formally choose among the Cox PH model, AFT model, and additive model simply using statistical tests. In practice, we may use methods described in Section 2.3.3 to check if the Cox model is appropriate. There are scenarios where multiple models can pass the check.

2.4 Remarks

In this chapter, we have presented the commonly adopted models and estimation and inference approaches for right censored data. Although the above sections include the majority of techniques studied in other textbooks, certain details (Fleming and Harrington, 1991) and excessive examples (Klein and Moeschberger, 2003) have been omitted due to space consideration. With parametric models for right censored data, estimation and inference can be accommodated using standard parametric likelihood theories. For readers with a basic knowledge of statistical estimation and inference, this should be straightforward. With semiparametric and nonparametric models, properties of most estimation and inference techniques can be established using martingale techniques, for which we provide a brief introduction in the section that follows. Compared to the data and model settings and analysis techniques to be described in the following chapters, right censored data and corresponding techniques have been well investigated, with routine analysis software available in R, STATA, SAS, MATLAB®, and others.

2.5 Theoretic Notes

All statistical procedures must come with a justification. Two necessary large sample properties are usually desirable: consistency and asymptotic normality, particularly for the estimates of parametric parameters. Consistency means convergence in probability and ensures that the estimator is accurate to the target. Asymptotic normality allows us to construct confidence intervals and to easily perform hypothesis testing.

For parametric methods, the proof of consistency and asymptotic normality follows from the well-known parametric likelihood theory (Shao, 1999). Essentially, attention is focused on the derivatives of the log-likelihood function with respect to the parameters. The first derivative of the log-likelihood function is also referred to as the *score* function, and setting the vector of score functions to zero leads to a set of estimating equations. The solutions are the maximum likelihood estimators of the parameters. The second derivative of the log-likelihood is useful too, and its negative expectation is called the *information* matrix. Now because the score vector is a sum of independent random variables, it is asymptotically normally distributed under regularity conditions. By using the functional mapping theory, we can obtain the asymptotic normality of the MLE as well. Straightforward calculation reveals that the expectation of covariance matrix of the MLE is the inverse of the information matrix.

Though it can be shown that the KM estimator is indeed a nonparametric MLE (NPMLE) by a clever construction (Kalbfleisch, 1978a), its asymptotic property cannot be easily justified by using the arguments in the previous paragraph. More generally, for nonparametric and semiparametric methods such as the KM estimator, NA estimator, log-rank test, and Cox regression, the proof of consistency and asymptotic normality for uncensored and right censored data is traditionally established by applying the counting process and martingale theory. The main technical tool is Rebolledo's martingale central limit theorem (Rebolledo, 1980). We supply a few preliminary notes related to this chapter in the following pages. The complete derivation requires a rather lengthy layout of mathematical elements including the definition of martingales and its properties, and is omitted in this book. See Fleming and Harrington (1991) for more details.

Empirical process theory (Kosorok, 2008) has become another school of theoretical tools that can be used to prove the asymptotic properties of nonparametric and semiparametric methods. The consistency of empirical distribution function such as (2.1) is justified by the Glivenko-Cantelli theorem, and all the functions with such a consistent property belong to the Glivenko-Cantelli class of functions. The asymptotic normality of the empirical distribution function is justified by the Donsker theorem, and all the functions with the property of being asymptotically normal belong to the Donsker class of functions. Therefore, in order to show the consistency and asymptotic normality of the nonparametric/semiparametric estimators such as those given in this chapter, we may only argue that the estimators are functions of the empirical process and the functions belong to the Glivenko-Cantelli and Donsker classes. The verification of the class membership requires some mathematical skills on functional analysis and combination theory, such as computation of the bracketing number and evaluation of Hadamard derivatives of functions (Kosorok, 2008). However, many important results have been established, and most users of the empirical process theory can rely on such results to establish the Glivenko-Cantelli and Donsker properties of the nonparametric/semiparametric estimators. Therefore, the arguments based on empirical process theory turn out to be much more succinct than those based on counting process and martingale theory. Empirical process theory is relevant with right censored data. However, with counting process and martingale techniques, it is not irreplaceable. With interval censored data, empirical process theory can be the only applicable tool. More relevant discussions are available in Chapter 3.

Nonparametric MLE. We assume the same settings given before equation (2.2) and show that the KM estimator is an NPMLE. Define $n_0 = n, d_0 = 0, t_0 = 0$, and denote n_{k+1} to be the number of remaining subjects beyond the largest failure time t_k. In fact, the outcome at each failure moment t_j ($j \leq k$) may be regarded as a binomial random variate with a failure probability $S(t_j-) - S(t_j)$. Since we have d_j failures and $n_{j+1} - n_j + d_j$ censored subjects between $[t_j, t_{j+1})$, the likelihood function is formed as

$$\mathcal{L} = \prod_{j=0}^{k} \{S(t_j-) - S(t_j)\}^{d_j} \prod_{i=1}^{n_{j+1}-n_j+d_j} S(c_{ji}), \qquad (2.65)$$

where c_{ji} is the censoring time for the ith subject censored in $[t_j, t_{j+1})$. It is obvious that the maximizer of (2.65) is a discontinuous function with jumps at each failure time t_j, and $S(c_{ji})$ is maximized by taking $S(c_{ji}) = S(t_j)$. Therefore, the estimation amounts to finding the survival function for a discrete random variable. Further, we notice that for a discrete failure time

$$S(t_j-) = \prod_{i=1}^{j-1}[1 - h(t_i)], \qquad (2.66)$$

$$S(t_j) = \prod_{i=1}^{j}[1 - h(t_i)], \qquad (2.67)$$

where the hazard $h(t)$ is simply the conditional failure probability $P(T = t | T \geq t)$. Substituting these expressions into (2.65), we have

$$\mathcal{L} = \prod_{j=0}^{k} h(t_j)^{d_j} \{1 - h(t_j)\}^{n_j-d_j}, \qquad (2.68)$$

and obtain the MLE as $\hat{h}(t_j) = d_j/n_j$. This immediately leads to the KM estimator.

Partial Likelihood. In a sample of size n, the ith subject has a time T_i to failure and a time C_i to censoring. We observe $Y_i = \min(T_i, C_i)$, $\Delta_i = I(T_i \leq C_i)$, and covariates $\mathbf{X}_i$. Under conditional independent censoring, up to a constant, a complete likelihood function of the data $\{Y_i = y_i, \Delta_i = \delta_i, \mathbf{X}_i = \mathbf{x}_i\}$ is given by

$$\mathcal{L} = \prod_{i=1}^{n} f(y_i|\mathbf{x}_i)^{\delta_i} S(y_i|\mathbf{x}_i)^{1-\delta_i}$$

$$= \prod_{i=1}^{n} h(y_i|\mathbf{x}_i)^{\delta_i} S(y_i|\mathbf{x}_i),$$

where $f(t|\mathbf{x})$, $S(t|\mathbf{x})$, and $h(t|\mathbf{x})$ are the conditional density function, conditional survival function, and conditional hazard function of T given $\mathbf{x}$, respectively. Under the Cox model (2.29), we may write the likelihood as

$$\mathcal{L} = \prod_{i=1}^{n} \left[h_0(y_i) \exp(\mathbf{x}_i^T \boldsymbol{\beta}) \right]^{\delta_i} \exp\{-H_0(y_i) \exp(\mathbf{x}_i^T \boldsymbol{\beta})\},$$

where $H_0(\cdot)$ is the cumulative baseline hazard function.

Following Breslow's argument (Cox, 1975), we consider the so-called least informative nonparametric modeling for $H_0(\cdot)$, in which $H_0(t)$ has a possible jump h_j at the observed failure time t_j $(j = 1, \cdots, m)$. More precisely, we write $H_0(t) = \sum_{j=1}^{m} h_j I(t_j \leq t)$. Then we have

$$H_0(y_i) = \sum_{j=1}^{m} h_j I(i \in R(t_j)). \tag{2.69}$$

Using (2.69), we can obtain a profile log-likelihood for $\mathbf{h} = (h_1, \cdots, h_m)$ given $\boldsymbol{\beta}$ as

$$l(\mathbf{h}|\boldsymbol{\beta}) = \sum_{j=1}^{m} \log(h_j) + \sum_{i=1}^{n} \delta_i \mathbf{x}_i^T \boldsymbol{\beta}$$

$$- \sum_{i=1}^{n} \left[\sum_{j=1}^{m} h_j I(i \in R(t_j)) \exp(\mathbf{x}_i^T \boldsymbol{\beta}) \right]. \tag{2.70}$$

Taking the derivative with respect to h_j and setting it to be zero, we obtain that

$$\hat{h}_j = \left[\sum_{i \in R(t_j)} \exp(\mathbf{x}_i^T \boldsymbol{\beta}) \right]^{-1}. \tag{2.71}$$

Replacing h_j with $\hat{h}_j$ in (2.70), we then obtain the log-partial likelihood, the logarithm of the partial likelihood (2.30), after dropping a constant term $-m$.

Counting Process. Using notations introduced earlier, we may define the observed counting process for the ith subject to be

$$N_i(t) = I(Y_i \leq t, \Delta_i = 1). \tag{2.72}$$

The counting process $\{N_i(t), t \geq 0\}$ counts the number of events on the subject in the interval $(0, t]$ by jumping from 0 to 1 at the observed event time. We denote $d N_i(t) = N_i\{(t + dt)-\} - N_i(t-)$ to be the number of events in the interval $[t, t + dt)$. The actual underlying counting process $\tilde{N}_i(t) = I(T_i \leq t)$ for the failure time T is related to $N_i(t)$ via

$$N_i(t) = \int_0^t A_i(u) \, d\tilde{N}_i(t), \tag{2.73}$$

where $A_i(t) = I(Y_i \geq t)$ is the *at-risk* process for the ith subject, indicating whether the subject is still being followed at time t. This function registers a jump from 1 to 0 when follow-up ends due to event or censoring.

We now define the *counting process martingale* as

$$M_i(t) = N_i(t) - \int_0^t A_i(u) h_i(u) \, du, \qquad t \geq 0, \tag{2.74}$$

where $h_i(t)dt = dH_i(t) = P(T_i \in [t, t+dt] | T_i \geq t)$. Using the above definitions, we can easily verify that

$$E\{dM_i(t) | \mathcal{F}_{t-}\} = E\{dN_i(t) - A_i(t)h_i(t)dt | \mathcal{F}_{t-}\} = 0 \tag{2.75}$$

for all t, where $\mathcal{F}_{t-}$ is the *filtration* or history and mathematically equivalent to the σ–field generated by the observed data up to but not including t. Rearranging (2.74), we obtain a decomposition of $N_i(t)$ as

$$N_i(t) = \int_0^t A_i(u) h_i(u) \, du + M_i(t), \tag{2.76}$$

where the first term on the right hand side is usually termed as the *compensator*. This representation draws an analogy between the martingale and the regression residual, since the compensator plays a role as the systematic component of $N_i(t)$.

There are well-established probability theories for the square integrable martingale (see for example Chapter 4 of Durrett, 2005) and they can be readily used to validate the inferences for survival data analysis after adopting the counting process notations. See Andersen and Gill (1982), Fleming and Harrington (1991), and a few other excellent texts for more details.

Equation (2.74) also suggests a way for model diagnostics, as discussed in Section 2.3.3. In fact, when covariates are present and the Cox PH model is fitted to the data, we can estimate the compensator process for the ith subject by

$$A_i(t) \hat{H}(t; \mathbf{x}_i) = -A_i(t) e^{\mathbf{x}_i^T \hat{\boldsymbol{\beta}}} \log\{\hat{S}_0(t)\}. \tag{2.77}$$

The Cox-Snell residual for the subject, given in (2.50), is simply the martingale process evaluated at the observed y_i.

2.6   Exercises

1. TRUE or FALSE?

   (a) The Kaplan-Meier estimator is an unbiased estimator of the survival function.

(b) In survival analysis, if a subject is still alive at the end of the study, the censoring time is considered independent of the survival time.

(c) A random variable X follows an inverse Gaussian distribution if X^{-1} follows a Gaussian distribution.

(d) The Cox-Snell residuals can be used to check the goodness-of-fit of a parametric regression model.

2. Give short answers to the following questions.

(a) Briefly describe the meaning of consistency.

(b) What is a partial likelihood?

(c) What is a score function?

(d) What is AIC and how do we use it in practice?

(e) How will you check the PH assumption?

3. Verify that in the absence of censoring, the KM estimator (2.2) reduces to the empirical survival function (2.1). Show that Greenwood's formula (2.3) reduces to the usual binomial variance estimate in this case.

4. For the following small data set of survival times: 3, 4, 5+, 6, 6+, 8+, 11, 14, 15, 16+, where "+" indicates a right censored survival time, do the following:

(a) Find the KM estimate of the survival function and its variance.

(b) Use the above KM estimate to get an estimate and its variance of the cumulative hazard function.

(c) Find the Nelson-Aalen estimate of the cumulative hazard function and its variance.

(d) Find an estimate and its variance of the survival function using the Nelson-Aalen estimate obtained in the above step.

5. Simulation is usually used in statistical research to assess the finite sample performance of various new estimators and procedures. It is easy to simulate random samples from a specified distribution using R.

(a) Use the function rexp to generate a random sample of size $n = 100$ from the exponential distribution with unit mean. Denote them by **C**.

(b) Use the function rmvnorm to generate a random sample of size $n = 100$ from the three-dimensional normal distribution with mean zero and identity covariance. Denote them by **X**.

(c) Use the function `rweibull` to generate a random sample of size $n = 100$ from a Weibull distribution whose hazard is given by the following PH model

$$h(t) = h_0(t) \exp\{\mathbf{X}^T \boldsymbol{\beta}\}, \tag{2.78}$$

where $h_0(t) = t/2$, $\mathbf{X}$ is the same as that simulated in part (b) and $\boldsymbol{\beta} = (1, 1.5, 2)^T$. Denote the simulated times by $\mathbf{T}$.

6. For the simulated samples from the previous problem, treat $\mathbf{C}$ as random censoring times and $\mathbf{T}$ as the true survival times. Create the actually observed data by evaluating $\mathbf{Y} = \min(\mathbf{C}, \mathbf{T})$ and $\boldsymbol{\Delta} = (\mathbf{C} > \mathbf{T})$. Fit a Cox PH model with R function `coxph` using the simulated data $\mathbf{Y}, \boldsymbol{\Delta}$, and $\mathbf{X}$. Inspect the bias of the regression coefficient estimates.

7. Repeat the previous two problems but with a larger sample size $n = 400$. Compare your findings.

8. One right censored observation consists of $(min(T, C), I(T \le C), Z)$, where T, C, and Z denote the event time, censoring time, and covariate, respectively. Consider the following small data set (8, 1, 3), (7, 1, 4), (9, 0, 5), and (10, 1, 6). Assume the Cox model. Denote β as the unknown regression coefficient.

   (a) Write the partial likelihood of β.

   (b) Plot the log partial likelihood of β in $[-8, 3]$, and see if this function is concave.

   (c) Find $\hat{\beta}$ that maximizes the log partial likelihood function by hand, and calculate the second-order derivative of the log partial likelihood function at $\hat{\beta}$.

   (d) Use software to fit the Cox model. Compare results from (c).

9. Patients suffering from chronic active hepatitis rapidly progress to an early death from liver failure. A new treatment has been suggested, and a clinical trial is designed to evaluate its effect. Specifically, the new treatment is expected to increase the survival probability at five years from 0.38, the value under the current standard practice, to 0.63. Take type I error $\alpha = 0.05$ and power $1 - \beta = 0.85$. What is the sample size for the required death if the treatment allocation ratio is 4 : 6 for the treatment arm versus the control arm?

10. Interpret the following counting process notations: $dN_i(t)$, $dN_i(t) - A_i(t)h_i(t)dt$, $\int_0^t \frac{dN_i(u)}{\sum_{j=1}^n A_j(u)}$, $\int_0^t \frac{\sum_{i=1}^n dN_i(u)}{\{\sum_{j=1}^n A_j(u)\}^2}$.

Chapter 3

Analysis of Interval Censored Data

In the first two chapters, we focused on the analysis of right censored data, which is the most typical censored survival data in biomedical studies. In addition to right censoring, the loss of information in a follow-up study may produce other types of incomplete data structure. We devote this chapter to the investigation of interval censored data, which may arise in many practical studies and pose a distinct challenge to data analysts. It should be noted that many aspects of interval censored data analysis, for example parametric methods, are very similar to those for right censored data. To avoid redundancy, such aspects will be briefly mentioned. Unlike some of the existing textbooks on survival analysis, this chapter aims to offer a more rigorous account of some theoretical aspects of interval censored data. The theoretical study of interval censored data can be more challenging than with right censored data, mainly because martingale techniques are no longer applicable, and advanced empirical processes techniques have to be adopted. We recognize the importance of the theoretical investigation, at the same time, we also acknowledge that it can be difficult for less theory-oriented readers. For those who are more interested in applications, you may selectively choose sections within the chapter and also refer to other textbooks, in particular Sun (2006). For those who are only interested in right censoring, it is also possible to completely skip this chapter. The notations in this chapter are mostly self-contained in order to be consistent with published papers in this area.

We first want to clarify that we talk about interval censoring in a narrow sense. Loosely speaking, an exact observation of failure time discussed in the previous chapters can be regarded as an interval censored data point when the status changes from event-absent to event-present. In practice, medical observations are taken continuously on subjects hourly, daily, weekly, or monthly, and an event has to be observed between two adjacent examination time points. When the two end points are close relative to the overall follow-up period, such an interval censored data point is routinely treated as an exact observation. What we discuss in this chapter will be interval censored data where the interval length cannot be neglected, and the data are not reducible to exact failure time records.

Early examples of interval censored data arose in demographic applications, with a common version occurring in studies of the distribution of the age at weaning in various settings (Diamond et al., 1986; Grummer-Strawn, 1993). Here, that is, the event time of interest is the time of weaning. However, such an event is usually not accurately documented. Quite often what is available is the *weaning status (that is, weaning or not)* at the time of observation. The event time is right censored, if weaning has not occurred at the time of observation. However, for a subject that has experienced weaning at the time of observation, the event time is only known to lie in the interval with the left end being time zero and the right end being the time of observation.

Another area of application of interval censored data is in carcinogenicity testing when a tumor under investigation is occult (Gart et al., 1986). Consider a typical animal oncological study. The event of interest is the development time of tumor under a certain "treatment" condition, for example, radiation at a certain dose. However, the tumor development time cannot be accurately observed. Often an animal has to be sacrificed in order to determine its status of tumor. Thus, what is available is the time of sacrifice and tumor status (event indicator) at that time. Here the interval censoring nature of data is similar to that described above.

The next example arises naturally in the study of infectious diseases, particularly when infection is an unobservable event, that is, one with often no or few clinical indications. The prototypical example is infection with the human immunodeficiency virus (HIV). A representative study is the California partner study of HIV infection (Jewell and Shiboski, 1990; Shiboski, 1998). The most straightforward partner study occurs when HIV infection data are collected on both partners involved in a long-term sexual relationship. These partnerships are assumed to include a primary infected individual who has been infected via an external source, and a susceptible partner who has no other means of infection other than sexual contact with the index case. Here, the event time of interest is the time from infection of the infected case to the infection of the susceptible partner. As the occurrence of infection has no clinically observable indication, the event time cannot be accurately observed. Its information is only accessible via medical examinations. At the time of examination, there are three possibilities for the event status. The first is that infection has already occurred by the time of the first examination, that is, infection occurs between time zero and the first examination time. The second is that infection happens between two examinations. And, the third is that infection has not happened by the last examination. In any case, the event time is not accurately observable and only known to lie in an interval.

Interval censored data also appear in the estimation of the distribution of age at incidence of a nonfatal human disease for which the exact incidence time is usually unknown, although accurate diagnostic tests for prevalent disease are available. When a cross-sectional sample of a given population receives such a diagnostic test, for each subject, the event can be determined to happen prior to or after the diagnostic time. When a longitudinal sample is available, the event time may be known to lie between two adjacent diagnostic times.

Interval censoring is a well-investigated area of survival analysis. We intend to cover some important methodological results in this chapter. As is clear from the previous chapters, there are usually two main aspects of survival data analysis. The first is the modeling of event time distribution, for which a large number of parametric, nonparametric, and semiparametric models have been developed. The second aspect is the estimation and inference procedure, followed by rigorous theoretical justifications.

For data with no censoring, right censored data, and interval censored data, usually it is assumed that censoring and the event are (conditionally) independent. Under the independence condition, different types of data have no difference in the event-generating mechanisms. Thus, the same models as described in Chapter 2 can be used to model interval censored data in this chapter. In particular, we may notice that parametric models play a relatively more important role for this type of data than for right censored data. Since interval censoring causes a more severe problem of missing information, non-parametric modeling purely based on empirical samples may suffer a lack of efficiency and generate less useful inference. On the other hand, analysis under an acceptable parametric assumption may yield simple and interpretable modeling results for interval censored data.

Because of the major differences in censoring schemes, the estimation and inference procedures are more complicated, and their justifications differ from what we present in Chapter 2. In fact, if all the observations in a sample of size n are interval censored, the nonparametric maximum likelihood estimate (NPMLE) of the failure time distribution can only approach the true distribution function at the $n^{1/3}$ rate, which is considerably slower than the usual $n^{1/2}$ convergence rate. The implication of this asymptotic result is that a finite-sample confidence interval can be wider, and such imprecise estimation results may be less convincing to determine the underlying distribution. Fortunately, several fine-tuned procedures whose details come later in this chapter may more or less ameliorate this problem.

The remainder of this chapter is organized as follows. We first provide descriptions and examples of case I and case II interval censored data. Parametric models are then described, followed by estimation and inference under nonparametric modeling. We then consider semiparametric models under case I and case II interval censoring separately. Special attention is devoted to the study of a Cox model under case I interval censoring, which may serve as the prototype for the study of other semiparametric models. Some technical details are provided in the Appendix (Section 3.8).

3.1   Definitions and Examples

3.1.1   Case I Interval Censored Data

Denote T as the event time of interest. Denote C as the examination or observation time (censoring). Under case I interval censoring, one observation consists of

$$(C, \delta = I(T \leq C)),$$

where $I(\cdot)$ is the indicator function. That is, the only information about the event time T is whether it has occurred before C or not. In the literature, case I interval censored data has also been commonly referred to as *current status data*. The two terminologies have no difference in implication.

A representative example is provided by Hoel and Walburg (1972), which described a lung tumor study involving 144 male RFM mice. The mice were randomly assigned to either a germ-free (GE) or conventional environment

TABLE 3.1: Death Times in Days for 144 Male RFM Mice with Lung Tumors

Group	Status	Death Times
CE	Tumor	381, 477, 485, 515, 539, 563, 565, 582, 603, 616, 624, 650, 651, 656, 659, 672, 679, 698, 702, 709, 723, 731, 775, 779, 795, 811, 839
	No tumor	45, 198, 215, 217, 257, 262, 266, 371, 431, 447, 454, 459, 475, 479, 484, 500, 502, 503, 505, 508, 516, 531, 541, 553, 556, 570, 572, 575, 577, 585, 588, 594, 600, 601, 608, 614, 616, 632, 632, 638, 642, 642, 642, 644, 644, 647, 647, 653, 659, 660, 662, 663, 667, 667, 673, 673, 677, 689, 693, 718, 720, 721, 728, 760, 762, 773, 777, 815, 886
GE	Tumor	546, 609, 692, 692, 710, 752, 773, 781, 782, 789, 808, 810, 814, 842, 846, 851, 871, 873, 876, 888, 888, 890, 894, 896, 911, 913, 914, 914, 916, 921, 921, 926, 936, 945, 1008
	No tumor	412, 524, 647, 648, 695, 785, 814, 817, 851, 880, 913, 942, 986

(CE). The purpose of this study was to compare the time from the beginning of the study until the time to observe a tumor. Case I interval censoring occurs, as lung tumors are nonlethal and cannot be observed before death in these RFM mice. The data, consisting of group assignment, death time, and tumor status at death, are shown in Table 3.1.

3.1.2 Case II Interval Censored Data

With case II interval censored data, it is only known that T has occurred either within some random time interval, or before the left end point of the time interval, or after the right end point of the time interval. More precisely, denote U and V as the two examination times. The observed data consists of

$$(U, V, \delta_1 = I(T \leq U), \delta_2 = I(U < T \leq V), \delta_3 = I(V < T < \infty)). \qquad (3.1)$$

In the above formulation, there are two examination times. A natural generalization of case II interval censoring is case k interval censoring where there are $k(> 2)$ examination times per subject.

We now consider a more general interval censoring scheme. Suppose that

$$0 < Y_{i1} < \ldots < Y_{in_i} < \infty$$

are the ordered examination times for the ith subject, $i = 1, \ldots, n$. Denote $\mathbf{Y_i} = (Y_{i1}, \ldots, Y_{in_i})$. Let T_i be the ith subject's unobservable event time. Computationally, it is convenient to reduce the general interval censoring to case II interval censoring by considering the following three possibilities: (i) the

TABLE 3.2: Observed Time Intervals (Months) for the Time to Breast Retraction of Early Breast Cancer Patients

Group	Observed Time Intervals
RT	(45,], (25, 37], (37,], (4, 11], (17, 25], (6, 10], (46,], (0, 5], (33,], (15,], (0, 7], (26, 40], (18,], (46,], (19, 26], (46,], (46,], (24,], (11, 15], (11, 18], (46,], (27, 34], (36,], (37,], (22,], (7, 16], (36, 44], (5, 12], (38,], (34,] (17,], (46,], (19, 35], (46,], (5, 12], (9, 14], (36, 48], (17, 25], (36,], (46,], (37, 44], (37,], (24,], (0, 8], (40,], (33,]
RCT	(8, 12], (0, 5], (30, 34], (16, 20], (13,], (0, 22], (5, 8], (13,], (30, 36], (18, 25], (24, 31], (12, 20], (10, 17], (17, 24], (18, 24], (17, 27], (11,], (8, 21], (17, 26], (35,], (17, 23], (33, 40], (4, 9], (16, 60], (33,], (24, 30], (31,], (11,], (15, 22], (35, 39], (16, 24], (13, 39], (15, 19], (23,], (11, 17], (13,], (19, 32], (4, 8], (22,], (44, 48], (11, 13], (34,], (34,], (22, 32], (11, 20], (14, 17], (10, 35], (48,]

event has occurred by the first examination. Denote $U_i = Y_{i1}$, and let $V_i = Y_{i2}$. Furthermore, let $\delta_{1i} = I(T_i \leq U_i)$ and $\delta_{2i} = I(U_i < T_i \leq V_i)$. Then $\delta_{1i} = 1$ and $\delta_{2i} = 0$; (ii) T_i is known to be bracketed between a pair of examination times (Y_{iL}, Y_{iR}), where Y_{iL} is the last examination time proceeding T_i, and Y_{iR} is the first examination time following T_i. Denote $U_i = Y_{iL}$ and $V_i = Y_{iR}$. Define δ_{1i} and δ_{2i} as in (i). Then $\delta_{1i} = 0$ and $\delta_{2i} = 1$; (iii) At the last examination, the event has not occurred. Then $\delta_{1i} = \delta_{2i} = 0$. Following the above formulation, the effective observations are

$$\{(U_i, V_i, \delta_{1i}, \delta_{2i}), \ i = 1, \ldots, n\}. \tag{3.2}$$

Computationally, the general interval censoring scheme can be reduced to case II interval censoring. The estimation approaches described in this chapter for case II interval censoring are applicable to general interval censoring. However, the asymptotic distributional results do *not* carry over to the general case although they can be extended to case k interval censoring with a fixed k.

Consider the following example of case II interval censored data. A retrospective study conducted between 1976 and 1980 in Boston on early breast cancer patients was presented in Finkelstein and Wolfe (1985). The data consist of 94 patients who were given either radiation therapy alone (RT, sample size 46) or radiation therapy plus adjuvant chemotherapy (RCT, sample size 48). The goal of the study was to compare the two treatments with respect to their cosmetic effects. In the study, patients were expected to visit the clinic every four to six months. However, the actual visit times varied across and within patients. At visits, physicians evaluated the cosmetic appearance of the patients including breast retraction, a response that has a negative impact on overall cosmetic appearance. In the data set (Table 3.2), there are 38 patients who did not experience breast retraction during the study, thus having right censored observations denoted by the intervals with no finite right end points. For the other patients, the observations are intervals, representing the time periods during which breast retraction occurred. The intervals are given by the last clinic visit time at which breast retraction had not yet occurred and the first clinic visit time at which breast retraction was detected.

Remarks. The interval censored data defined in this section are "standard." Another type of interval censored data, which is relatively rare compared with case I and case II interval censored data, is *doubly interval censored data* (De Gruttola and Lagokos, 1988; Sun, 1995). Here, the event time $T = T_2 - T_1$, that is, it is defined as the difference of two event times. Instead of observing T_1 and T_2 accurately, it is only known that

$$T_1 \in (U_1, V_1], \quad T_2 \in (U_2, V_2].$$

That is, both ends are interval censored. Consider for example the HIV partner study. Here T_2 is the infection time of the susceptible partner, and T_1 is the infection time of the infected case. As described above, T_2 is usually interval censored. Because of the same rationale, T_1 can also be interval censored, leading to doubly interval censored data. We refer to other publications, for example Chapter 8 of Sun (2006), for descriptions of methods for doubly interval censored data.

As shown in Kalbfleisch and Lawless (1985) and Sun and Wei (2000), interval censored data can also be embedded into the bigger framework of *panel count data*. With panel count data, the quantity of interest is the *number or count* of a certain event. Consider, for example, cancer recurrence after treatment. A patient may experience multiple events of recurrence. One way of modeling the recurrence process is to focus on the time between two consecutive recurrences. See Section 4.2.4 in Chapter 4. An alternative approach is to focus on the count of recurrences at a specific time. Interval censoring may happen if the exact time of recurrence is not available. For example, in many observational studies, the record only contains the recurrence status at a prespecified time (for example, the last day of a calendar year). The interval censored survival data we describe above is a special case of panel count data, with the maximum number of event equal to one. Chapter 9 of Sun (2006) provides brief discussions on panel count data.

3.2 Parametric Modeling

The simplest way of modeling interval censored data is to use parametric models. For data with no censoring and right censored data, we describe commonly adopted parametric models in Chapter 2. The main difference between right censored data and interval censored data is on the censoring mechanism, not the data generating mechanism. Thus, the same set of parametric models as described in Chapter 2 can be adopted here. Particularly, commonly adopted parametric survival models for interval censored data include the following.

Exponential. The exponential model assumes a constant hazard $h(t) = \lambda$. Thus, the corresponding survival function is $S(t) = \exp(-\lambda t)$, and the density is $f(t) = \lambda \exp(-\lambda t)$. See Section 2.1.2 for more details.

Weibull. Under the two-parameter Weibull model, the hazard function is $h(t) = \lambda \gamma (\lambda t)^{\gamma-1}$ with $\lambda, \gamma > 0$. The cumulative hazard function is $H(t) = (\lambda t)^{\gamma}$, and the survival function is $S(t) = \exp(-(\lambda t)^{\gamma})$. The logarithm of the Weibull hazard is a linear function of the logarithm of time with intercept $\gamma \log(\lambda) + \log(\gamma)$ and slope $\gamma - 1$. Thus, the hazard is rising if $\gamma > 1$, constant if $\gamma = 1$, and declining if $\gamma = 1$. Also see Section 2.1.2 for more details.

Gompertz-Makeham. The Gompertz distribution is characterized by the fact that the logarithm of the hazard is linear in t, so $h(t) = \exp(\alpha + \beta t)$ and is thus closely related to the Weibull distribution where the logarithm of the hazard is linear in $\log(t)$. As a matter of fact, the Gompertz distribution is a log-Weibull distribution. This distribution provides a remarkably close fit to adult mortality in contemporary developed nations. Also see Section 2.1.2 for more details. The Gompertz-Makeham distribution has a hazard function given by

$$h(t) = \alpha - \beta e^{-\gamma t},$$

with $\alpha > 0$. This hazard function is slightly more general than that of the Gompertz distribution since it allows the offset at $t = 0$ to be parameterized.

Gamma. The gamma distribution with parameters λ and k, denoted as $\Gamma(\lambda, k)$, has density

$$f(t) = \frac{\lambda (\lambda t)^{k-1} \exp(-\lambda t)}{\Gamma(k)}$$

and survival function $S(t) = 1 - I_k(\lambda t)$, where $I_k(x)$ is the incomplete gamma function defined as $I_k(x) = \int_0^x \lambda^{k-1} \exp(-x) \, dx / \Gamma(k)$. There is no closed-form expression for the survival function. There is no explicit formula for the hazard function either, but this may be computed easily as the ratio of the density to the survival function. R has a function **pgamma** that computes the cumulative distribution function and survival function. This function calls k the shape parameter and $1/\gamma$ the scale parameter. The gamma hazard increases monotonically if $k > 1$, from a value of 0 at the origin to a maximum of λ, is constant if $k = 1$, and decreases monotonically if $k < 1$, from ∞ at the origin to an asymptotic value of λ. When $k = 1$, the gamma distribution reduces to the exponential distribution.

Generalized Gamma. The generalized gamma distribution adds a scale parameter in the expression for $\log(T)$ so that $\log(T) = \alpha + \sigma W$, where W

has a generalized extreme value distribution with parameter k. The density of the generalized gamma distribution can be written as

$$f(t) = \frac{\lambda p (\lambda t)^{pk-1} \exp(-(\lambda t)^p)}{\Gamma(k)}$$

where $p = 1/\sigma$ and $\lambda = \exp(-\alpha)$. The generalized gamma includes the gamma ($\sigma = 1$), the Weibull ($k = 1$), and the exponential ($k = \sigma = 1$) distributions as special cases. It also includes the log-normal distribution as a special limiting case when $k \to \infty$.

Log-Normal. T has a log-normal distribution if $\log(T) = \alpha + \sigma W$ where W has a standard normal distribution. The hazard function of the log-normal distribution increases from 0 to reach a maximum and then decreases monotonically, approaching 0 at $t \to \infty$. The survival function can be written as

$$S(t) = 1 - \Phi\left(\frac{\log(t) - \alpha}{\sigma}\right).$$

Log-Logistic. T has a log-logistic distribution if $\log(T) = \alpha + \sigma W$, where W has a standard logistic distribution. The survival function is $S(t) = \frac{1}{1+(\lambda t)^p}$, where $\alpha = -\log(\lambda)$ and $p = 1/\sigma$. The hazard function is $h(t) = \frac{\lambda p (\lambda t)^{p-1}}{1+(\lambda t)^p}$. The hazard function is monotone decreasing from ∞ if $p < 1$, monotone increasing from λ if $p = 1$, and similar to the log-normal if $p > 1$.

Generalized F. Kalbfleish and Prentice (1980) considered the more general case where $\log(T) = \alpha + \sigma W$, and W is distributed as the logarithm of an F variate (which adds two more parameters) whose density is given by

$$f(w) = \frac{(m_1/m_2)^{m_1} \exp(m_1 w)(1 + m_1 e^w/m_2)^{-(m_1+m_2)}}{\Gamma(m_1)\Gamma(m_2)/\Gamma(m_1 + m_2)}.$$

It is interesting to note that this distribution includes all of the above distributions as special cases. If $(m_1, m_2) = (1, 1)$, we have the log-logistic distribution; $m_2 = \infty$ gives the generalized gamma distribution. This observation enables a nested-model testing approach to examine the adequacy of different parametric forms.

For estimation and inference, consider a sample of n i.i.d. subjects. For subject $i = 1, \ldots, n$, denote T_i as the survival time of interest, and we observe

$$\{(L_i, R_i], \boldsymbol{Z}_i\},$$

where $L_i < T_i \leq R_i$ and $\boldsymbol{Z}_i$ denotes the covariate vector that may be associated with survival. Note that case I interval censored data can also be rewritten into the above form. The strategy for introducing covariate effects into parametric survival modeling has been discussed in Section 2.3. Assume that T_is have a parametric survival function $S(t, \beta)$, where β denotes the unknown regression parameters. When $L_i < R_i$, that is, no event time is accurately observed, the unnormalized log-likelihood function is

$$\ell(\beta) = \sum_{i=1}^{n} \log(S(L_i, \beta) - S(R_i, \beta)).$$

With parametric survival models, once the log-likelihood function is properly specified, under most cases, standard likelihood-based estimation and inference can be conducted. Here the estimate $\hat{\beta}$ is the maximizer of $\ell(\beta)$. Under fairly mild regularity conditions (Shao, 1999), $\hat{\beta}$ is uniquely defined and a consistent estimate of β_0, the true regression parameter. In addition, it is asymptotically normal and fully efficient, with asymptotic variance equal to the inverse of the Fisher information matrix. In practical data analysis, the variance estimate can be obtained using a plug-in approach, replacing β_0 with its estimate in the Fisher information matrix. Inference can also be based on the maximum likelihood theory. Hypothesis testing of $H_0 : \beta = \beta_0$ can be based on the score test, where the test statistic is asymptotically χ^2 distributed under the Null. Comparing two nested models can be conducted using a likelihood ratio test, where the test statistic is also asymptotically χ^2 distributed. As likelihood-based estimation and inference with parametric models are rather straightforward, we do not pursue further discussions.

As an example, consider the breast cancer data provided in Table 3.2. In this example there is only one covariate—the treatment group indicator. Consider the exponential model, where the event times satisfy exponential distributions with hazards λ_1 and λ_2 in the RT and RCT groups, respectively. To introduce the covariate effect, define $\beta = \log(\lambda_2/\lambda_1)$. This reparameterization mimics the Cox proportional hazards model. Under the exponential model, the MLEs are

$$\hat{\lambda}_1 = 0.0163 \ (0.0036), \quad \hat{\beta} = 0.7416 \ (0.2769),$$

where numbers in the "()" are the estimated standard errors. Hypothesis testing of $H_0 : \beta = 0$ (i.e., equal hazard for the two groups) produces a test statistic 2.6782, leading to a p-value 0.0074. This suggests that patients in the RCT group have a significantly higher risk of developing breast retraction. The same data can also be analyzed using other parametric models. In particular, fitting "bigger" models, for example, the Weibull model, can help assess whether the exponential model is sufficient. We leave this task to the readers.

3.3 Nonparametric Modeling

Parametric models are intuitive and easy to implement. However, they may suffer from model misspecification. Nonparametric modeling can be more flexible and provide a more accurate description of data. For data with no censoring and right censored data, the Kaplan-Meier (KM) approach described in Chapter 2 provides the nonparametric maximum likelihood estimate (NPMLE) of the survival function. Under mild regularity conditions, the KM estimate is point-wise $n^{1/2}$ consistent and asymptotically normally distributed. Martingale theory can be applied to rigorously establish such results. With interval censored data, the KM approach is no longer applicable, as the exact event times are unknown, and so the martingale structure cannot be straightforwardly constructed. In the following two sections, we investigate the nonparametric estimation of survival function (or equivalently distribution function) for case I and case II interval censored data separately.

3.3.1   Case I Interval Censoring

Consider n i.i.d. subjects under case I interval censoring. Denote $\{T_i : i = 1, \ldots, n\}$ as the unobservable event times of interest, satisfying survival function $S(t)$. The observed data consist of

$$\{(C_i, \delta_i) : i = 1, \ldots, n\},$$

where C_i is the censoring time and δ_i is the event indicator for subject i. Assume that the censoring distribution does not depend on S. Then, up to a constant not depending on S, the likelihood function is

$$L(S) = \prod_{i=1}^{n} [S(C_i)]^{1-\delta_i} [1 - S(C_i)]^{\delta_i}.$$

Denote $\{s_j : j = 1, \ldots, m\}$ as the unique ordered statistics of $0 \cup \{C_i : i = 1, \ldots, n\}$. To accommodate ties, define $r_j = \sum_{i=1}^{n} \delta_i I(C_i = s_j)$, the number of events at time s_j, and $n_j = \sum_{i=1}^{n} I(C_i = s_j)$, the number of observations at time s_j. The likelihood function can be rewritten as

$$L(S) = \prod_{j=1}^{m} [S(s_j)]^{n_j - r_j} [1 - S(s_j)]^{r_j} = \prod_{j=1}^{m} [F(s_j)]^{r_j} [1 - F(s_j)]^{n_j - r_j},$$

where $F = 1 - S$ is the distribution function.

As the likelihood function depends on S and F only through their values at s_js, we can only estimate the survival/distribution functions at those discrete time points. The distribution function is nondecreasing. Robertson et al. (1998) show that maximizing L with respect to F is equivalent to minimizing

$$\sum_{j=1}^{m} n_j \left[\frac{r_j}{n_j} - F(s_j) \right]^2$$

subject to the nondecreasing constraint of F. Such an optimization problem has been referred to as *isotonic regression*. The max-min approach of isotonic regression shows that the NPMLE of F is

$$\hat{F}(s_j) = \max_{u \leq j} \min_{v \geq j} \frac{\sum_{l=u}^{v} d_l}{\sum_{l=u}^{v} n_l}.$$

Computationally, $\hat{F}$ can also be obtained as follows: (1) Order the censoring times $C_{(1)} \leq C_{(2)} \leq \ldots \leq C_{(n)}$. Denote the corresponding event indicators as $\delta_{(1)}, \ldots, \delta_{(n)}$; (2) Plot $(i, \sum_{j=1}^{i} \delta_{(j)}), i = 1, \ldots, n$; (3) Construct the Greatest Convex Minorant (GCM) G^* of the points in Step (2) (Huang, 1996); (4) $\hat{F}(C_{(i)})$ =left-derivative of G^* at i, $i = 1, \ldots, n$. The above two approaches both only involve simple calculations and can be realized easily. Brief discussion on software implementation is provided later in this chapter. It is easy to see that the estimate of $F(S)$ differs vastly from the KM estimate for right censored data.

We now describe the asymptotic properties of the estimate. Denote G as the distribution function of the censoring time C. Assume that T and C are independent. The scenario with discrete F and/or G is relatively simpler and less interesting. Here we focus on the scenario where both F and G are continuous. Further assume that the support of F is dominated by that of G. Under some mild regularity conditions, Groeneboom and Wellner (1992) establish the uniform consistency by showing that

$$P\left\{ \lim_{n \to \infty} \sup_{t \in \mathbb{R}^+} \left| \hat{F}(t) - F(t) \right| \to 0 \right\} = 1.$$

This consistency result also holds for case II interval censored data but under different technical conditions.

If we further assume that T and C are both bounded by a positive constant $M < \infty$, following Huang and Wellner (1995), it can be proved that

$$\|\hat{F} - F\|_G^2 = \int_0^M \left[\hat{F}(t) - F(t) \right]^2 dG(t) = O_p(n^{-2/3}).$$

That is, the NPMLE has convergence rate $n^{1/3}$, which is considerably slower than the $n^{1/2}$ rate for the KM estimate with right censored data. The intuition is that with interval censoring, there is an excessive "loss of information", leading to a slower convergence rate. Theoretically, it can be proved that without making stronger assumptions, this rate cannot be improved. The rigorous proof demands constructing a sharp bound for the entropy (which measures the "size" of an infinite dimensional functional space) of the functional set composed of bounded, monotone functions. In addition to the L_2 norm consistency, Schick and Yu (2000) also establish consistency under the L_1 and other norms.

Now consider a fixed time point t_0 where $0 < F(t_0), G(t_0) < 1$. Assume that F and G have density functions f and g, respectively. In addition, assume that $f(t_0), g(t_0) > 0$. Groeneboom and Wellner (1992) prove that as $n \to \infty$,

$$\frac{1}{2} n^{1/3} \left[\frac{2g(t_0)}{f(t_0)F(t_0)[1 - F(t_0)]} \right]^{1/3} [\hat{F}(t_0) - F(t_0)] \to \Omega, \tag{3.3}$$

in distribution. Here, Ω is the last time when a standard Brownian motion minus the parabola $\eta(t) = t^2$ reaches its peak. This local asymptotic distribution differs in its presentation and complexity from that with right censored data. In practice, the above result does not suggest an easy way to construct point-wise confidence interval.

3.3.2 Case II Interval Censoring

Under case II interval censoring, the joint density of a single observation $X = (U, V, \delta_1, \delta_2)$ is

$$p(x) = F(u)^{\delta_1} [F(v) - F(u)]^{\delta_2} (1 - F(v))^{\delta_3} h(u, v), \tag{3.4}$$

where $\delta_3 = 1 - \delta_1 - \delta_2$ and $h(u, v)$ is the joint density of (U, V). Assume independent censoring. With n i.i.d. observations, up to a constant, the unnormalized log-likelihood function is

$$\ell(F) = \sum_{i=1}^{n} \{\delta_{1i} \log F(U_i) + \delta_{2i} \log(F(V_i) - F(U_i)) + \delta_{3i} \log(1 - F(V_i))\}. \tag{3.5}$$

Iterative Convex Minorant Approach. Under case II interval censoring, there is no closed form expression for $\hat{F}$. Let P_n denote the empirical measure based on the n observations. For any distribution function F and $t \geq 0$, define

$$\begin{aligned}
W_F(t) &= \int_{u \leq t} \left\{ \frac{\delta_1}{F(u)} - \frac{\delta_2}{F(v) - F(u)} \right\} d\mathrm{P}_n \\
&\quad + \int_{v \leq t} \left\{ \frac{\delta_2}{F(v) - F(u)} - \frac{\delta_3}{1 - F(v)} \right\} d\mathrm{P}_n \\
G_F(t) &= \int_{u \leq t} \left\{ \frac{\delta_1}{(F(u))^2} + \frac{\delta_2}{(F(v) - F(u))^2} \right\} d\mathrm{P}_n \\
&\quad + \int_{v \leq t} \left\{ \frac{\delta_2}{(F(v) - F(u))^2} + \frac{\delta_3}{(1 - F(v))^2} \right\} d\mathrm{P}_n
\end{aligned}$$

and

$$V_F(t) = W_F(t) + \int_{[0,t]} F(s) dG(s).$$

Let J_{1n} be the set of examination times U_i such that T_i either belongs to

$[0, U_i]$ or $(U_i, V_i]$, and let J_{2n} be the set of examination times V_i such that T_i either belongs to $(U_i, V_i]$ or (V_i, ∞). Furthermore, let $J_n = J_{1n} \cup J_{2n}$, and let $T_{(j)}$ be the jth order statistic of the set J_n.

Let $T_{(1)}$ correspond to an observation U_i such that $\delta_{1i} = 1$, and let the largest order statistic $T_{(m)}$ correspond to an observation V_i such that $\delta_{1i} = \delta_{2i} = 0$. Then $\hat{F}$ is the NPMLE of F if and only if $\hat{F}$ is the left derivative of the convex minorant of the self-induced cumulative sum diagram formed by the points

$$(G_{\hat{F}}(T_{(j)}), V_{\hat{F}}(T_{(j)})), \quad j = 1, \ldots, m$$

and $(0, 0)$. With this property, Groeneboom (1991) introduces the *iterative convex minorant algorithm* to compute $\hat{F}$. This approach starts with a sensible initial estimate, which can be obtained by treating interval censored data as right censored data and computing the KM estimate. Then the estimate is repeatedly updated by computing the left derivative of the convex minorant. This process is iterated until convergence. Advancing from Groeneboom (1991), Jongbloed (1998) proposes modification of the iterative convex minorant algorithm and shows that this modified algorithm always converges.

It can be verified that the NPMLE $\hat{F}$ satisfies the self-consistency equation

$$
\begin{aligned}
\hat{F} &= E_{\hat{F}}[\tilde{F}(t)|U_i, V_i, \delta_{1i}, \delta_{2i}, i = 1, \ldots, n] \\
&= \frac{1}{n} \sum_{i=1}^{n} \delta_{1i} \frac{\hat{F}(U_i \wedge t)}{\hat{F}(U_i)} + \delta_{2i} \frac{\hat{F}(V_i \wedge t) - \hat{F}(U_i \wedge t)}{\hat{F}(V_i) - \hat{F}(U_i)} + \delta_{3i} \frac{\hat{F}(t) - \hat{F}(V_i \wedge t)}{1 - \hat{F}(V_i)},
\end{aligned}
$$

where $\tilde{F}$ is the (unobservable) empirical distribution function of $T_1, \ldots, T_n$, and $x \wedge y = \min(x, y)$. The above results suggest ways to verify whether a function is NPMLE, however, not a simple solution to compute it. In addition, it should be noted that this equation does not sufficiently characterize the NPMLE. That is, there may exist other distribution functions different from $\hat{F}$ that satisfy the above equation.

Turnbull's Approach. Here, we present an alternative approach, which also has no closed form and is based on an iterative procedure.

To construct the estimator, let $0 = \tau_0 < \tau_1 < \ldots < \tau_m$ be a grid of time points which includes all the points U_i and V_i for $i = 1, \ldots, n$. For the ith observation, define the weight α_{ij} to be 1 if the interval (τ_{j-1}, τ_j) is contained in the interval $(U_i, V_i]$ and 0 otherwise. The weight α_{ij} indicates whether the event which occurs in the interval $(U_i, V_i]$ could have occurred at τ_j. An initial estimate $\hat{S}(\tau_j)$ of the survival function at τ_j is made. Then the Turnbull's algorithm proceeds as follows: (1) Compute the probability of an event occurring at time τ_j by $p_j = \hat{S}(\tau_{j-1}) - \hat{S}(\tau_j)$ for $j = 1, \ldots, m$; (2) Estimate the number of events occurring at τ_j by $d_j = \sum_{i=1}^{n} \frac{\alpha_{ij} p_j}{\sum_{k=1}^{m} \alpha_{ik} p_k}$ for $j = 1, \ldots, m$; (3) Compute the estimated number at risk at time τ_j by $Y_j = \sum_{k=j}^{m} d_k$; (4) Compute the KM (product limit) estimate using the pseudo data generated in Steps (2) and (3); (5) Repeat Steps (2)–(4) until convergence.

Asymptotic Properties. The asymptotic properties of NPMLE are similarly established as those for case I interval censored data. Specifically, denote H_1 and H_2 as the distribution functions of U and V, respectively. Assume that the support of F is dominated by that of $H_1 + H_2$, and T and V are both bounded by a positive constant $M < \infty$. Geskus and Groeneboom (1996) prove that

$$||\hat{F} - F||_{H_i}^2 = \int_0^M \left[\hat{F}(t) - F(t) \right]^2 dH_i(t) = O_p(n^{-2/3})$$

for $i = 1, 2$. That is, the estimate is $n^{1/3}$ consistent.

Denote f, h_1, h_2 as the density functions of F, H_1, and H_2, respectively. Assume that (a) $f(t) \geq a_0 > 0$ for all $t \in (0, M)$ and a constant a_0; (b) h_1 and h_2 are continuous with $h_1(t) + h_2(t) > 0$ for all $t \in (0, M)$; and (c) $P(V - U > \xi) = 1$ for a positive constant $\xi > 0$. Then under some mild regularity conditions, at a fixed time $t_0 \in (0, M)$,

$$\frac{1}{2} n^{1/3} \left[\frac{2c(t_0)}{f(t_0)} \right]^{1/3} \left[\hat{F}(t_0) - F(t_0) \right] \to \Omega,$$

where Ω has the same definition as with case I interval censored data and

$$c(t_0) = \frac{h_1(t_0)}{F(t_0)} + \frac{h_2(t_0)}{1 - F(t_0)} + \int_{t_0}^M \frac{h(t_0, x)}{F(x) - F(t_0)} \, dx + \int_0^{t_0} \frac{h(x, t_0)}{F(t_0) - F(x)} \, dx.$$

Among the conditions needed for the validity of the above results, condition (c) is worth further attention. This condition assumes that the lengths of the time intervals are bounded away from zero. Under this condition, no event time is accurately observed.

The above $n^{1/3}$ convergence rate and nonstandard asymptotic distribution may be inconvenient in practice. Unfortunately, without making stronger assumptions on either the data or model or both, the asymptotic results cannot be further improved. Now consider an alternative scenario with a different assumption on data. Modifying model assumptions is discussed later in this chapter. Consider the scenario where among the n event times, n_1 are accurately observed, and the rest are interval censored (Huang, 1999b). Such a scenario can be realistic in certain biomedical studies. Further assume that as $n \to \infty$, n_1/n, the fraction of accurately observed event times, is bounded away from zero. Then it can be proved that at a fixed time $t_0 \in (0, M)$

$$n^{1/2} \left[\hat{F}(t_0) - F(t_0) \right] \to Z_1(t_0),$$

where Z_1 is a normally distributed random variable. That is, the convergence rate can be improved to $n^{1/2}$. Here, the variance of Z_1 is given by the information lower bound for the estimate of F. Under other assumptions on the event time and censoring distributions, convergence rate between $n^{1/3}$ and $n^{1/2}$ can also be obtained. For example, Groeneboom and Wellner (1992) study the scenario where $U = V$ has a sufficient mass and show that $\hat{F}$ has $(n \log n)^{1/3}$ convergence rate. For more relevant discussions, we refer readers to Banerjee and Wellner (2005), Goodall et al. (2004), and Hudgens (2005).

3.3.3 Estimation of Linear Functionals

Beyond the distribution (survival) function itself, its linear functional, which is defined as

$$K(F) = \int c(t)\, d\, F(t)$$

where $c(t)$ is a given function, is often of interest. Many statistical quantities of interest can be written as linear functionals of the distribution function. Important examples include the moments of T and distribution quantiles. Assume that the supports of event and censoring times are bounded.

With case I interval censored data, further assume that the support of T is dominated by that of C, and G has a density function g. Under mild regularity conditions, it can be shown that

$$\sqrt{n}\left[K(\hat{F}) - K(F)\right] \to N(0, \sigma^2) \tag{3.6}$$

in distribution as $n \to \infty$, where

$$K(\hat{F}) = \int c(t)\, d\, \hat{F}(t),$$

$$\sigma^2 = \int_0^M \frac{F(t)(1 - F(t))}{g(t)} \left[c^{(1)}(t)\right]^2 dt$$

and $c^{(1)}(t)$ is the first-order derivative of $c(t)$. In practice, it is usually difficult to verify the nonsingularity of the variance, and it needs to be assumed that $0 < \sigma < \infty$. We refer to Huang and Wellner (1995) for detailed proof. The above result states that despite the $n^{1/3}$ convergence rate for $\hat{F}$, the linear functionals can still be consistently estimated at the usual $n^{1/2}$ rate and achieve the asymptotic normality.

With case II interval censored data, the $n^{1/2}$ consistency and asymptotic normality may still hold, although the asymptotic variance does not have a simple analytic form. In addition, special attention needs to be paid to the assumptions on censoring distribution. We omit the detailed discussions in this book. See Geskus and Groeneboom (1996, 1999) for more information.

3.3.4 Smoothed Estimation

With both case I and case II interval censored data, the NPMLEs of the distribution functions have $n^{1/3}$ convergence rate (if $V - U$ is bounded away

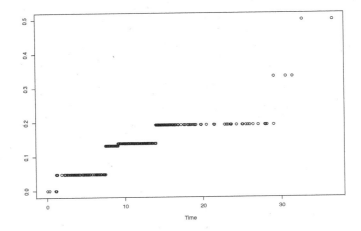

FIGURE 3.1: A representative plot of the distribution function estimate with case I interval censored data, sample size ~ 300.

from zero for case II interval censored data). In addition, they are only estimable at the discrete $\{s_j : j = 0, \ldots, m\}$. In Figure 3.1, we show a representative plot of the distribution function estimate with case I interval censored data (sample size ~ 300). The estimate is a step function with a very small number of "jumps." Such an estimate can be difficult to interpret or use in downstream analysis.

Many biological processes are "smooth," making it reasonable to assume differentiability conditions on F. One possible assumption, which has been commonly made for nonparametric smoothed estimation, is that F belongs to the Sobolev space indexed by the order of derivative $s_0 (\geq 1)$. With a slight abuse of notation, for both case I and case II interval censored data, denote $(U_i, V_i]$ as the interval that contains the event time T_i for the ith subject in a sample of size n. Under the smoothness assumption, consider the penalized estimate

$$\hat{F} = argmax \left\{ \sum_{i=1}^{n} \frac{1}{n} \log[F(V_i) - F(U_i)] - \lambda^2 \int [F^{(s_0)}(t)]^2 dt \right\},$$

where $\lambda > 0$ is a data-dependent tuning parameter, and $F^{(s_0)}(t)$ is the s_0th order derivative of F. Here the integration is over the support of U and V.

With smooth nonparametric parameters, penalized estimation has been commonly adopted under the framework of spline. We refer to Wahba (1990) for thorough discussions of penalized spline estimation. The tuning parameter λ controls the smoothness of the estimate. In particular, when $\lambda \to \infty$, $\hat{F}^{(s_0)} \to 0$, and the estimate is "too smooth." On the other hand, when λ is too small, the estimate may be too wiggly and poses a threat of overfitting. In practice, λ can be chosen via for example cross-validation or generalized cross-validation. Under mild regularity conditions, when $n \to \infty$, it can be proved that if $\lambda^2 = O(n^{-\frac{2s_0}{2s_0+1}})$,

$$n^{\frac{s_0}{2s_0+1}}(\hat{F}(t_0) - F(t_0)) \to Z(t_0), \tag{3.7}$$

at a specific time point t_0, where $Z(t_0)$ is a $O_p(1)$ random variable. A commonly made assumption is that $s_0 = 2$ (cubic spline), leading to the $n^{2/5}$ convergence rate, which is considerably faster than the $n^{1/3}$ rate. It is obvious that if stronger smoothness assumptions can be made, the estimate has a faster convergence rate. In addition, it can be proved that

$$\int [\hat{F}^{(s_0)}(t)]^2 dt = O_p(1). \tag{3.8}$$

That is, the estimate is not "too smooth" or "too wiggly." It is at the same level of smoothness as the original parameter.

In numerical studies, $\hat{F}$ can be obtained via the basis function expansion approach (Wahba, 1990). Denote $\{B_1(t), \ldots, B_k(t)\}$ as a set of k basis functions, where k grows with n as $n \to \infty$. Specification of the basis functions has been well investigated in Wahba (1990) and can be realized using many existing software packages (for example, the bs function for B-spline in R package splines). Once we make the basis expansion $F = \sum_{j=1}^{k} \alpha_j B_j(t)$ with unknown regression parameters $\{\alpha_1, \ldots, \alpha_k\}$, the nonparametric optimization problem becomes a parametric one. Note that the parametric optimization needs to be conducted under the constraint that $\hat{F}$ is nonnegative and nondecreasing.

When the smoothness assumption is not made, we focus on the estimation of distribution function. The density function estimate is not directly obtainable. When the distribution function is assumed to be s_0th-order differentiable, if $s_0 \geq 2$, then the density function is $(s_0 - 1)$th-order differentiable. Its estimate can be obtained by taking the first-order derivative of $\hat{F}$. It can be shown that the estimate of the density function is consistent and has convergence rate $n^{\frac{s_0-1}{2(s_0-1)+1}}$.

Under the smoothness assumption, an alternative estimation approach is the kernel estimation. We refer to Pan (2000b) and Braun et al. (2005) for the development of kernel estimation approaches with interval censored data. Loosely speaking, if similar smoothness assumptions are made, then penalization and kernel approaches lead to the same convergence rate. We return to the discussion on functional estimation in Section 4.1.

3.3.5 Software Implementation

Compared with right censored data, software for interval censored data analysis is less developed. Among the available packages, fitdistrplus can fit parametric distributions to interval censored data based on likelihood approaches. The `Icens` package, which is available from `Bioconductor` and written by Gentleman and Vandal, provides several ways to compute the NPMLE under various censoring and truncation schemes. The `R` package `MLEcens` computes the NPMLE of distribution function for bivariate interval censored data, which is not discussed in this chapter but nevertheless can be important in some biomedical studies (we refer the readers to Sun, 2006, Chapter 7 for more discussions). The `icfit` function in package `interval` can also compute the NPMLE. The package `dcens` has been developed by Julia, Gomez and Cortes for estimation with doubly censored data. For the Turnbull approach, research `R` code and examples are available from Suely Ruiz Giolo's Web site *http://www.est.ufpr.br/rt/suely04a.htm*. The `ICE` package can estimate the hazard function. For some other tasks, for example, estimation of linear functionals and smoothed estimation, to the best of our knowledge, there is no well-written software.

As a simple demonstration, consider the `cosmesis` data set which is available in the `Icens` package. The first few lines of data are as follows

```
    L   R Trt
1  45 100   0
2   6  10   0
3   0   7   0
4  46 100   0
```

where the interval $(L, R]$ contains the unobserved event time, and `Trt` is the treatment group indicator. An example of using the `EMICM` function (hybrid EM iterative convex minorant) to compute the NPMLE for the `Trt=0` group, as provided by the package authors, proceeds as follows

```
data(cosmesis)
csub1 <- subset(cosmesis, subset=Trt==0, select=c(L, R))
EMICM(csub1)
```

3.4   Two-Sample Comparison

In many biomedical studies, particularly clinical trials, it is of great interest to compare the survival experiences of two different groups of subjects. A representative example of different groups is different treatment regimens, such as a placebo control group versus an active treatment group. With right censored data, comparison of the survival functions of two or more groups can be conducted using the log-rank test introduced in Section 2.2.1. The theoretical validity of the log-rank test is based on the martingale theory, which does not extend straightforwardly to interval censored data. In the following two sections, we discuss comparing the survival functions of two groups with interval censored data. The approaches can be readily generalized to K-sample problems as well.

3.4.1 Case I Interval Censoring

Denote S_1 and S_2 as the survival functions of two groups respectively. The goal is to test the hypothesis $H_0 : S_1(t) = S_2(t)$ for all t. Consider the scenario that *subjects in the two groups have the same censoring distribution.* This assumption is reasonable in, for example, controlled clinical trials, and needs to be verified on a case-by-case basis.

Rank-Based Approach. Denote the observed data as $\{(C_i, \delta_i, Z_i) : i = 1, \ldots, n\}$, where $Z_i = 0$ or 1 is the group membership indicator for subject i. Denote $s_1 < \ldots < s_k$ as the distinct ordered statistics of $\{C_1, \ldots, C_n\}$ and $\hat{S}$ as the NPMLE of $\hat{S}$ constructed using all observations under H_0. The estimate of distribution function is $\hat{F} = 1 - \hat{S}$. Define $n_j = \sum_i \delta_i I(C_i = s_j)$ as the number of subjects that have censoring times equal to s_j and have experienced events. Define $l_j = \sum_i I(C_i = s_j)$ as the number of subjects with censoring times equal to s_j. As described in Section 3.3,

$$\hat{F}(s_j) = \max_{r \leq j} \min_{s \geq j} \frac{\sum_{v=r}^{s} n_v}{\sum_{v=r}^{s} l_v}.$$

Consider the test statistic

$$U = \frac{1}{\sqrt{n}} \sum_{i=1}^{n} Z_i(\delta_i - \hat{F}(C_i)), \tag{3.9}$$

which is the sum of the *observed* minus the *expected* number of events over all subjects. This formulation has been motivated by the form of the log-rank statistic for right censored data. Under H_0, Sun and Kalbfleisch (1996) show that as $n \to \infty$, U is asymptotically normally distributed with mean zero. The asymptotic variance can be consistently estimated by

$$\hat{\sigma}^2 = \hat{\sigma}_z^2 \sum_{i=1}^{n} \frac{1}{n}[\delta_i - \hat{F}(C_i)]^2, \tag{3.10}$$

where $\hat{\sigma}_z^2$ is the sample variance of Z_is. With the asymptotic normality result, the rest of the hypothesis testing procedure is standard.

Survival-Based Approach. Consider G, the distribution function of C. Denote $\hat{G}$ as the NPMLE of G. With case I interval censored data, if we view censoring as a type of "event," then its times are all accurately observed. Thus $\hat{G}$ can be obtained using the standard empirical estimation approach, that is, adapting equation (2.1) from Chapter 2. It is $n^{1/2}$ consistent and point-wise asymptotically normally distributed. Denote $\hat{S}_1$ and $\hat{S}_2$ as the NPMLE of S_1 and S_2 constructed using observations in the two groups separately. Denote $n_1 = \sum_i I(Z_i = 0)$ and $n_2 = \sum_i I(Z_i = 1)$. Assume that as $n \to \infty$, $n_1/n \to p$, a constant bounded away from 0 and 1. Consider the test statistic

$$U = \int_0^\tau [\hat{S}_2(t) - \hat{S}_1(t)] \, d\hat{G}(t), \tag{3.11}$$

where τ is a constant satisfying $S(\tau) > 0$ and S is the common survival function under H_0. This test statistic has been motivated by the Kolmogorov-Smirnov test and the weighted Kaplan-Meier estimator. Under the null, as $n \to \infty$, U is asymptotically normally distributed with mean zero. The asymptotic variance can be consistently estimated by

$$\hat{\sigma}^2 = \frac{n^2}{n_1 n_2} \int_0^\tau \hat{S}(t)[1 - \hat{S}(t)] d\hat{G}(t). \tag{3.12}$$

Here $\hat{S}$ is the NPMLE of S constructed using all observations. U may then be used to test the difference between two survival curves over the period $[0, \tau]$.

Wilcoxon-Type Approach. Under the null, the event times in the two groups have the same distribution. In addition, it has been assumed that the censoring times have the same distribution. Thus, δ_is should be i.i.d., which motivates the following Wilcoxon-type statistic

$$U = \frac{1}{\sqrt{n}} \sum_{i=1}^n \sum_{j=1}^n (Z_i - Z_j)(\delta_i - \delta_j). \tag{3.13}$$

It can be proved that under the null, as $n \to \infty$, U is asymptotically normally distributed with mean zero. Its asymptotic variance can be consistently estimated with

$$\hat{\sigma}^2 = \frac{1}{n} \sum_{i=1}^n \left(Z_i - \frac{\sum_{j=1}^n Z_j}{n} \right)^2 \delta_i^2. \tag{3.14}$$

Remarks. Beyond the above three approaches, there are other approaches that can be used for two-sample comparison (e.g., Lim and Sun, 2003; Zhang et al., 2003). These three approaches are singled out because of their intuitive format and computational simplicity. Although the three approaches have simple, intuitive forms, because of the interval censoring nature of data, their theoretical validity is highly nontrivial (Sun and Kalbfleisch, 1996; Anderson and Ronn, 1995).

With all the three approaches, a critical assumption is that the censoring times in different groups have the same distribution. In controlled clinical trials, randomization usually leads to the validity of this assumption. In observational studies, this assumption may or may not hold. In Sun (2006), an approach that can accommodate different censoring distributions for different groups is described. A limitation of this approach is that the dependence between the censoring time and group membership needs to be specified. In Sun (2006), the Cox proportional hazards model for the censoring time is assumed, whose validity needs to be verified on a case-by-case basis.

3.4.2   Case II Interval Censoring

Consider case II interval censored data. For convenience of notation, we rewrite the observed data as

$$\{(U_i, V_i], Z_i : i = 1, \ldots, n\}.$$

Here the event time $T_i \in (U_i, V_i]$, and Z_i is the group indicator. Denote S_1 and S_2 as the survival functions for the two groups, respectively. The goal is to test $H_0 : S_1(t) = S_2(t)$ for all t. Again it is assumed that all subjects have the same censoring distribution.

Rank-Based Approach. Denote $\hat{S}$ as the NPMLE of the common survival function S estimated using all observations under H_0 and $s_1 < \ldots < s_m$ as the ordered distinct $\{U_i, V_i : i = 1, \ldots, n\}$ where $\hat{S}$ has jumps. Denote s_{m+1} as the time point after which $\hat{S} = 0$. For $i = 1, \ldots, n$ and $j = 1, \ldots, m + 1$, define $\alpha_{ij} = I(s_j \in (U_i, V_i])$. In addition, define $\delta_i = 0$ if the observation on T_i is right censored ($V_i = \infty$) and 1 otherwise. Define $\rho_{ij} = I(\delta_i = 0, U_i \geq s_j)$, which is equal to 1 if T_i is right censored and is at-risk at time s_j-. Under the null, $\hat{S}$ is a consistent estimate of S. Consider the estimate of *the overall observed failure*

$$d_j = \sum_{i=1}^{n} \delta_i \frac{\alpha_{ij}[\hat{S}(s_j-) - \hat{S}(s_j)]}{\sum_{u=1}^{m+1} \alpha_{iu}[\hat{S}(s_u-) - \hat{S}(s_u)]}. \tag{3.15}$$

for $j = 1, \ldots, m$. The estimate of *the overall number at-risk* is

$$n_j = \sum_{r=j}^{m+1} \sum_{i=1}^{n} \delta_i \frac{\alpha_{ir}[\hat{S}(s_r-) - \hat{S}(s_r)]}{\sum_{u=1}^{m+1} \alpha_{iu}[\hat{S}(s_u-) - \hat{S}(s_u)]} + \sum_{i=1}^{n} \rho_{ij}. \tag{3.16}$$

Following a similar strategy, for groups $l = 1$ and 2, respectively, the estimated observed failure and at-risk numbers are

$$d_{jl} = \sum_{i=1}^{n} I(Z_i = l)\delta_i \frac{\alpha_{ij}[\hat{S}(s_j-) - \hat{S}(s_j)]}{\sum_{u=1}^{m+1} \alpha_{iu}[\hat{S}(s_u-) - \hat{S}(s_u)]}. \tag{3.17}$$

and

$$n_{jl} = \sum_{r=j}^{m+1} \sum_{i=1}^{n} I(Z_i = l)\delta_i \frac{\alpha_{ir}[\hat{S}(s_r-) - \hat{S}(s_r)]}{\sum_{u=1}^{m+1} \alpha_{iu}[\hat{S}(s_u-) - \hat{S}(s_u)]} + \sum_{i=1}^{n} I(Z_i = l)\rho_{ij}.$$

Recall that with right censored data, the log-rank statistic is defined as the sum of differences between observed and expected numbers of failure. It is worth noting that under right censoring, the above estimates simplify to the numbers of observed failure and at-risk at a specific time point. Motivated by the right censored data log-rank statistic, we consider the statistic

$$\boldsymbol{U}_r = (U_{r1}, U_{r2})^T$$

where

$$U_{rl} = \sum_{j=1}^{m} \left(d_{jl} - \frac{n_{jl}d_j}{n_j} \right), \quad l = 1, 2.$$

It can be proved that under the null, as $n \to \infty$, $\boldsymbol{U}_r$ is mean zero and asymptotically normally distributed. However, there is a lack of simple analytic form for the asymptotic variance.

For variance estimation, Zhao and Sun (2004) developed a multiple imputation based approach. With an interval censored observation, the exact event time is unknown and can be viewed as a "missing" measurement. Imputation approaches target at filling in "reasonable" values of the missing time. Denote B as the number of imputations. In practice, $10 \le B \le 20$ imputations are usually sufficient. An imputation-based inference approach proceeds as follows. For $b = 1, \ldots, B$:

1. For $i = 1, \ldots, n$:

    (a) if $\delta_i = 0$, let $T_i^{(b)} = U_i$ and $\delta_i^{(b)} = 0$;

    (b) if $\delta_i = 1$, randomly sample $T_i^{(b)}$ from the conditional probability function

    $$f_i(s) = P(T_i^{(b)} = s) = \frac{\hat{S}(s-) - \hat{S}(s)}{\hat{S}(U_i) - \hat{S}(V_i)}$$

    and $\delta_i^{(b)} = 1$. Here s takes values over all $s_j \in (U_i, V_i]$.

    This step "converts" interval censored data into right censored data through imputation. In particular, all right censored observations are kept unchanged. For those interval censored observations with finite right bounds, we fill in possible values by sampling from their conditional distributions (conditional on that they belong to known intervals).

2. With the right censored data, construct the log-rank statistic $U^{(b)}$. This procedure has been described in Section 2.2.1. Denote $V^{(b)}$ as the estimated variance of $U^{(b)}$ (see Section 2.2.1).

With the above procedure, we now have B log-rank statistics $\{U^{(b)} : b =$

$1, \ldots, B\}$ for the imputed data and the corresponding variance estimates $\{V^{(b)} : b = 1, \ldots, B\}$. Compute

$$
\boldsymbol{V}_{r1} = \sum_{b=1}^{B} \frac{1}{B} V^{(b)}
$$

$$
\boldsymbol{V}_{r2} = \left(1 + \frac{1}{B}\right) \frac{\sum_{b=1}^{B} [U^{(b)} - \sum_{k=1}^{B} U^{(k)}/B][U^{(b)} - \sum_{k=1}^{B} U^{(k)}/B]^{T}}{B - 1}.
$$

The variance of $\boldsymbol{U}_r$ can be estimated by

$$
\boldsymbol{V}_r = \boldsymbol{V}_{r1} + \boldsymbol{V}_{r2}.
$$

Here, the variance has two components. The first is the variance for right censored data, while the second reflects the variation caused by multiple imputations. Similar variance estimates have been extensively adopted in the multiple imputation literature. Hypothesis testing can be conducted based on the asymptotic normality under the null and the variance estimate.

Survival-Based Approach.   As with case I interval censored data, a survival-based approach adopts a test statistic that is the integration of difference in survival functions over time. Particularly, consider the statistic

$$
U_s = \int_0^{\tau} W(t)[\hat{S}_1(t) - \hat{S}_2(t)]dt,
$$

where τ is less than or equal to the largest observation time. $W(t)$ is the weight function and possibly data-dependent. It can be used to emphasize difference within a certain range. For example $W(t) = 1$ gives equal weight to all time points, whereas $W(t) = I(a < t < b)$ focuses the comparison on the time interval (a, b). Choosing a proper weight function follows the same strategy as for right censored data.

Consider the scenario where T has a continuous distribution and its support is dominated by $[0, \tau]$. Denote n_1, n_2 as the number of subjects in groups 1 and 2 respectively. Consider the observed data

$$
\{(U_i, V_i, \delta_{1i}, \delta_{2i}, \delta_{3i}) : i = 1, \ldots, n\},
$$

where $\delta_{1i} = I(T_i \leq U_i)$, $\delta_{2i} = I(U_i < T_i \leq V_i)$, and $\delta_{3i} = I(V_i < T_i)$. Denote $H(U, V)$ as as the joint distribution function of U and V, and $H_1(U)$ and $H_2(V)$ as the marginal distribution functions of U and V, respectively. In addition, denote h, h_1 and h_2 as the corresponding density functions.

Assume that as $n \to \infty$, $n_1/n \to p$ with p bounded away from 0 and 1. Also assume that $W(t)$ is a deterministic function with a finite variation. Consider the rescaled test statistic

$$
U_s = \sqrt{\frac{n_1 n_2}{n}} \int_0^{\tau} W(t)[\hat{S}_1(t) - \hat{S}_2(t)]dt. \tag{3.18}
$$

Under the null, Fang et al. (2002) prove that U_s is asymptotically normally distributed. Some regularity conditions are needed for this asymptotic result. Particularly, it is assumed that $V_i - U_i$s are uniformly bounded away from zero. That is, there is no accurately observed event time.

Unfortunately, the asymptotic variance does not have a simple form. It can be obtained as follows. Consider $\hat{S}$, the NPMLE of the common survival function S computed using all observations, and $\hat{F} = 1 - \hat{S}$. Denote $0 < s_1 < \ldots < s_m < \tau$ as the time points at which $\hat{S}$ and $\hat{F}$ have jumps. Denote $a_j = \hat{F}(s_j)$ for $j = 1, \ldots, m$ and $\phi_{\hat{F}}$ as the solution to the following Fredholm integral equation

$$\phi_{\hat{F}}(t) = d_{\hat{F}}(t) \left[W(t) - \int_0^\tau \frac{\phi_{\hat{F}}(t) - \phi_{\hat{F}}(t')}{|\hat{F}(t) - \hat{F}(t')|} h^*(t', t) dt' \right]$$

where

$$d_{\hat{F}}(t) = \frac{\hat{F}(t)[1 - \hat{F}(t)]}{h_1(t)[1 - \hat{F}(t)] + h_2(t)\hat{F}(t)}$$

and $h^*(t', t) = h(t', t) + h(t, t')$. $\phi_{\hat{F}}$ is absolutely continuous with respect to $\hat{F}$ and a step function with jumps at the s_js.

Denote $\hat{H}$, $\hat{H}_1$, and $\hat{H}_2$ as the NPMLEs of H, H_1, and H_2, respectively. Of note, as the censoring times are accurately observed, such estimates can be straightforwardly obtained and are $n^{1/2}$ consistent and point-wise asymptotically normal. Define $y_j = \phi_{\hat{F}}(s_j)$,

$$\Delta_j(h_r) = \int_{s_j}^{s_{j+1}} h_r(t) dt \approx \int_{s_j}^{s_{j+1}} d\hat{H}_r(t)$$

$$\Delta_{jl}(h) = \int_{s_j}^{s_{j+1}} \int_{s_l}^{s_{l+1}} h(u, v) dv du \approx \int_{s_j}^{s_{j+1}} \int_{s_l}^{s_{l+1}} d\hat{H}(u, v)$$

and

$$d_j = \frac{a_j(1 - a_j)}{\Delta_j(h_1)(1 - a_j) + \Delta_j(h_2)a_j}$$

for $j, l = 1, \ldots, m$ and $r = 1, 2$. It can be proved that $\boldsymbol{y} = (y_1, \ldots, y_m)'$ is the unique solution to the linear equations

$$y_j \left[d_j^{-1} + \sum_{l<j} \frac{\Delta_{lj}(h)}{a_j - a_l} + \sum_{l>j} \frac{\Delta_{jl}(h)}{a_l - a_j} \right] = \Delta_j(W) + \sum_{l<j} \frac{\Delta_{lj}(h)}{a_j - a_l} y_l + \sum_{l>j} \frac{\Delta_{jl}(h)}{a_l - a_j} y_l$$

for $j = 1, \ldots, m$. In addition, define

$$\tilde{\theta}_{\hat{F}}(u, v, \delta_1, \delta_2) = -\delta_1 \frac{\phi_{\hat{F}}(u)}{\hat{F}(u)} - \delta_2 \frac{\phi_{\hat{F}}(v) - \phi_{\hat{F}}(u)}{\hat{F}(v) - \hat{F}(u)} + \delta_3 \frac{\phi_{\hat{F}}(v)}{1 - \hat{F}(v)}$$

and

$$\hat{Q}(u, v, \delta_1, \delta_2) = \sum_{i=1}^{n} I(U_i \leq u, V_i \leq v, \delta_{1i} = \delta_1, \delta_{2i} = \delta_2)$$

$$\Delta\hat{H}(U_i, V_i)\hat{F}^{\delta_{1i}}(U_i)[\hat{F}(V_i) - \hat{F}(U_i)]^{\delta_{2i}}[1 - \hat{F}(V_i)]^{1-\delta_{1i}-\delta_{i2}}.$$

Here, $\hat{Q}$ is the empirical distribution function of $(U, V, \delta_1, \delta_2)$. Fang et al. (2002) show that a consistent estimate of the asymptotic variance of U_s is

$$\int \tilde{\theta}_{\hat{F}}^2(u, v, \delta_1, \delta_2)d\hat{Q}(u, v, \delta_1, \delta_2).$$

Inference can then be based on the asymptotic normality and the above variance estimate.

Remarks. With the rank- and survival-based approaches described above, it has been assumed that the event time is continuously distributed. Sun (2006) describes an approach that can accommodate discrete event time. The test statistic can be viewed as a discretization of the survival-based test statistic for continuous survival time described above. In addition, with continuously distributed survival time, there are other survival-based approaches. We refer to Sun (2006) for more detailed discussions. With right censored data, log-rank-based approaches have dominated in popularity and can be realized using many software packages. The test statistics and variance estimates are not even hard to compute manually. With interval censored data, the story is quite different. We have to strive more to evaluate the test statistics and their variances. In general, different approaches present different advantages in computation and interpretation. It is less conclusive which approach should be recommended as the most suitable for a practical data set.

3.4.3   Software Implementation

In the R library `interval`, the `ictest` function has been developed to conduct two- or multiple-sample comparisons. This function implements log-rank and Wilcoxon type tests, with multiple options (for example, exact or asymptotic test). For more details, we refer readers to the library manual. Consider the `cosmesis` data set again. In the `interval` package, this data set is referred to as `bcos`, with variable names "left, right, treatment." An example of using the `ictest` function to conduct two-sample comparison, as provided by the package authors, proceeds as follows

```
## perform a log-rank-type test using the permutation form of
the test data(bcos)
testresult<-ictest(Surv(left,right,type="interval2")~treatment,
scores="log-rank1",data=bcos)
testresult
```

```
## perform a Wilcoxon rank sum-type test
## using asymptotic permutation variance
left<-bcos$left
right<-bcos$right
trt<-bcos$treatment
## save time by using previous fit
ictest(left,right,trt, initfit=testresult$fit, method="pclt",
scores="wmw")
```

In addition, three generalized log-rank tests and a score test for interval-censored data are implemented in the `glrt` package.

3.5 Semiparametric Modeling with Case I Interval Censored Data

In the above sections, we have discussed parametric and nonparametric modeling of interval censored data. The most impressive advantage of parametric modeling is its simplicity. However, parametric models are usually restricted and may suffer from model misspecification. Nonparametric modeling can provide a more honest description of data. However, it does not have lucid interpretations and cannot easily accommodate covariate effects. In practical data analysis, semiparametric models, as a bridge between parametric and nonparametric models, can be preferred. In this section, we investigate semiparametric models for case I interval censored data. We provide detailed investigation of the Cox model because of its popularity and, more importantly, because other semiparametric models can be studied in a very similar manner (note that this is quite different from right censored data). We also study the additive risk model, which provides a useful alternative to the Cox model. With this model, we consider two approaches. The first approach takes advantage of the special form of the likelihood function and is computationally simple, however, not fully efficient. The second approach, on the other hand, can achieve full efficiency at the price of more complicated estimating equations. We also very briefly discuss other semiparametric models including the proportional odds model and the AFT (accelerated failure time) model.

Beyond the event time T and censoring time C, assume that a length-d vector of covariate Z is available. Under case I interval censoring, one observation consists of

$$(C, \delta = I(T \leq C), Z).$$

Assume n i.i.d. observations.

3.5.1 Cox Proportional Hazards Model

For simplicity of notation, assume that $d = 1$ (i.e., there is only one covariate). The estimation and inference procedure described below can be straightforwardly extended to the multiple covarites case. The Cox model assumes that the conditional cumulative hazard function

$$\Lambda(t|Z) = \exp(\beta Z)\Lambda(t), \tag{3.19}$$

where β is the unknown regression coefficient and $\Lambda(t)$ is the baseline cumulative hazard function.

With right censored data, the profile likelihood objective function depends on β only and takes a parametric form, which significantly simplifies the computation. With interval censored data, in general, we cannot profile out the nonparametric baseline function. Thus, we have to jointly consider β and Λ in estimation and inference.

Assume that T and C are conditionally independent given Z and that the joint distribution of (C, Z) does not involve β or Λ. For a single observation $X = (C, \delta, Z)$, up to a constant, the log-likelihood function is

$$l(\beta, \Lambda) = \delta \log(1 - \exp(-\Lambda(C)\exp(\beta Z))) - (1 - \delta)\exp(\beta Z)\Lambda(C).$$

With n i.i.d. observations, let P_n denote the empirical measure. Consider the maximum likelihood estimate

$$(\hat{\beta}, \hat{\Lambda}) = argmax \mathrm{P}_n l(\beta, \Lambda).$$

Let $\{C_{(1)}, \ldots, C_{(n)}\}$ be the order statistics of $C_1, \ldots, C_n$. Denote $\delta_{(i)}$ and $Z_{(i)}$ as the event indicator and covariate corresponding to $C_{(i)}$. Following arguments similar to those for the estimation of F, Λ only has jumps at C_is. Here we take $\hat{\Lambda}(t)$ as a right-continuous increasing step function. In addition, we specify that $\hat{\Lambda}(t) = 0$ for $t < C_{(1)}$. $\hat{\Lambda}(t)$ for $t > C_{(n)}$ can be specified as any constant of convenience, for example $\hat{\Lambda}(C_{(n)})$. We now have $\hat{\Lambda}(t)$ completely specified based on $\hat{\Lambda}(C_{(i)})$s.

Assume that $\delta_{(1)} = 1$ and $\delta_{(n)} = 0$. This assumption does not have any real effect on the downstream analysis. Since if $\delta_{(1)} = 0$ or $\delta_{(n)} = 1$, then we can have $\hat{\Lambda}(C_{(1)}) = 0$ or $\hat{\Lambda}(C_{(n)}) = \infty$. So in the log-likelihood function $\mathrm{P}_n \ell(\hat{\beta}, \hat{\Lambda})$, the terms associated with $\delta_{(1)} = 0$ or $\delta_{(n)} = 1$ are zero. That is, they have no contribution to $\mathrm{P}_n \ell(\hat{\beta}, \hat{\Lambda})$.

The MLE $(\hat{\beta}, \hat{\Lambda})$ has the following properties

$$\sum_{i=1}^{n}\left\{\delta_i \frac{\exp(-\hat{\Lambda}(C_i)\exp(\hat{\beta}Z_i))}{1 - \exp(-\hat{\Lambda}(C_i)\exp(\hat{\beta}Z_i))} - (1 - \delta_i)\right\} \times \hat{\Lambda}(C_i)\exp(\hat{\beta}Z_i)Z_i = 0,$$

$$\sum_{j \geq i}\left\{\delta_{(j)} \frac{\exp(\hat{\beta}Z_{(j)})\exp(-\exp(\hat{\beta}Z_{(j)})\hat{\Lambda}(C_{(j)}))}{1 - \exp(-\exp(\hat{\beta}Z_{(j)})\hat{\Lambda}(C_{(j)}))} - (1 - \delta_{(j)})\exp(\hat{\beta}Z_{(j)})\right\} \leq 0,$$

for $i = 1, \ldots, n$,

$$\sum_{i=1}^{n} \left\{ \delta_{(i)} \frac{\exp(\hat{\beta} Z_{(i)}) \exp(-\exp(\hat{\beta} Z_{(i)}) \hat{\Lambda}(C_{(i)}))}{1 - \exp(-\exp(\hat{\beta} Z_{(i)}) \hat{\Lambda}(C_{(i)}))} - (1 - \delta_{(i)}) \exp(\hat{\beta} Z_{(i)}) \right\} \hat{\Lambda}(C_{(i)}) = 0.$$

The first equation can be established simply by taking the first-order derivative of $\mathrm{P}_n l$ with respect to β and evaluated at $\hat{\beta}$. The last two equations can be proved following Proposition 1.1 of Groeneboom and Wellner (1992). Together, these three equations can lead to the following computational algorithm.

3.5.1.1 Computation

With the Cox model, there are two unknown parameters β and Λ. In principle, it is possible to maximize over both of them simultaneously. In practice, consider the following iterative algorithm

(a) Initialize $\hat{\beta} = 0$ (component-wise if $d > 1$).

(b) With the current estimate $\hat{\beta}$, compute the estimate $\hat{\Lambda}$ by maximizing $\mathrm{P}_n l(\hat{\beta}, \Lambda)$ with respect to Λ under the constraint that Λ is a right continuous, nonnegative, nondecreasing step function.

(c) With the current estimate $\hat{\Lambda}$, compute the estimate $\hat{\beta}$ by maximizing $\mathrm{P}_n l(\beta, \hat{\Lambda})$ with respect to β.

(d) Repeat Steps (b) and (c) until convergence.

With the above algorithm, we iteratively maximize over β and Λ. Convergence can be concluded if the difference between two consecutive estimates is smaller than a predefined cut-off (say, 0.01). Maximizing over β is a parametric problem and can be realized using the Newton-Raphson and related algorithms. A more challenging step is (b), which involves a constraint optimization. Step (b) can be achieved using the "pool-adjacent-violators" (PAV) algorithm, which has been designed for generic isotonic regression problems. For reference, see Robertson et al. (1998). In R, the `isotone` function in the `DDHFm` library conducts PAV, although under a much simpler setting. With the PAV algorithm, the number of nonlinear equations that need to solved is of the order at least $O(n)$, which may lead to a high computational cost with moderate to large n. An alternative, computationally more affordable approach proceeds as follows. First, we define the processes W_Λ, G_Λ, and V_Λ as

$$W_\Lambda(c) = \int_{c'=0}^{c} \left\{ I(t \le c') \frac{\exp(-\exp(\beta z) \Lambda(c'))}{1 - \exp(-\exp(\beta z) \Lambda(c'))} - I(t > c') \exp(\beta z) \right\} dQ_n(t, c', z)$$

$$G_\Lambda(c) = \int_{c'=0}^{c} I(t \le c') \frac{\exp(2\beta z) \exp(-\exp(\beta z) \Lambda(c'))}{[1 - \exp(-\exp(\beta z) \Lambda(c'))]^2} dQ_n(t, c', z)$$

and

$$V_\Lambda(c) = W_\Lambda(c) + \int_0^c \Lambda(c')dG_\Lambda(c'), \ c \geq 0.$$

Here, Q_n is the empirical measure of the (unobservable) points $\{(T_i, C_i, Z_i), i = 1, \ldots, n\}$.

Huang (1996) shows that for any fixed β, $\hat{\Lambda}$ maximizes $P_n l(\beta, \Lambda)$ if and only if $\hat{\Lambda}$ is the left derivative of the greatest convex minorant of the "self-induced" cumulative sum diagram, consisting of the points

$$\{G_{\hat{\Lambda}}(C_{(j)}), V_{\hat{\Lambda}}(C_{(j)})\}, \ j = 1, \ldots, n$$

and the origin (0, 0).

The above result suggests an iterative procedure to compute $\hat{\Lambda}$ for any fixed $\hat{\beta}$. First we initialize the estimate by treating interval censored data as right censored data. Suppose that $\hat{\Lambda}^k$ is obtained at the kth iteration. Then $\hat{\Lambda}^{k+1}$ is computed as the left derivative of the convex minorant of the cumulative sum diagram, consisting of the points

$$\{G_{\hat{\Lambda}^k}(C_{(j)}), V_{\hat{\Lambda}^k}(C_{(j)})\}, \ j = 1, \ldots, n$$

and the origin (0,0). This procedure iterates until convergence.

The above computational algorithm is proposed in Huang (1996). Several followup studies have adopted modifications of this algorithm and shown its effectiveness. Alternative computational algorithms are described in Sun (2006).

In R, the `intcox` function in the `intcox` package computes the Cox model estimate using the iterative convex minorant algorithm. Application of this function is very similar to that of `coxph` for right censored data.

3.5.1.2   Asymptotic Results

The Cox model is semiparametric. Establishing its asymptotic estimation properties is nontrivial. As martingale theories—which are the main technical tools with right censored data—are no longer applicable, empirical process techniques need to be employed. Below we describe the asymptotic estimation results with the Cox model. For more general theories underlying such results, we refer to van der Vaart (2000) and van der Vaart and Wellner (1996). We first make the following assumptions.

(A1) The true parametric regression coefficient $\beta_0 \in \Theta$, which is a bounded subset of $\mathbb{R}^d$. Usually it is also assumed that β_0 is an interior point of Θ.

(A2) (a) The covariate Z has bounded support. (b) For any $\tilde{\beta} \neq \beta_0$, the probability $P(\tilde{\beta}Z \neq \beta_0 Z) > 0$.

(A3) $F(0) = 0$. Let $\tau_F = inf\{t : F(t) = 1\}$. The support of C is an interval $[l_C, u_C]$ and $0 \leq l_C \leq u_C < \tau_F$. That is, the support of T dominates that of C.

(A3)* Everything is the same as (A3) except that it is now assumed that $0 < l_C \leq u_C < \tau_F$.

(A4) The true cumulative baseline hazard function Λ_0 has a strictly positive derivative on $[l_C, u_C]$, and the joint distribution $G(c, z)$ of (C, Z) has a bounded second-order (partial) derivative with respect to c.

The boundedness assumptions are made to control the "size" of the space of likelihood function, which can be measured by entropy (see the Appendix, Section 3.8). It is worth noting that the boundedness assumptions are for a theoretical purpose. In practice, the actual bounds may remain unknown. Making such assumptions seems to have little impact on practical data analysis.

With parametric models, the MLEs of the parametric parameters are usually $n^{1/2}$ consistent and asymptotically normally distributed. A necessary condition is that the *Fisher information (matrix) is component-wise finite and not singular.* Such a condition, although not completely trivial, can be easily satisfied with most parametric models. With semiparametric models, such a condition is nontrivial. We refer to Groeneboom and Wellner (1992) for more general investigations of the information calculation with semiparametric models. For the Cox model with case I interval censored data, the information calculation is as follows. More details are provided in the Appendix (Section 3.8).

Denote

$$Q(c, \delta, z) = \delta \frac{S(c|z)}{1 - S(c|z)} - (1 - \delta)$$

and

$$O(c|z) = E[Q^2(C, \delta, Z)|C = c, Z = z] = \frac{S(c|z)}{1 - S(c|z)}.$$

Suppose that assumptions (A2) and (A3)* are satisfied. Huang (1996) shows that

(a) The efficient score function for β is

$$l_\beta^* = \exp(\beta Z)Q(c, \delta, z)\Lambda(c) \left\{ z - \frac{E((Z \exp(2\beta Z))O(C|Z)|C = c)}{E((\exp(2\beta Z))O(C|Z)|C = c)} \right\}.$$

(b) The Fisher information for β is

$$I(\beta) = E[l_\beta^*]^{\otimes 2} = E \left\{ R(C, Z) \left[Z - \frac{E(ZR(C, Z)|C)}{E(R(C, Z)|C)} \right]^{\otimes 2} \right\},$$

where $a^{\otimes 2} = aa^T$ for $a \in \mathbb{R}^d$ and $R(C, Z) = \Lambda^2(C|Z)O(C|Z)$.

As can be seen from the above formulation, the Fisher information does not have a simple form. Usually it is hard to derive simple sufficient conditions under which the Fisher information is nonsingular. Rather, the nonsingularity needs to be *assumed*.

Suppose that assumptions (A1), (A2), and (A3) are satisfied. Then the estimation consistency result can be summarized as follows:

$$\hat{\beta} \to \beta_0 \quad a.s.,$$

and if $c \in [l_C, u_C]$ is a continuity point of F,

$$\hat{\Lambda}(c) \to \Lambda_0(c) \quad a.s.$$

Moreover, if F is continuous, then

$$\sup_{c \in [l_C, u_C]} |\hat{\Lambda}(c) - \Lambda_0(c)| \to 0, \quad a.s.$$

The second step is to establish the rate of convergence. As established earlier in this chapter, under nonparametric modeling, the rate of convergence is $n^{1/3}$. With semiparametric models, there are additional parametric parameters. Intuitively, the rate of convergence of the nonparametric parameter estimates cannot be faster than $n^{1/3}$. Then the question becomes whether a $n^{1/3}$ rate can be achieved.

Assume that the support of the censoring time C is finite and is strictly contained in the support of the event time T. Under this assumption, Λ_0 is bounded away from 0 and ∞ on $[l_C, u_C]$. Recall that the true parameter value β_0 is assumed to be in a bounded set. Under this assumption and the consistency of $\hat{\beta}$, we are able to focus on a bounded set for the estimation of β_0. With the boundedness of Λ_0 and consistency of $\hat{\Lambda}$, we restrict our estimation of Λ_0 to the following class of functions:

$$\Phi = \{\Lambda : \Lambda \text{ is increasing and } 0 < 1/M \leq \Lambda(c) \leq M < \infty \text{ for all } c \in [l_C, u_C]\},$$

where M is a positive constant. In practice, a very large value of M can be chosen so that this assumption has little impact. In theoretical development, this assumption is crucial and can significantly reduce the complexity.

Define the distance d on $\mathbb{R}^d \times \Phi$ as

$$d((\beta_1, \Lambda_1), (\beta_2, \Lambda_2)) = |\beta_1 - \beta_2| + ||\Lambda_1 - \Lambda_2||,$$

where $|\beta_1 - \beta_2|$ is the Euclidean distance in $\mathbb{R}^d$,

$$||\Lambda_1 - \Lambda_2||_2 = \left[\int (\Lambda_1(c) - \Lambda_2(c))^2 dQ_C(c)\right]^{1/2},$$

and Q_C is the marginal probability measure of the censoring time C. The rate of convergence result can be summarized as follows.

Assume that assumptions (A1), (A2), and (A3)* are satisfied. Then

$$d((\hat{\beta}, \hat{\Lambda}), (\beta_0, \Lambda_0)) = O_p(n^{-1/3}).$$

The above result shows that the nonparametric parameter estimate can achieve the $n^{1/3}$ convergence rate. Considering the $n^{1/3}$ convergence rate under nonparametric modeling, we conclude that this rate for $\hat{\Lambda}$ cannot be further improved without making stronger assumptions. For semiparametric models, van der Vaart and Wellner (1996) establish that it is possible to "separate" the parametric parameter estimation from nonparametric parameter estimation. Thus, for the parametric parameter estimate, it is still possible to achieve the $n^{1/2}$ convergence rate and asymptotic normality, despite the slow convergence rate of the nonparametric parameter estimate. However, such a result is nontrivial and needs to be established on a case-by-case basis. For the present model, we have the following asymptotic distribution result.

Suppose that β_0 is an interior point of Θ and that assumptions (A1), (A2), (A3)* and (A4) are satisfied. In addition, assume that the Fisher information matrix $I(\beta)$ evaluated at β_0 is nonsingular. Then

$$n^{1/2}(\hat{\beta} - \beta_0) \to_d N(0, I(\beta_0)^{-1}).$$

The above result shows that $\hat{\beta}$ is asymptotically efficient in the sense that any regular estimator has asymptotic variance matrix no less than that of $\hat{\beta}$. For more details on the asymptotic efficiency of parametric parameter estimation in semiparametric models, we refer the readers to Bickel et al. (1993).

3.5.1.3   Inference

The Cox model has two unknown parameters. Λ describes the baseline hazard of the whole cohort. The estimate $\hat{\Lambda}$ only has $n^{1/3}$ convergence rate and its asymptotic distribution is not normal. In contrast, the parameter β describes the difference between subjects with different characteristics (covariates). The estimate $\hat{\beta}$ is $n^{1/2}$ consistent and asymptotically normally distributed. Thus, in practice, inference on $\hat{\beta}$ is of more interest.

In what follows, we describe two inference approaches. The first is a "plug-in" approach. This approach involves first deriving the analytic form of the Fisher information matrix. Then each component in the Fisher information is separately, consistently estimated. Under some regularity conditions, the overall estimate of the information matrix (and hence asymptotic variance) can be consistent. This approach has been extensively adopted with parametric models and simple semiparametric models. The second approach is a bootstrap approach. Compared with the plug-in approach, the bootstrap approach is computationally more expensive. However, as it does not demand the exact form of the Fisher information, it can be potentially more broadly applicable.

The Plug-In Approach.   For the Cox model with case I interval censoring, this approach is proposed in Huang (1996) and proceeds as follows. In the expression of $I(\beta)$, $R(c,z)$ is defined as

$$
\begin{aligned}
R(c,z) &= \Lambda^2(c|z)O(c|z) \\
&= \exp(2\beta z)\Lambda^2(c)\frac{\exp(-\exp(\beta z)\Lambda(c))}{1-\exp(-\exp(\beta z)\Lambda(c))}.
\end{aligned}
$$

Denote

$$
\hat{R}(c,z) = \exp(2\hat{\beta}z)\hat{\Lambda}^2(c)\frac{\exp(-\exp(\hat{\beta}z)\hat{\Lambda}(c))}{1-\exp(-\exp(\hat{\beta}z)\hat{\Lambda}(c))}.
$$

Let $\mu_1(c) = E(R(C,Z)|C=c)$ and $\mu_2(c) = E(ZR(C,Z)|C=c)$. Then if we can obtain reasonable estimates $\hat{\mu}_1$ and $\hat{\mu}_2$ for $\mu_1(c)$ and $\mu_2(c)$, we can estimate $I(\beta)$ by

$$
\hat{I}(\beta) = \frac{1}{n}\sum_{i=1}^{n}\left\{\hat{R}(C_i,Z_i)\left[Z_i - \frac{\hat{\mu}_2(C_i)}{\hat{\mu}_1(C_i)}\right]^{\otimes 2}\right\}.
$$

When Z is a continuous covariate vector, $\mu_1(c) = E(R(C,Z)|C=c)$ can be approximated by $E(\hat{R}(C,Z)|C=c)$. Then we can estimate $E(\hat{R}(C,Z)|C=c)$ by nonparametric regression approaches, which usually involve smoothing such as kernel estimation. For details, we refer to Stone (1977). When Z is a categorical covariate, the nonparametric smoothing procedure does not work well because of the discrete nature of the values of $\hat{R}(c,z)$. Consider the simplest case where Z is binary. Assume that $P(Z=1) = \gamma$ and $P(Z=0) = 1-\gamma$. Then

$$
E(ZR(C,Z)|C=c) = R(c,1)P(Z=1|C=c) = R(c,1)f_1(c)\gamma/f(c),
$$

and

$$
\begin{aligned}
E(R(C,Z)|C=c) &= R(c,1)P(Z=1|C=c) + R(c,0)P(Z=0|C=c) \\
&= \frac{R(c,1)f_1(c)\gamma}{f(c)} + \frac{R(c,0)f_0(c)(1-\gamma)}{f(c)},
\end{aligned}
$$

where $f_1(c)$ is the conditional density of C given $Z=1$, $f_0(c)$ is the conditional density of C given $Z=0$, and $f(c)$ is the marginal density of C. Note that we only need to estimate the ratio of the above two conditional expectations. $f(c)$ will cancel when we take the ratio of the two conditional expectations. γ can be estimated by the total number of subjects in the treatment group with $Z=1$ divided by the sample size. Let $\hat{f}_1(c)$ be a kernel density estimator of f_1 and $\hat{f}_0$ be a kernel density estimator of f_0. Then a natural estimator of $E(ZR(C,Z)|C=c)/E(R(C,Z)|C=c)$ is

$$
\hat{\mu}(c) = \frac{\hat{R}(c,1)\hat{f}_1(c)\hat{\gamma}}{\hat{R}(c,1)\hat{f}_1(c)\hat{\gamma} + \hat{R}(c,0)\hat{f}_0(c)(1-\hat{\gamma})}.
$$

Here, $\hat{R}(c,z)$ is defined above. With a proper choice of the bandwidth and kernel in the estimation of $f_1(c)$ and $f_0(c)$, the above estimator is consistent. Hence, a reasonable estimator of $I(\beta)$ is

$$\hat{I}(\hat{\beta}) = \frac{1}{n} \sum_{i=1}^{n} \{\hat{R}(C_i, Z_i)(Z_i - \hat{\mu}(C_i))^2\}.$$

In the special case when C and Z are independent, the above nonparametric smoothing is not necessary. In this case, we have

$$I(\beta) = E\left\{ R(C,Z) \left[Z - \frac{E_Z(ZR(C,Z))}{E_Z(R(C,Z))} \right]^{\otimes 2} \right\},$$

where E_Z is the expectation with respect to Z. We can simply estimate $I(\beta)$ by

$$\hat{I}(\hat{\beta}) = \frac{1}{n} \sum_{i=1}^{n} \left\{ \hat{R}(C_i, Z_i) \left[Z_i - \frac{\sum_{j=1}^{n} Z_j \hat{R}(C_j, Z_j)}{\sum_{j=1}^{n} \hat{R}(C_j, Z_j)} \right]^{\otimes 2} \right\}.$$

With case I interval censored data, the Cox and a few other semiparametric models have closed-form Fisher information matrices. Thus it is possible to employ the plug-in approach. It is worth noting that with those models, the plug-in approach involves multiple smoothing estimation, which, theoretically and computationally, is nontrivial. Numerically, our limited experience suggests that the plug-in approach tends to work well when the dimensionality of Z is low and when the distribution of Z is simple. With more complicated models, for example, the Cox model with partly linear covariate effects, the Fisher information matrices do not have closed forms. Below we describe a bootstrap approach. Compared with the plug-in approach, it has the advantage of not demanding a closed form of the Fisher information matrix, not involving smoothing estimation, and being more broadly applicable.

The Weighted Bootstrap. Recall that the MLE is defined as

$$(\hat{\beta}, \hat{\Lambda}) = argmax\frac{1}{n} \sum_{i=1}^{n} l(\beta, \Lambda; D_i),$$

where $D_i = (C_i, \delta_i, Z_i)$ denotes the ith observation.
Consider the *weighted MLE*, which is defined as

$$(\tilde{\beta}, \tilde{\Lambda}) = argmax\frac{1}{n} \sum_{i=1}^{n} w_i \times l(\beta, \Lambda; D_i),$$

where w_is are the independent continuously distributed positive weights satisfying $E(W) = var(W) = 1$. Ma and Kosorok (2005) show that conditional on the observed data, $\tilde{\beta} - \hat{\beta}$ has the same asymptotic variance as $\hat{\beta} - \beta_0$.
The above result suggests the following inference procedure.

(a) Generate n i.i.d. random weights. A large number of distributions are positive and satisfy the mean and variance requirements. A simple distribution, which has been adopted in multiple practical data analysis, is the exp(1) distribution.

(b) Compute the weighted MLE $\tilde{\beta}$.

(c) Repeat Steps (a) and (b) B (for example $B=1000$) times. Compute the sample variance of the weighted MLEs, which provides a reasonable estimate of the variance of $\hat{\beta}$.

The rationale of the weighted bootstrap is as follows. It is well known that the ordinary nonparametric bootstrap can be viewed as a weighted estimation, with the weights having the multinomial distribution. With the nonparametric bootstrap, some observations may show up multiple times in the weighted estimation, while some others may not be included at all (i.e., having zero weights). The weighted bootstrap described above shares a similar strategy as the nonparametric bootstrap. The key difference is that *all observations have nonzero weights and contribute to the weighted estimation.* Intuitively, the weighted bootstrap can be more "smooth" than the nonparametric bootstrap. This seemingly small difference makes properties of the weighted bootstrap much easier to establish.

The weighted bootstrap only involves computing the weighted MLEs, which can be realized using the same algorithm and computer code as for the ordinary MLE. It is worth noting that no smoothing estimation is involved in this procedure. Another advantage of the weighted bootstrap is its generic applicability. Examining Ma and Kosorok (2005) suggests that for many semiparametric models, particularly including *all* of the semiparametric models for case I and case II interval censored data studied in this chapter, the weighted bootstrap is valid without making additional assumptions. On the negative side, the weighted bootstrap involves a large number of weighted estimation and can be computationally expensive.

Consider the lung tumor data provided in Table 3.1. Introduce the covariate Z, with $Z = 1$ corresponding to the conventional mice and $Z = 0$ corresponding to the germ-free mice. Assume the Cox model. The estimation procedure described above leads to the estimate

$$\hat{\beta} = -0.55 \ (0.29),$$

where the number in "()" is the estimated standard error using the plug-in approach. The weighted bootstrap approach with exp(1) weights leads to a similar estimated standard error. With the asymptotic normality, we can construct the 95% confidence interval for β_0 as (-1.12, 0.01). The p-value for testing $\beta_0 = 0$ is 0.054. Thus, under the Cox model, there is no significant difference between the two groups.

In the above sections, we have considered the standard Cox model and the maximum likelihood estimation, which is the "default" estimation approach.

In the following sections, we consider several useful extensions, including different estimation procedures and alternative models.

3.5.1.4 Least Squares Estimation

The likelihood-based approach is fully efficient, with the asymptotic variance equal to the inverse of the Fisher information matrix. An alternative estimation approach, which is not fully efficient but more closely mimics linear regression analysis, is the least squares approach. Note that this estimation approach is not necessarily simpler than the likelihood approach. However, it does provide further insight into the Cox model and its properties under interval censoring.

Consider the estimate

$$(\hat{\beta}_{LSE}, \hat{\Lambda}_{LSE}) = argmin_{\beta, \Lambda} \sum_{i=1}^{n} (1 - \delta_i - \exp(-\exp(\beta Z_i)\Lambda(C_i)))^2.$$

This estimate has been investigated in Ma and Kosorok (2006b). It has been motivated by the fact that when the event and censoring are independent, $E(\delta) = E(1 - \exp(-\exp(\beta Z)\Lambda(C)))$. Of note, although the objective function has a simpler form than the likelihood function, the least squares approach is *not* computationally simpler, because of the presence of Λ.

With a finite sample, properties of the least squares estimate are similar to those for the MLE. For computation, results can be developed in a parallel manner as described in the above sections. Because of the significant overlap, we will not repeat those results. Examining the proofs of asymptotic properties described in the Appendix (Section 3.8) as well as the Bibliography suggests that such results are applicable to general M-estimation, particularly including the least squares estimation. First, assumptions similar to those made for the MLE are needed. Following the proofs for the MLE, we can show that

$$||\hat{\Lambda}_{LSE} - \Lambda_0||_2 = O_p(n^{-1/3})$$

and

$$n^{1/2}(\hat{\beta}_{LSE} - \beta_0) \to N(0, I_{LSE}^{-1}).$$

in distribution. That is, the convergence rate and asymptotic distribution results are quite similar to those for the MLE. The key difference is that I_{LSE} *is no longer the Fisher information matrix*. To calculate the asymptotic variance, we first make the following notations:

$$m_1 = 2Ze^{\beta Z}\Lambda \exp(-e^{\beta Z}\Lambda)(1 - \delta - \exp(-e^{\beta Z}\Lambda)),$$
$$m_2[a] = 2e^{\beta Z}\exp(-e^{\beta Z}\Lambda)(1 - \delta - \exp(-e^{\beta Z}\Lambda))a,$$
$$L(\beta, \Lambda) \equiv 2ze^{\beta Z}\exp(-e^{\beta Z}\Lambda)$$
$$((1 - \Lambda e^{\beta Z})(1 - \delta - \exp(-e^{\beta Z}\Lambda)) + \Lambda e^{\beta Z}\exp(-e^{\beta Z}\Lambda)),$$
$$N(\beta, \Lambda) \equiv -2e^{2\beta Z}\exp(-e^{\beta Z}\Lambda)\left[1 - \delta - 2\exp(-e^{\beta Z}\Lambda)\right], \quad \text{and}$$
$$A^* = E(L(\beta, \Lambda)|C)/E(N(\beta, \Lambda)|C).$$

Here, the definition of the a function is similar to that in the Appendix (Section 3.8) for the MLE. We also set

$$
\begin{aligned}
m_{11} = \ & 2Z^2 \Lambda e^{\beta Z} \exp(-e^{\beta Z}\Lambda) \\
& \times \left[(1 - \Lambda e^{\beta Z})(1 - \delta - \exp(-e^{\beta Z}\Lambda)) + \Lambda e^{\beta Z} \exp(-e^{\beta Z}\Lambda) \right]
\end{aligned}
$$

and $m_{21}[a] = L(\beta, \Lambda)a$. Then

$$
I_{LSE} = \{ \mathrm{P}(m_{11} - m_{21}[A^*]) \}^{-1} \mathrm{P}[m_1 - m_2[A^*]]^2 \{ \mathrm{P}(m_{11} - m_{21}[A^*]) \}^{-1}.
$$

This result can be obtained using the projection of score function and sandwich variance estimation approach. More details on this calculation are provided in Ma and Kosorok (2006b). In general, it is expected for I_{LSE} to be smaller than or equal to the Fisher information matrix. Thus, the least squares estimate would have a larger variance than the MLE. For inference of $\hat{\beta}_{LSE}$, it is possible to develop a plug-in approach following Huang (1996). In addition, the weighted bootstrap is directly applicable.

3.5.1.5　Smooth Estimation of Cumulative Baseline Hazard

In the above sections, estimate of Λ is specified to be a step function. Asymptotically, it is only $n^{1/3}$ consistent. Such an estimate can be difficult to interpret or use in downstream analysis. A similar problem has been discussed for the estimate of the distribution function F under nonparametric modeling. In this section, we consider the scenario where Λ_0 is assumed to be a smooth function. Such an assumption is often sensible in practical data analysis. As with nonparametric modeling, we describe the smoothness of Λ_0 by assuming that it belongs to a certain Sobolev space.

First consider the scenario that $\Lambda \in \mathbb{S}_{s_0}$, the Sobolev space indexed by the *known* order of derivative s_0. A commonly made assumption is that $s_0 = 2$, which corresponds to the space of cubic splines. Consider the penalized MLE (PMLE) defined as

$$
(\hat{\beta}, \hat{\Lambda}) = argmax \left\{ \mathrm{P}_n l(\beta, \Lambda) - \lambda^2 J^2(\Lambda) \right\}.
$$

Consider the reparameterization

$$
\Lambda(t) = \int_0^t \lambda(u)du = \int_0^t \exp(a(u))du.
$$

$\Lambda(t)$ is a nonnegative, nondecreasing function, leading to a constraint optimization problem and may demand special computational algorithms. With the reparameterization, there is no constraint on the function a. Consider the smoothness penalty

$$
J^2(\Lambda) = I(s_0 = 1) \int (\Lambda^{(1)}/\Lambda)^2 dc + I(s_0 \neq 1) \int [a^{(s_0 - 1)}(c)]^2 dc.
$$

λ is a data-dependent tuning parameter and can be determined via cross-validation (Wahba, 1990). $\Lambda^{(1)}$ is the first-order derivative of Λ, $a^{(s_0-1)}$ is the $s_0 - 1$th order derivative of a, and the integration is over the support of C.

The tuning parameter λ controls the degree of smoothness. For example as $\lambda \to \infty$, $\hat{\Lambda}$ becomes flat. Asymptotically, it can be proved that if $\lambda^2 = O(n^{-\frac{2s_0}{2s_0+1}})$, then

$$\hat{\Lambda} \in \mathbb{S}_{s_0}.$$

That is, the estimate has "a proper degree of smoothness." In addition,

$$n^{-\frac{s_0}{2s_0+1}}(\hat{\Lambda}(c) - \Lambda(c)) = O_p(1), \qquad (3.20)$$

at a specific time point c. With $s_0 > 1$, a convergence rate faster than $n^{1/3}$ can be obtained. In addition, if a stronger assumption on smoothness can be made, then a faster convergence rate can be achieved. The intuition is that a stronger smoothness assumption leads to a smaller functional space measured with entropy for the unknown nonparametric parameter, resulting in more accurate estimation. It is also worth noting that with a finite s_0, the convergence rate is always slower than $n^{1/2}$. That is, a parametric convergence rate cannot be achieved under a nonparametric assumption.

When investigating the asymptotic properties of $\hat{\beta}$, we first note that imposing the smoothness assumption on Λ_0 has little impact on the Fisher information calculation. The difference is that the perturbation and projection are made in the spaces corresponding to $\mathbb{S}_{s_0}$. Under the assumption of nonsingular Fisher information, it can be proved that as $n \to \infty$

$$n^{1/2}(\hat{\beta} - \beta) \to N(0, I^{-1}(\beta)) \qquad (3.21)$$

in distribution. For detailed proofs of the asymptotic properties and discussions on the computational aspect, we refer to the readers to Ma and Kosorok (2006b).

In the above development, it has been *assumed* that s_0, the degree of smoothness, is known *a priori*. The resulted convergence rate of $\hat{\Lambda}$ heavily depends on this assumption. The most commonly made assumption is that $s_0 = 2$, which is mainly for theoretical and computational convenience. In practice, it is usually difficult to verify this assumption. If in fact $s_0 > 2$, but an assumption of $s_0 = 2$ is made, then the resulting estimate has a suboptimal convergence rate. Even though in most practical cases $s_0 = 2$ may suffice, in terms of methodology and theory, it is still of interest to pursue the best possible convergence rate.

We now consider an alternative approach with a weaker assumption, which postulates that

$$\Lambda_0 \in \mathbb{S}_{s_0}, \text{ the Sobolev space indexed by the } unknown \ s_0.$$

Furthermore, it is assumed that

$$\exists \text{ a } known \text{ integer } s_{max}, \text{ s.t., } 1 \leq s_0 \leq s_{max}$$

Under this revised condition, we assume that the order of smoothness cannot be over s_{max}, however, the exact order is unknown. In practice, to be cautious, a large value of s_{max} can be used. Now that s_0 is unknown, the above approach is not directly applicable. Ma and Kosorok (2006b) describe a penalization approach, which can *adaptively* determine the order of smoothness and obtain the optimal convergence rate.

Consider the revised PMLE

$$(\hat{\beta}, \hat{\Lambda}) = argmax_{1 \leq s \leq s_{max}} argmax \left\{ P_n l(\beta, \Lambda) - \lambda_s^2 \left[J_s^2(\Lambda) + \lambda_0^2 \right] \right\}.$$

Here λ_s is a data-dependent tuning parameter.

$$J_s^2(\Lambda) = I(s = 1) \int (\Lambda^{(1)}/\Lambda)^2 dc + I(s \neq 1) \int [a^{(s-1)}(c)]^2 dc,$$

where notations have similar definitions as described above. λ_0 is a *model-dependent* constant. Theoretically, the existence and determination of λ_0 have been established in Ma and Kosorok (2006b). In a simulation study, it is shown that fixing $\lambda_0 = 1$ can generate satisfactory results under a variety of scenarios.

With this revised approach, the penalized MLE is computed for each $s \ (= 1, \ldots, s_{max})$ separately, and the one that leads to the largest value of the penalized objective function is chosen as the final estimate. Compared with the ordinary penalized estimation, an extra model-dependent constant λ_0 is introduced, which plays an important role in controlling the size of the nonparametric parameter estimate.

To establish the asymptotic properties, assume that

(a) $\lambda_s^2 = O_p(n^{-2s/(2s+1)})$;

(b) $J_s^2(\hat{\Lambda}) = o_p(n^{1/6})$.

Assumption (a) is the same as that with the approach described above. Under most smoothing spline settings, it is possible to show that $J_s(\hat{\Lambda}) = O_p(1)$. Thus, assumption (b) is achievable under most reasonable circumstances. As $n \to \infty$, under mild regularity conditions, it can be proved that

$$||\hat{\Lambda} - \Lambda_0||_2^2 = \int (\hat{\Lambda}(c) - \Lambda_0(c))^2 dc = O_p(n^{-2s_0/(2s_0+1)}).$$

This result suggests that the revised PMLE can *adaptively* "find" the proper order of smoothness and achieve the optimal convergence rate, which marks a significant improvement over the approach described before. The detailed proof is available in Ma and Kosorok (2006b). As it is rather technical and lacks statistical insight, we do not provide it here. For $\hat{\beta}$, it can be proved that it is $n^{1/2}$ consistent, asymptotically normal, and efficient. For inference with β, the plug-in approach and weighted bootstrap are applicable.

We note that there exist functions that belong to $\mathbb{S}_s$ for any s, for example, the exponential function. In this case, for any prespecified s_{max}, $\|\hat{\Lambda} - \Lambda_0\|_2^2 = O_p(n^{-2s_{max}/(2s_{max}+1)})$. That is, the estimate achieves the best convergence rate allowed by the assumption. In this case, the approach described above does not have full adaptivity. Wahba and Wendelberger (1980) and Ma and Kosorok (2006b) suggest that in practice, taking s_{max}=5 or 6 yields satisfactory results in most situations.

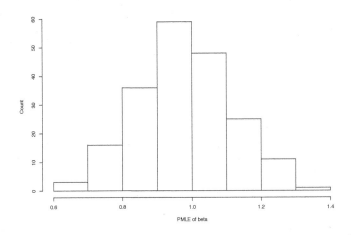

FIGURE 3.2: Simulation study: histogram of $\hat{\beta}$ over 200 replicates.

We use a small simulation study to compare the performance of MLE (without the smoothness assumption) and PMLE (under the smoothness assumption). For simplicity, we consider only a one-dimensional covariate. The event times are generated from the Cox model with $\beta_0 = 1$, the covariate $Z \sim Unif[-1.3, 1.3]$, and $\Lambda_0(t) = \int_0^t \exp(u^2/32 - 1)du$. The censoring times are assumed to be exponentially distributed and independent of the event times. The observations are limited to the time interval $[0.2, 5]$. We simulate 200 realizations for a sample size equal to 400. In Figure 3.2, we provide the histogram of $\hat{\beta}$, showing that the marginal distribution appears to be Gaussian. For inference, we adopt the weighted bootstrap approach with $\exp(1)$ weights. The empirical 95% confidence intervals have coverage ratio 0.955, based on 200 bootstraps for each dataset. In Figure 3.3, the plot of PMLE $\hat{\Lambda}$ versus time and corresponding point-wise 95% confidence intervals shows satisfying coverage. It can be seen that *the PMLE has tighter confidence intervals for estimating Λ_0*, which justifies the value of making stronger smoothness assumptions and conducting penalized estimation.

To the best of our knowledge, there is still no user-friendly software implementing the approaches described in this section. In practice, researchers will have to select a proper set of basis functions, transform the nonparametric op-

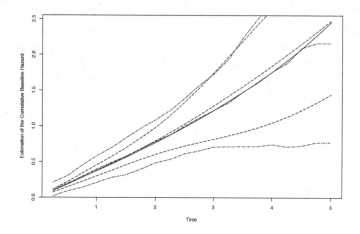

FIGURE 3.3: Simulation study: plot of the PMLE and MLE of Λ_0 versus time. The solid line is the true cumulative baseline hazard. The dashed lines are the PMLE and corresponding point-wise 95% confidence intervals. The dot-dashed lines are the MLE and corresponding 95% confidence intervals.

timization problem into a parametric one, and solve the parametric problem using for example Newton approaches.

3.5.1.6   Partly Linear Covariate Effects

Under the standard Cox model, it is assumed that the covariate effect is linear. In practice, such an assumption may limit the applicability of the Cox model, as many practical biological processes are nonlinear. As an extension of the standard Cox model, Ma and Huang (2005) consider a Cox model with *partly linear covariate effect*. A similar model has been previously considered for right censored data.

Assume that for each subject, two covariates (Z, X) are measured. Extension to more covariates may demand advancement in computation but is theoretically trivial. Consider the Cox model

$$\lambda(t|Z, X) = \lambda(t) \exp(\beta Z + f(X)),$$

where notations have similar implications as with the simple Cox model. $f(X)$ is the *unknown, nonparametric* covariate effect. Intuitively, the space of unspecified nonparametric functions is "too large." Some constraint is needed for estimation. Considering that most biological processes are smooth, we assume that

$$f \in \mathbb{S}_{s_0},$$

the Sobolev space indexed by the known order of derivative s_0. Such an assumption naturally leads to the penalized MLE defined as

$$(\hat{\beta}, \hat{\Lambda}, \hat{f}) = argmax \left\{ P_n l(\beta, \Lambda, f) - \lambda^2 J^2(f) \right\}.$$

Here, $\lambda^2 J^2(f)$ is the penalty on smoothness as in standard penalized spline studies.

There are three unknown parameters β, Λ, and f, which can be difficult to maximize simultaneously. Computationally, consider the following iterative algorithm:

(a) Initialize $\hat{\beta}=0$ and $\hat{f} = 0$.

(b) With fixed $\hat{\beta}$ and $\hat{f}$, compute $\hat{\Lambda} = argmax P_n l(\hat{\beta}, \Lambda, \hat{f})$. This can be achieved via multiple approaches, including, for example, the PAV algorithm. In addition, a cumulative sum diagram approach similar to that for the standard Cox model can be derived. For details, refer to Ma and Huang (2005).

(c) With fixed $\hat{\beta}$ and $\hat{\Lambda}$, compute $\hat{f} = argmax \left\{ P_n l(\hat{\beta}, \hat{\Lambda}, f) - \lambda^2 J^2(f) \right\}$. This is a "standard" penalized estimation problem. In practice, it is achieved by making basis expansion of f and transforming a nonparametric optimization problem into a parametric one. We refer to Wahba (1990) for more details.

(d) With fixed $\hat{\Lambda}$ and $\hat{f}$, compute $\hat{\beta} = argmax P_n l(\beta, \hat{\Lambda}, \hat{f})$. This can be achieved using the Newton-Raphson algorithm.

(e) Repeat Steps (b)–(d) until convergence. In practice, convergence can be concluded if the difference between two consecutive estimates is smaller than a predefined cut-off.

Existing software can be modified to execute each step of the above algorithm. However, there is still no software ready to carry out the complete process.

Now consider the asymptotic properties of the PMLE. As with other partly linear regression problems, f is only identifiable up to a constant. For identifiability, assume that $E(f(X)) = 0$ and absorb the mean into Λ. Beyond assumptions made with the standard Cox model, also assume that covariate X has bounded support.

When establishing the asymptotic results, the first step is to compute the Fisher information matrix and establish its nonsingularity. In the Appendix (Section 3.8), it is shown that for the simple Cox model, computation of the efficient score function involves projecting the score function of β onto the space of the score functions of Λ. Under the partly linear Cox model, computation of the efficient score function involves the following three steps:

(a) Project the score function for β onto the space generated by the score functions of Λ.

(b) Project the score function for f onto the space generated by the score functions of Λ.

(c) Project the space generated in Step (a) onto the space generated in Step (b).

We refer to Ma and Huang (2005) for more details on this calculation and to Sasieni (1992) for a general investigation of the nonorthogonal projection approach.

Further assume that $\lambda = O_p(n^{-1/3})$. Then, the consistency and convergence rate result can be summarized as

$$d((\hat{\beta}, \hat{\Lambda}, \hat{f}), (\beta_0, \Lambda_0, f_0)) = O_p(n^{-1/3}),$$

where

$$
\begin{aligned}
d((\beta_1, \Lambda_1, f_1), (\beta_2, \Lambda_2, f_2)) &= |\beta_1 - \beta_2| + ||\Lambda_1 - \Lambda_2||_2 \\
&\quad + \left\{ \int (f_1(X) - f_2(X))^2 dQ(X) \right\}^{1/2}
\end{aligned}
$$

and $Q(X)$ is the marginal distribution function of X. As described in the previous sections, without making additional assumptions, $\hat{\Lambda}$ cannot have a convergence rate faster than $n^{1/3}$. The above result shows that the optimal convergence rate for $\hat{\Lambda}$ can in fact be achieved. Because of this rate, the convergence rate of $\hat{f}$ is "slowed down" from the ordinary $n^{\frac{s_0}{2s_0+1}}$ rate (as under standard penalized spline estimation settings) to the $n^{1/3}$ rate. The assumption on λ is made to match this rate. Ma and Huang (2005) suggest that the convergence rate of $\hat{f}$ cannot be further improved. On the other hand, it is worth noting that, when it is reasonable to assume that $\Lambda_0 \in \mathbb{S}_{s_0}$, a double-penalization approach can be employed. This approach imposes penalization on both Λ and f and can achieve the $n^{\frac{s_0}{2s_0+1}}$ convergence rate.

For the asymptotic distribution of $\hat{\beta}$, it can be proved that

$$n^{1/2}(\hat{\beta} - \beta_0) \to N(0, I^{-1}(\beta_0))$$

in distribution as $n \to \infty$, where $I(\beta_0)$ is the Fisher information matrix. That is, despite the slow convergence rate of $\hat{\Lambda}$ and $\hat{f}$, particularly the suboptimal rate of $\hat{f}$, $\hat{\beta}$ is still $n^{1/2}$ consistent, asymptotically normal, and fully efficient. For inference, as the form of the Fisher information matrix is much more complicated than that of the standard Cox model, a plug-in approach, although in principle still feasible, can be extremely difficult. However, the weighted bootstrap approach is still applicable.

3.5.1.7 Estimation via Multiple Imputation

An alternative estimation approach can be based on multiple imputation. Before introducing the detailed approach, let's look into the differences between right and case I interval censored data. Observations with $\delta = 0$ (i.e., right censored observations) are quite similar under both censoring schemes: it is only known that the event times are larger than the censoring times. The key difference arises from observations with $\delta = 1$. Under right censoring, observations with $\delta = 1$ have *exact* event times. However, under case I interval censoring, if an observation has $\delta = 1$, it is only known that $T \in (0, C]$.

Imputation is a generic approach for dealing with missing data. We refer to Rubin (1987) for a thorough development of imputation methods. With case I interval censored data, when $\delta = 1$, the exact event time can be viewed as "missing." Imputation targets at "filling in" a missing event time with *plausible values*. If we can somehow fill in all the interval-censored event times, then case I interval censored data can be transformed into right censored data, for which statistical methodologies and software are very well developed. Pan (2000a) and several follow-up studies develop multiple imputation approaches for the Cox model under general interval censoring schemes. Below we consider the special case with case I interval censored data, and focus on estimation and inference with β.

We use the same notations as in the previous sections. Consider n i.i.d. observations $\{(C_i, \delta_i, Z_i) : i = 1, \ldots, n\}$ under the Cox model $\lambda(t|Z) = \lambda(t) \exp(\beta Z)$. Again for simplicity of notation, assume that Z is one-dimensional. Here we consider two multiple imputation approaches and note that other imputation approaches may also be applicable here.

Poor Man's Multiple Imputation. This approach is proposed in Wei and Tanner (2000). We use the superscript (i) and subscript (k) to represent the ith iteration and kth imputed data set, respectively. This approach proceeds as follows:

1. Initialize $\hat{\beta}$ and $\hat{\Lambda}$. A reasonable initial estimate for β is $\hat{\beta} = 0$. A possible initial estimate for Λ is the KM estimate, treating the interval censored data as if it were right censored.

2. With the current estimate of $\hat{\beta}$ and $\hat{\Lambda}$, generate m sets of possibly right censored observations $\{T_{(1)}, \delta_{(1)}, Z\}, \ldots, \{T_{(m)}, \delta_{(m)}, Z\}$ as follows. For each observation $(C_j, \delta_j, Z_j), j = 1, \ldots, n$.

   (a) if $\delta_{(k),j} = 0$, take $T_{(k),j} = C_j$ and $\delta_{(k),j} = 0$.

   (b) if $\delta_{(k),j} = 1$, sample $T_{(k),j}$ from the survival function $\exp(-\hat{\Lambda} \exp(\hat{\beta} Z_j))$ *conditional* on that $T_{(k),j} \leq C_j$.

3. Use each $\{T_{(k)}, \delta_{(k)}, Z\}$ (which is a set of right censored data) to fit a Cox model to obtain an estimate $\hat{\beta}_{(k)}^{(i)}$ and its covariance estimate $\hat{\Sigma}_{(k)}^{(i)}$.

4. Based on $\{T_{(k)}, \delta_{(k)}, Z\}$ and $\hat{\beta}_{(k)}^{(i)}$, calculate $\hat{\Lambda}_{(k)}^{(i)}$ following the approach described in Section 2.3.

5. Compute

$$\hat{\beta}^{(i+1)} = \frac{1}{m} \sum_{k=1}^{m} \hat{\beta}_{(k)}^{(i)}$$

$$\hat{\Lambda}^{(i+1)} = \frac{1}{m} \sum_{k=1}^{m} \hat{\Lambda}_{(k)}^{(i)}$$

$$\hat{\Sigma}^{(i+1)} = \frac{1}{m} \sum_{k=1}^{m} \hat{\Sigma}_{(k)}^{(i)} + \left(1 + \frac{1}{m}\right) \frac{\sum_{k=1}^{m} [\hat{\beta}_{(k)}^{(i)} - \hat{\beta}^{(i+1)}]^2}{m-1}.$$

6. Repeat Steps 2–5 until convergence. The estimates at convergence are the final estimates.

In Step 1, we choose reasonable initial values. Under mild regularity conditions, the overall objective function is globally concave. Thus, the choice of initial values may affect the number of iterations needed for convergence, but not convergence itself. In Step 2, we "convert" case I interval censored data into right censored data. Particularly, all right censored observations (with $\delta = 0$) are kept unchanged. For an observation with $\delta = 1$, we *randomly sample* the event time from the current estimation of the survival function. When sampling, we limit to the interval $[0, C]$ to be consistent with the original observation. We refer to Gelman et al. (2003) for available techniques on sampling. Estimation and other analysis can be conducted with the imputed right censored data. The estimates in Step 5 are taken as the average from multiple imputed data sets. The variance estimate has two parts. The first part is the average of the variances for the imputed right censored data. The second part reflects the variation caused by imputation. An inflation factor $1/m$ is used when estimating the between-imputation variance to take account of a finite number of imputations. In practice, $10 \leq m \leq 20$ is usually sufficient to generate reasonable estimates.

Asymptotic Normal Multiple Imputation.   There are studies showing that the Poor Man's approach may underestimate the true variability when the missingness is severe (here this means that a lot of observations have $\delta = 1$), whereas the asymptotic normal approach can be more accurate (Wei and Tanner, 2000). The asymptotic normal approach is based on the result that $\hat{\beta}$ is asymptotically normally distributed. It can be implemented by modifying two steps in the Poor Man's approach as follows

(a) In Step 5, the posterior of the regression coefficient is approximated by a mixture of normal distributions

$$g^{(i+1)}(\beta) = \frac{1}{m} \sum_{k=1}^{m} N(\hat{\beta}_{(k)}^{(i)}, \hat{\Sigma}_{(k)}^{(i)}).$$

(b) In Step 2, first sample m times from $g^{(i)}(\beta)$ to obtain $\hat{\beta}^{(i)}_{(k)}$, $k = 1, \ldots, m$. Then for each $k(= 1, \ldots, m)$ and each observation with $\delta_j = 1$ $(j = 1, \ldots, n)$, sample $T_{(k),j}$ from the survival function $\exp(-\hat{\Lambda}\exp(\hat{\beta}Z_j))$ *conditional on* that $T_{(k),j} \leq C_j$. Keep right censored observations.

Other steps remain the same.

The most significant advantage of the imputation approaches is the computational simplicity. As can be seen from the above algorithms, no special algorithm is needed for estimation or inference. The computational cost can be considerably lower than that with MLE. On the negative side, the asymptotic properties of imputation estimates are difficult to establish. For example even the establishment of the convergence of the algorithms is highly nontrivial. A rule of thumb is that if an EM algorithm converges, then the corresponding multiple imputation algorithms converge for the same problem. For models with a tolerable computational complexity, we recommend likelihood-based approaches, as they have more lucid asymptotic properties.

Remarks. In the above sections, we have devoted considerable effort to estimation and inference with the Cox model. The Cox model is by far the most popular semiparametric survival model and warrants a detailed investigation. More importantly, study of the Cox model can serve as a prototype for other semiparametric models. With right censored data, the profile likelihood approach for the Cox model is very much model specific and cannot be easily extended to other models. In contrast, with interval censored data, the estimation and inference approaches and even asymptotic properties for the Cox model can be readily extended to other semiparametric models with only minor modifications.

3.5.2 Additive Risk Model

Consider the additive risk model, which provides a useful alternative to the Cox model. For simplicity of notation, again assume only one covariate. Under the additive risk model, the conditional hazard function is

$$\lambda(t|Z) = \lambda(t) + \beta Z,$$

where notations have similar implications as with the Cox model. The additive risk model describes a different aspect of the association between the failure time and covariates than the Cox model, and is more plausible than the latter in many applications (Lin and Ying, 1994b; Lin et al., 1998). It is noted that, unlike the Cox model, the additive risk model does not belong to the family of transformation models.

In what follows, we describe two estimation approaches. The first approach is developed in Lin et al. (1998). It takes advantage of the special form of the likelihood function and is computationally simple. However, it is not fully efficient, and in some cases the estimates may have rather high variability. The second approach was developed in Martinsussen and Scheike (2002). As argued in the end of the last section, with the additive risk model, it is possible to follow the same strategy as with the Cox model and develop fully efficient maximum likelihood estimate. The approach in Martinsussen and Scheike (2002) follows such a strategy. In terms of presentation, it takes advantage of the closed-form of the efficient score function and, in some sense, shares similarity with estimating equation-based approaches. Assume independent censoring. It is noted that both approaches described below can be extended to accommodate dependent censoring. We refer to the original publications for such extensions.

Lin, Oakes, and Ying's Approach.   In the development of this approach, considerable martingale and counting process techniques are used. Assume n i.i.d. observations $\{(C_i, \delta_i = I(C_i \leq T_i), Z_i) : i = 1, \ldots, n\}$. Note that here the *definition of δ_i is "reversed" for notational simplicity.*

For $i = 1, \ldots, n$, define the counting process $N_i(t) = \delta_i I(C_i \leq t)$, which jumps by one unit when subject i is observed at time t and still has not experienced the event. The process $N_i(t)$ is censored when subject i is observed and found to have experienced the event. Subject i is at risk at time t if and only if $N_i(t)$ is not censored and has not taken its jump by time t, which happens if and only if $C_i \geq t$. Consider the hazard $dH_i(t)$ of a jump of $N_i(t)$ for a subject that has not been observed. Two things have to happen for $dN_i(t) = 1$: (i) $C_i = t$; (ii) subject i has not experienced the event by time t. Denote the hazard of censoring as $d\Lambda_c(t)$. Under the additive risk model, event (ii) has the conditional probability

$$P(T_i \geq t | Z_i) = \exp(-\Lambda(t) - \beta Z_i^*(t)),$$

where $\Lambda(t)$ is the cumulative baseline hazard function and $Z_i^*(t) = \int_0^t Z_i ds$. We use the integration notation so that the approach can be easily extended to accommodate time-dependent covariates. The product of the two probabilities for events (i) and (ii) is

$$dH_i(t) = \exp(-\beta' Z_i^*(t)) dH_0(t), \tag{3.22}$$

where $dH_0(t) = \exp(-\Lambda(t)) d\Lambda_C(t)$.

It is noted that the above equation takes the familiar form of a Cox model. Let $Y_i(t) = I(C_i \geq t)$. Lin et al. (1998) prove that the compensated counting processes

$$M_i(t) = N_i(t) - \int_0^t Y_i(s) \exp(-\beta' Z_i^*(s)) dH_0(s), \quad i = 1, \ldots, n$$

are martingales with respect to the σ-filtration

$$\mathcal{F} = \sigma\{H_i(s), Y_i(s), Z_i : s \leq t, i = 1, \ldots, n\}.$$

Thus, one can conduct estimation and inference by applying the partial likelihood principle to model (3.22) with data $\{(N_i, Y_i, Z_i^*) : i = 1, \ldots, n\}$. Specifically, the partial likelihood score function and observed information matrix for β are

$$U(\beta) = \sum_{i=1}^{n} \int_0^\infty \left\{ Z_i^*(t) - \frac{S^{(1)}(\beta, t)}{S^{(0)}(\beta, t)} \right\} dN_i(t)$$

$$I(\beta) = \sum_{i=1}^{n} \int_0^\infty \left\{ \frac{S^{(2)}(\beta, t)}{S^{(0)}(\beta, t)} - \frac{S^{(1)}(\beta, t)^{\otimes 2}}{S^{(0)}(\beta, t)^2} \right\} dN_i(t),$$

where

$$S^{(k)}(\beta, t) = \sum Y_j(t) \exp(-\beta' Z_j^*(t)) Z_j^*(t)^{\otimes k}, \quad k = 0, 1, 2.$$

Consider the estimate defined as

$$U(\hat{\beta}) = 0.$$

By the counting-process martingale theories (Anderson and Gill, 1982), it can be shown that

$$n^{1/2}(\hat{\beta} - \beta) \to N(0, \Omega^{-1})$$

in distribution, where $\Omega^{-1} = \lim_{n \to \infty} n^{-1} I(\hat{\beta})$.

The above asymptotic results can serve as the basis for estimation and inference on the parametric regression parameters. It is obvious that the estimation procedure is computationally simple and does not demand any special software. In fact, one can just input the data $\{(C_i, \delta_i, Z_i^*) : i = 1, \ldots, n\}$ into standard software for fitting the Cox model with right censored data. It is noted that the distribution of the censoring time for $N_i(t)$ also involves β so that the censoring times are informative about β. This implies that $\hat{\beta}$ may not be fully efficient. Such a concern motivates the following fully efficient approach.

Martinussen and Scheike's Approach. Assume that the distribution of censoring time C is absolutely continuous with hazard function $\alpha(t)$. Assume that T and C are independent given covariate Z. Consider n i.i.d. observations $\{(C_i, \delta_i = I(C_i \leq T_i), Z_i) : i = 1, \ldots, n\}$. Define the following two counting processes

$$N_{1i}(t) = \delta_i I(C_i \leq t), \quad N_{2i}(t) = (1 - \delta_i) I(C_i \leq t).$$

The process N_{1i} jumps by one when subject i is observed and event-free. N_{2i} jumps by one when subject i is observed and has experienced the event. Further define $N_i(t) = N_{1i}(t) + N_{2i}(t)$. Define $Y_i(t) = I(t \leq C_i)$ and $\mu(t) = \alpha(t) \exp \left\{ -\int_0^t \lambda(s) ds \right\}$.

Under the additive risk model, up to an additive constant not depending on λ and β, the log-likelihood function is

$$l(\beta, \Lambda) \quad = \quad \sum_i \delta_i \log(\exp(-\Lambda(C_i) - \beta' Z_i^*))$$

$$+ (1 - \delta_i) \log(1 - \exp(-\Lambda(C_i) - \beta' Z_i^*)), \qquad (3.23)$$

where $Z_i^* = \int Z_i ds$. Under the assumption of independent censoring, the efficient score for β is

$$i_\beta^* = \int \left\{ Z^* - E\left(Z^* Y \frac{p}{1-p} \right) E\left(Y \frac{p}{1-p} \right)^{-1} \right\} \left(\frac{p}{1-p} dN_2 - dN_1 \right),$$

where $p = p(t, \Lambda, \beta) = \exp(-\Lambda(t) - \beta Z^*)$. The empirical version of the efficient score leads to

$$U(\beta, \Lambda) = \sum_{i=1}^n \int \left(Z_i^* - \frac{S_1}{S_0} \right) \left(\frac{p_i}{1 - p_i} dN_{2i} - dN_{1i} \right), \qquad (3.24)$$

where

$$S_j(t) = \sum_i \frac{p_i}{1 - p_i} Y_i Z_i^{*\otimes j}.$$

Define the estimate $\hat{\beta}$ as the solution to the estimated empirical efficient score equation

$$U(\beta, \hat{\Lambda}) = 0,$$

where $\hat{\Lambda}$ is an estimate of Λ.

The empirical efficient score (3.24) evaluated at β_0 is a martingale, evaluated at infinity, as the compensated counting processes

$$M_{1i}(t) = N_{1i}(t) - \int_0^t Y_i \alpha p_i ds, \quad M_{2i}(t) = N_{2i}(t) - \int_0^t Y_i \alpha (1 - p_i) ds$$

are martingales. This result also holds for $U(\beta_0, \hat{\Lambda})$, if $\hat{\Lambda}$ is a predictable estimator of Λ.

The information for β_0 is

$$I(\beta_0) \quad = \quad E(i_{\beta_0}^*)^{\otimes 2}$$

$$= \quad E\left[\int \left\{ Z^* - E\left(Z^* Y \frac{p}{1-p} \right) E\left(Y \frac{p}{1-p} \right)^{-1} \right\}^{\otimes 2} \frac{p}{1-p} Y \alpha dt \right].$$

Under the assumption that $\hat{\Lambda} - \Lambda_0 = o_p(n^{-1/4})$ uniformly and some additional standard regularity conditions, Martinsussen and Scheike (2002) prove that as $n \to \infty$,

$$n^{1/2}(\hat{\beta} - \beta_0) \to N(0, I^{-1}(\beta_0))$$

in distribution. This estimate is thus fully efficient. Note that this approach demands a consistent estimate $\hat{\Lambda}$ as input. Several possibilities have been suggested in Martinsussen and Scheike (2002). As shown with the Cox model, it is usually possible to obtain an estimate $\hat{\Lambda}$ with the $n^{1/3}$ convergence rate. Thus, the requirement $\hat{\Lambda} - \Lambda_0 = o_p(n^{-1/4})$ can be fulfilled.

3.5.3   Alternative Semiparametric Models

In this subsection, we briefly introduce the proportional odds model and
AFT model. Establishing their asymptotic properties demands using empirical
process techniques extensively. Some of the proofs are similar to those for
the Cox model. Here, we only briefly state the information calculation and
distributional results, without discussing the details.

Proportional Odds Model.   Under the proportional odds model,

$$logit(F(t|Z)) = logit(F(t)) + \beta Z.$$

Here, $F(t|Z))$ is the conditional distribution function, and $F(t)$ is the baseline
distribution function. Again we assume one-dimensional covariate.

We parameterize the model in terms of the regression parameter and the
baseline log-odds function $\alpha(t) = logit(F(t))$. Then the joint density function
of a single observation is

$$p_{\beta,\alpha}(\delta, c, z) = \exp(\delta(\alpha(t) + \beta z))(1 + \exp(\alpha(t) + \beta z))^{-1}h(c, z).$$

$h(c, z)$ is the joint density function of censoring time and covariate. With
n i.i.d. observations, the empirical log-likelihood function, up to an additive
constant not depending on $(\beta, \alpha(t))$ is

$$\sum_{i=1}^{n}\{\delta_i(\alpha(C_i) + \beta Z_i) - \log(1 + \exp(\alpha(C_i) + \beta Z_i))\}.$$

Consider $(\hat{\beta}, \hat{\alpha}(t))$, the maximizer of the above objective function, under the
constraint that $\hat{\alpha}$ is a nondecreasing function.

To establish the asymptotic properties, the following assumptions are
needed: (A1) The event time is independent of the censoring time given the
covariates; (A2) The joint distribution of the censoring times and covariates
is independent of the parameters of interest; (A3) The distribution of Z is not
concentrated on any proper subspace of R^d; and Z is bounded; (A4) Denote
F_0 as the true value of $F(t)$. $F_0(0) = 0$. Let $\tau_{F_0} = inf\{t : F_0(t) = 1\}$. The
support of C is an interval $[\tau_0, \tau_1]$, with $0 < \tau_0 \leq \tau_1 < \tau_{F_0}$; and (A5) F_0 has
a strictly positive and continuous density on $[\tau_0, \tau_1]$. The joint distribution
function $H(C, Z)$ of (C, Z) has bounded second-order partial derivative with
respect to C.

Under the above assumptions, the efficient score function for β is

$$l_{\beta}^* = (\delta - E(\delta|C = c, Z = z))\left(z - \frac{E(Z var(\delta|C, Z)|C = c)}{E(var(\delta|C, Z)|C = c)}\right),$$

and the information for β is $I(\beta) = E(l_{\beta}^*)^2$.

Under assumptions (A1)-(A5), it can be proved that

$$n^{1/2}(\hat{\beta} - \beta_0) \to N(0, I^{-1}(\beta_0))$$

in distribution, as $n \to \infty$. In addition,

$$\left\{ \int [\hat{\alpha}(c) - \alpha_0(c)]^2 dG(c) \right\}^{1/2} = O_p(n^{-1/3}).$$

Here, $G(c)$ is the marginal distribution function of the censoring time. As the efficient score and Fisher information have relatively simple forms, inference on $\hat{\beta}$ can be conducted using a plug-in approach similar to that for the Cox model. In addition, the weighted bootstrap is valid.

AFT Model. Let T be the logarithm or another known transformation of the event time. Assume that the same transformation has been made for the observation time. Assume the model

$$T = \beta Z + \epsilon.$$

When the distribution of ϵ is known, this is a simple parametric model, and a likelihood-based estimation can be easily obtained. Consider the more difficult scenario with $\epsilon \sim F_0$ *unspecified*. Assume that ϵ is independent of Z and the censoring time. Then the density function of a single observation is

$$p_{\beta,F}(\delta, c, z) = F(c - \beta z)^\delta (1 - F(c - \beta z))^{1-\delta} h(c, z).$$

With n i.i.d. observations, up to a constant not depending on β and F, the log-likelihood function is

$$\ell(\beta, F) = \sum_{i=1}^n \delta_i \log F(C_i - \beta Z_i) + (1 - \delta_i) \log(1 - F(C_i - \beta Z_i)).$$

Consider $(\hat{\beta}, \hat{F})$, the maximizer of the above objective function under the constraint that $\hat{F}$ is a (sub)distribution function.

$(\hat{\beta}, \hat{F})$ can be computed using the following maximum profile likelihood approach:

(1) For any fixed β, maximize $\ell(\beta, F)$ with respect to F under the constraint that F is a distribution function. This step can be achieved using multiple isotonic regression techniques, for example, the PAV algorithm. The estimate may need to be truncated to ensure that it is a distribution function. Denote the resulting maximizer by $\hat{F}(\cdot; \beta)$.

(2) Substitute $\hat{F}(\cdot; \beta)$ back into $\ell(\beta, F)$, and obtain the profile likelihood $\ell(\beta, \hat{F}(\cdot; \beta))$. Then the maximum profile likelihood estimate $\hat{\beta}$ is the value of β that maximizes $\ell(\beta, \hat{F}(\cdot; \beta))$ (assuming that it exists).

(3) A natural estimate of F is $\hat{F}(\cdot; \hat{\beta})$.

In the calculation of Fisher information, let $k(s) = E(Z|C - \beta Z = s)$. Assume that $k(\cdot)$ is bounded. Then the efficient score for β is

$$\dot{l}_\beta^* = f(c - \beta z)[z - E(Z|C - \beta Z = u - \beta z)] \times \left[\frac{1 - \delta}{1 - F(c - \beta z)} - \frac{\delta}{F(c - \beta z)} \right].$$

The information for estimation of β is

$$I(\beta) = E\left\{ \frac{f(C - \beta Z)^2}{F(C - \beta Z)(1 - F(C - \beta Z))} [Z - E(Z|C - \beta Z)]^{\otimes 2} \right\}.$$

Consistency of $(\hat{\beta}, \hat{F}(\cdot, \hat{\beta}))$ can be proved as in Cosslett (1983). However, the distributional results of $\hat{\beta}$ have not been established under assumptions similar to those with the Cox model. It is not clear whether or not $\hat{\beta}$ has a $n^{1/2}$ convergence rate and asymptotic normality. The main technical difficulty is that $\ell(\beta, \hat{F}(\cdot, \beta))$ is not a smooth function of β. Our own limited simulation study suggests that the estimate $\hat{\beta}$ may still have asymptotic normality. More research is needed on this topic.

3.6 Semiparametric Modeling with Case II Interval Censored Data

In this section, we consider semiparametric models with case II interval censored data. The general strategy is similar to that with case I interval censored data. However, because of significant differences in censoring schemes, the assumptions needed can be different. In addition, there are also differences in computational approaches.

Denote the observed data as $X = (U, V, \delta_1, \delta_2, \delta_3, Z)$. Again for simplicity of notation, we consider the case with $d = 1$ (one-dimensional covariate). Then the joint density function for a single observation is

$$p(x) = F(u|z)^{\delta_1} [F(v|z) - F(u|z)]^{\delta_2} (1 - F(v|z))^{\delta_3} h(u, v, z),$$

where h is the joint density function of (U, V, Z). We make the following assumptions:

(A1) The event time is independent of the censoring times given the covariates.

(A2) The joint distribution of the censoring times and covariates is independent of the parameters of interest.

(A3) (a) The distribution of Z is not concentrated on any proper affine subspace of R^d (i.e., of dimension $d - 1$ or smaller). (b) Z is bounded.

These assumptions are identical to those made with case I interval censored data. Then up to a constant not depending on $F(\cdot|Z)$, the log-likelihood function is

$$l = \delta_1 \log F(U|Z) + \delta_2 \log[F(V|Z) - F(U|Z)] + \delta_3 \log(1 - F(V|Z)).$$

As with case I interval censored data, in what follows, we provide detailed investigation of the Cox model. We also briefly discuss the proportional odds model and AFT model. Beyond (A1)–(A3), we also make the following assumptions:

(A4) (a) There exists a positive constant η such that $P(V - U \geq \eta) = 1$; That is, there is no accurately observed event time. (b) The union of the support of U and V is contained in an interval $[\tau_0, \tau_1]$ with $0 < \tau_0 < \tau_1 < \infty$.

(A5) F_0, the true value of F, has strictly positive, bounded and continuous derivative on $[\tau_0, \tau_1]$.

(A6) The conditional density $g(u, v|z)$ of (U, V) given Z has bounded partial derivatives with respect to u and v. The bounds of these partial derivatives do not depend on z.

These assumptions are comparable to those made with case I interval censored data. Assume n i.i.d. observations.

3.6.1   Cox Proportional Hazards Model

Under the Cox model, the empirical log-likelihood function as a function of the regression parameter β and the cumulative baseline hazard Λ is

$$P_n l(\beta, \Lambda) = \frac{1}{n} \times \sum_{i=1}^{n} \{\delta_{1i} \log(1 - \exp(-\Lambda(U_i) \exp(\beta Z_i))) +$$

$$\delta_{2i} \log[\exp(-\Lambda(U_i) \exp(\beta Z_i)) - \exp(-\Lambda(V_i) \exp(\beta Z_i))] - \delta_{3i} \Lambda(V_i) \exp(\beta Z_i)\}.$$

Consider the MLE defined as $(\hat{\beta}, \hat{\Lambda}) = argmax P_n l(\beta, \Lambda)$, under the constraint that $\hat{\Lambda}$ is a nonnegative and nondecreasing function.

Computation.   As with case I interval censored data, we iteratively maximize over β and Λ until convergence. Maximization over β can be achieved using the Newton-Raphson algorithm, as the likelihood function is continuously differentiable. Below we describe an iterative convex minorant algorithm for maximizing over Λ. We use X to generically denote the observed data. Define

$$a_1(x; \Lambda) = \frac{\exp(\beta z)\exp(-\Lambda(u)\exp(\beta z))}{1 - \exp(-\Lambda(u)\exp(\beta z))}$$

$$a_2(x; \Lambda) = \frac{\exp(\beta z)\exp(-\Lambda(u)\exp(\beta z))}{\exp(-\Lambda(u)\exp(\beta z)) - \exp(-\Lambda(v)\exp(\beta z))}$$

$$a_3(x; \Lambda) = \frac{\exp(\beta z)\exp(-\Lambda(v)\exp(\beta z))}{\exp(-\Lambda(u)\exp(\beta z)) - \exp(-\Lambda(v)\exp(\beta z))}.$$

Let

$$W_\Lambda(t) = \sum_{i=1}^n \{\delta_{1i}a_1(X_i; \Lambda) - \delta_{2i}a_2(X_i, \Lambda)\}I(U_i \leq t)$$
$$+ \{\delta_{2i}a_3(X_i, \Lambda) - \delta_{3i}\exp(\beta Z_i)\}I(V_i \leq t)$$

$$G_\Lambda(t) = \sum_{i=1}^n \{\delta_{1i}a_1^2(X_i; \Lambda) + \delta_{2i}a_2^2(X_i, \Lambda)\}I(U_i \leq t)$$
$$+ \{\delta_{2i}a_3^2(X_i, \Lambda) + \delta_{3i}\exp(2\beta Z_i)\}I(V_i \leq t)$$

$$V_\Lambda(t) = W_\Lambda(t) + \int_{[0,t]} \Lambda(s)dG_\Lambda(s).$$

Then, $\hat{\Lambda}$ is the left derivative of the greatest convex minorant of the self-induced cumulative sum diagram formed by the points

$$(G_{\hat{\Lambda}}(Y_{(j)}), V_{\hat{\Lambda}}(Y_{(j)}))$$

and the origin $(0, 0)$. Here, $Y_{(j)}$s are the order statistics of the set $J_{1n} \cup J_{2n}$. J_{1n} is the set of examination times U_i such that T_i either belongs to $[0, U_i]$ or $(U_i, V_i]$, and J_{2n} is the set of examination times V_i such that T_i either belongs to $(U_i, V_i]$ or (V_i, ∞). With this result, $\hat{\Lambda}$ can be computed using a similar iterative algorithm as with case I interval censored data.

Information Calculation. As with case I interval censored data, information calculation is needed for establishing the asymptotic properties of $\hat{\beta}$. Let f and F be the density and distribution functions corresponding to Λ, and let f_s be a one-dimensional smooth curve through f, where the smoothness is with respect to s. Denote $a = \frac{\partial}{\partial s}\log f_s|_{s=0}$. Then $a \in L_2^0(F) = \{a : \int adF = 0 \text{ and } \int a^2 dF < \infty\}$. Let

$$h = \frac{\partial}{\partial s}\Lambda_s|_{s=0} = \frac{-\int_{\cdot}^\infty adF}{1 - F}.$$

It can be proved that $h \in L_2(F)$.

The score function for β is

$$\dot{l}_\beta(x) = \frac{\partial}{\partial \beta}l(x; \beta, \Lambda),$$

and the score operator for Λ is

$$i_\Lambda[a](x) = \frac{\partial}{\partial s} l(x; \beta, \Lambda_s)|_{s=0}.$$

The explicit forms of i_β and i_Λ can be obtained by carrying out the differentiation. The score operator i_Λ maps $L_2^0(F)$ to $L_2^0(P)$, where P is the joint probability measure of $(U, V, \delta_1, \delta_2, Z)$ and $L_0^2(P)$ is defined similarly as $L_0^2(F)$.

Let i_Λ^T be the adjoint operator of i_Λ. That is, for any $a \in L_2^0(F)$ and $b \in L_2^0(P)$,

$$< b, i_\Lambda[a] >_P = < i_\Lambda^T[b], a >_F,$$

where $< \cdot, \cdot >_P$ and $< \cdot, \cdot >_F$ are the inner products in $L_0^2(P)$ and $L_0^2(F)$, respectively. We need to find a_* such that $i_\beta - i_\Lambda[a_*]$ is orthogonal to $i_\Lambda[a]$ in $L_2^0(P)$. This amounts to solving the following normal equation

$$i_\Lambda^T i_\Lambda[a_*] = i_\Lambda^T i_\beta. \tag{3.25}$$

The value of i_Λ^T at any $b \in L_2^0(P)$ can be computed by

$$i_\Lambda^T b(t) = E_Z E(b(X)|T = t, Z).$$

Huang and Wellner (1997) show that under conditions (A1)–(A6), equation (3.25) has a unique solution. Moreover, $h_* = \frac{-\int a_* dF}{1-F}$ has a bounded derivative. The efficient score for β is

$$l_\beta^* = i_\beta - i_\Lambda[a_*],$$

and the information is

$$I(\beta) = E(l_\beta^*)^{\otimes 2}.$$

Asymptotic Properties.   Denote (β_0, Λ_0) as the true value of (β, Λ). Denote G as the joint distribution function of (U, V) and let G_1 and G_2 be the marginal distribution functions of U and V, respectively. Then, the consistency result can be summarized as follows:

$$\hat{\beta} \to \beta_0 \text{ and } \hat{\Lambda} \to \Lambda_0, \ G_1 + G_2 - \text{almost everywhere.}$$

Consider the following two special cases. If both U and V are discrete, then $\hat{\Lambda} \to \Lambda_0$ at all the mass points of U and V. If at least one of U and V has a continuous distribution whose support contains the support of the event time, then for any finite $M > 0$,

$$\sup_{0 \le t \le M} |\hat{\Lambda}(t) - \Lambda_0(t)| \to 0.$$

The convergence rate and asymptotic distribution results can be further established as follows. Denote the support of β as Θ. Assume that Θ is compact set of $\mathbb{R}^d$, and β_0 is an interior point of Θ. Then, under assumptions (A1)–(A6),

$$n^{1/2}(\hat{\beta} - \beta_0) \to_d N(0, I(\beta_0)^{-1})$$

in distribution as $n \to \infty$. That is, $\hat{\beta}$ is asymptotically normal and efficient. For $\hat{\Lambda}$, we have

$$||\hat{\Lambda} - \Lambda_0||_2 = O_p(n^{-1/3}),$$

which is similar to that for case I interval censored data. For detailed proofs, we refer to Huang and Wellner (1997) and references therein.

Inference.   Unlike with case I interval censored data, there is no closed-form expression for the asymptotic variance of $\hat{\beta}$. The plug-in approach can be infeasible. Here we introduce an alternative approach, which treats the semiparametric estimate as if it were parametric. This approach has been carefully developed for multiple models in Sun (2006).

Denote $\ell_d(\hat{\beta}, \hat{\Lambda})$ as the log-likelihood function of $\hat{\beta}$ and the *distinct* values of $\hat{\Lambda}$. Denote

$$\hat{\Sigma}_{11} = \frac{\partial^2}{\partial \hat{\beta}^2} \ell_d, \quad \hat{\Sigma}_{12} = \frac{\partial^2}{\partial \hat{\beta} \partial \hat{\Lambda}} \ell_d, \quad \hat{\Sigma}_{22} = \frac{\partial^2}{\partial \hat{\Lambda}^2} \ell_d,$$

and

$$\hat{\Sigma} = \left(\begin{array}{cc} \hat{\Sigma}_{11} & \hat{\Sigma}_{12} \\ \hat{\Sigma}_{12}^T & \hat{\Sigma}_{22} \end{array} \right).$$

Then the asymptotic variance of $\hat{\beta}$ can be estimated with

$$(\hat{\Sigma}_{11} - \hat{\Sigma}_{12}\hat{\Sigma}_{22}^{-1}\hat{\Sigma}_{12}')^{-1}.$$

When Λ is parametric (for example, has a finite number of nonzero mass points), the above estimate is exact. With nonparametric Λ, as shown above, $\hat{\Lambda}$ only has convergence rate $n^{1/3}$. The number of distinct values of $\hat{\Lambda}$ is expected to be much smaller than the sample size. Thus, in practical data analysis, we can use the above parametric approximation for the semiparametric model. This approach has been shown to have reasonable performance in numerical studies (Sun, 2006), although theoretical justification of its validity can be extremely difficult.

An alternative approach is the weighted bootstrap, whose validity does not require a closed-form Fisher information matrix. Here, implementation of the weighted bootstrap is the same as with case I interval censored data.

3.6.2   Proportional Odds Model

Under the proportional odds model (see Section 2.3.4), the conditional distribution function has the form

$$F(t|z) = \frac{\exp(\alpha(t) + \beta z)}{1 + \exp(\alpha(t) + \beta z)}.$$

Here, notations have similar definitions as with case I interval censored data. With the log-likelihood function specified in the beginning of this section, we consider the MLE based on n i.i.d. observations. Note that the estimation is carried out under the constraint that $\hat{\alpha}$ is a nondecreasing function. The estimate $(\hat{\beta}, \hat{\alpha}(t))$ can be computed in an iterative manner, as with the Cox model. An iterative convex minorant algorithm can be derived for maximization over $\alpha(t)$. Calculation of the Fisher information matrix can be conducted in a similar manner as for the Cox model. Details are provided in Huang and Wellner (1997). Denote $(\beta_0, \alpha_0(t))$ as the true value of $(\beta, \alpha(t))$. Asymptotic properties can be summarized as

$$||\hat{\alpha} - \alpha_0||_2 = O_p(n^{-1/3}), \tag{3.26}$$

and

$$n^{1/2}(\hat{\beta} - \beta_0) \to N(0, I(\beta_0)^{-1}) \tag{3.27}$$

in distribution as $n \to \infty$, where $I(\beta_0)$ is the Fisher information matrix. Inference of $\hat{\beta}$ can be conducted either by inverting the observed Fisher information matrix or using the weighted bootstrap approach, in a similar manner as described in the above sections.

3.6.3 AFT Model

Under the AFT model, up to a constant, the empirical log-likelihood function based on n i.i.d. observations is

$$\ell(\beta, F) = \sum_{i=1}^{n} \delta_{1i} \log F(U_i - \beta Z_i) + \delta_{2i} \log[F(V_i - \beta Z_i) - F(U_i - \beta Z_i)]$$
$$+ \delta_{3i} \log(1 - F(V_i - \beta Z_i)).$$

Consider the MLE $(\hat{\beta}, \hat{F}) = argmax\ell(\beta, F)$. Computation of this estimate can be realized using a maximum profile likelihood approach similar to that with case I interval censored data.

Calculation of the Fisher information is presented in Huang and Wellner (1997). It is proved that the efficient score and hence information matrix exist. It is noted that additional assumptions on the distribution of (U, V) and F function are needed for information calculation.

For asymptotic results, further assume that F_0, the unknown true value of F, is continuous. Then, it can be proved that

$$\hat{\beta} \to \beta_0,$$

and

$$\hat{F}(t; \hat{\beta}) \to F_0(t)$$

for all t except on a set with $G_1 + G_2$ measure zero.

The establishment of convergence rate and asymptotic distribution is very challenging. There is no result similar to that for the Cox model *unless stronger smoothness conditions are satisfied.* Intuitively, the main challenge is that the two parameters β and F cannot be fully "separated." The direct consequence is that $\hat{F}(\cdot, \beta)$ and the resulting profile likelihood function are not smooth functions of β. In addition, the estimation of the variance matrix cannot be obtained by inverting the observed information as $\ell(\beta, F)$ is not a twice-differentiable function of β. More theoretical investigations on this model will be needed in the future.

3.7 Discussions

Jianguo Sun's book (2006) provides an updated and authoritative account for interval censored data. Many topics covered in this chapter are in parallel with those in Sun (2006) but may be more refined with additional mathematical details. The key difference is that Sun (2006) presents the methodologies in a more intuitive, friendly manner, whereas in this chapter, we have been trying to be more rigorous and present the theoretical basis of such methods. For a few models and methods, we only provide very crude presentations with rather sketchy justifications. For those, we have pointed out the original publications as necessary references for more information. For first-time learners, it can be very helpful to combine reading this chapter with Sun (2006).

Compared with right censored data, the analysis of interval censored data is much more challenging. Many useful statistical techniques, especially the martingale theory, cannot be applied directly as the event times are completely "missing." In the study of asymptotic properties with semiparametric models, advanced empirical processes techniques are extensively used. Our limited experiences suggest that such techniques are inevitable. For more theory-oriented readers, we refer to Kosorok (2008), van der Vaart (2000), and van der Vaart and Wellner (1996) for the general theorems that have been used to establish the asymptotic properties presented in this chapter.

With both case I and case II interval censored data, this chapter has mostly focused on "standard" parametric and semiparametric models and likelihood-based estimation and inference. Topics not covered in this chapter but nevertheless of significant importance may include but are definitely not limited to the following:

(a) Bayesian estimation and inference, where it is assumed that the unknown parameters satisfy a certain prior distribution and the estimates are taken as the posterior mode/median/mean. As can be seen from this chapter, the computation of the maximum likelihood estimates can be very challenging and may demand special algorithms. Bayesian approaches with conjugate priors may provide a computationally more efficient solution. The trade-off is that the theoretical properties of Bayesian estimators may be difficult to establish. Bayesian estimation in survival analysis will be briefly examined in Section 4.4.

(b) Alternative inference methods. In this chapter, for inference of the parametric parameter estimates in semiparametric models, we introduce three inference methods: the plug-in approach, the weighted bootstrap, and the inversion of the observed information. All three methods, unfortunately, have limitations. Particularly, the plug-in approach is only applicable to models with closed-form, simple Fisher information. Smoothed estimation may need to be conducted, which may incur considerable computational cost and is theoretically nontrivial. The weighted bootstrap approach can be computationally expensive, depending on optimization algorithms. And, the inversion approach is a parametric approximation to a semiparametric problem. It is not applicable to all models, and its asymptotic validity may be difficult to establish. There are other inference methods, including, for example, the profile sampler, piggyback bootstrap (Kosorok, 2008), nonparametric bootstrap, and others. However, they all suffer high computational cost and have other drawbacks. The inference aspect with interval censored data is still being actively studied.

(c) Model diagnostics. With right censored data, there is a family of well-developed methods that can systematically carry out model diagnostics (for example, examination of martingale residuals in the Cox model; see Section 2.3.3). With interval censored data, model diagnostics techniques are less developed. In Chapter 5, we introduce general diagnostic approaches based on the ROC (receiver operating characteristics) curves, which can provide a partial solution to this problem. In Li and Ma (2010), the authors adopt a bootstrap approach for model diagnostics under interval censoring. This approach is asymptotically valid. However, limited numerical studies suggest that it is not very sensitive to certain distributions and may demand a substantially large sample size for a reasonably good performance in practice.

(d) Implementation. In this chapter, we have mostly focused on the development of methodologies and asymptotic properties. For parametric models, computation of the maximum likelihood estimates and inference can be achieved with simple coding. For nonparametric and semiparametric models, computation of the estimates is nontrivial. For a few models, we briefly introduce implementation using R. In SAS, there are macros available for computing the NPMLE, and conducting the generalized log-rank test of Zhao and Sun (2004) and a few other simple tasks. For more complicated models, for example, those with nonparametric covariate effects, there is no well developed software.

3.8 Appendix

In this chapter we have described maximum likelihood estimation with the Cox model under case I interval censored data. Results on consistency, convergence rate and asymptotic distribution have been provided. Here in this Appendix we provide some technical details for establishing those results. We refer to Huang (1996), Ma and Huang (2005), and van der Vaart (2000) for full details. It is worth noting that *most of the following asymptotic development can be extended to other semiparametric models under case I interval censoring.*

Information Calculation. For simplicity of notation, consider the simple scenario with $Z \in \mathbb{R}$ $(d = 1)$. The general case can be proved in a similar manner. We first compute the score functions for β and F. Note that for information calculation, it is easier to formulate the Cox model using the baseline distribution function F as opposed to the cumulative hazard Λ. Theoretically speaking, different formulations are strictly equivalent.

The score function for β is simply the derivative of the log-likelihood with respect to β, that is

$$\dot{l}_\beta = z \exp(\beta z) \Lambda(c) \left[\delta \frac{S(c|z)}{1 - S(c|z)} - (1 - \delta) \right].$$

Suppose that $\mathbb{F}_0 = \{F_\eta, |\eta| < 1\}$ is a regular parametric subfamily of $\mathbb{F} = \{F : F << \mu, \mu = \text{Lebesgue measure}\}$. Set $(\partial/\partial\eta) \log f_\eta(t)|_{\eta=0} = a(t)$. Then, $a \in L_2^0(F)$ and $(\partial/\partial\eta) S_\eta(t)|_{\eta=0} = \int_t^\infty a dF$. The score operator for f is

$$\dot{l}_f(a) = \exp(\beta z) \frac{\int_c^\infty a dF}{S(c)} \left[-\delta \frac{S(c|z)}{1 - S(c|z)} + (1 - \delta) \right].$$

With $Q(c, \delta, z) = \delta \frac{S(c|z)}{1 - S(c|z)} - (1 - \delta)$, we have

$$\begin{aligned}
\dot{l}_\beta &= z \exp(\beta z) \Lambda(c) Q(c, \delta, z), \\
\dot{l}_f(a) &= -\exp(\beta z) \frac{\int_c^\infty a dF}{S(c)} Q(c, \delta, z).
\end{aligned}$$

With assumptions (A2) and (A3)*, $\dot{l}_\beta$ is square integrable. For any $a \in L_2(F)$, $\dot{l}_f(a)$ is square integrable. To calculate the efficient score $\dot{l}_\beta^*$ for β, we need to find a function a_* such that

$$\dot{l}_\beta - \dot{l}_f a_* \perp \dot{l}_f a,$$

for all $a \in L_2^0(F)$. That is

$$E(\dot{l}_\beta - \dot{l}_f a_*)(\dot{l}_f a) = 0.$$

We note that

$$E(\dot{l}_\beta - \dot{l}_f a_*)(\dot{l}_f a) = -E_C \left\{ \frac{\int_C^\infty a dF}{S(C)} E\left[\exp(2\beta Z)O(C|Z)\left[Z\Lambda(C) + \frac{\int_C^\infty a_* dF}{S(C)} \right]|C\right] \right\}.$$

Let

$$E\left[\exp(2\beta Z)O(C|Z)\left[Z\Lambda(C) + \frac{\int_C^\infty a_* dF}{S(C)} \right]|C\right] = 0.$$

Thus, a_* is determined by

$$\int_c^\infty a_* dF = -\frac{\Lambda(c)S(c)E(Z\exp(2\beta Z)O(C|Z)|C=c)}{E(\exp(2\beta Z)O(C|Z)|C=c)}.$$

Note that a_* is only determined on the support of C. $\dot{l}_f a_*$ is a square integrable function with expectation 0. So the efficient score function for β is

$$l_\beta^* = \exp(\beta z)Q(c,\delta,z)\Lambda(c)\left\{ z - \frac{E(Z\exp(2\beta Z)O(C|Z)|C=c)}{E(\exp(2\beta Z)O(C|Z)|C=c)} \right\}.$$

The Fisher information matrix is then $I(\beta) = E(l_\beta^{*2})$. *Assume* that the Fisher information is positive-definite and component-wise bounded.

Consistency.   The first step is to establish the identifiability of the model. Under the Cox model and the boundedness assumptions, the log-likelihood function is continuously differentiable. In addition, assumption (A2b) postulates that for any $\tilde{\beta} \neq \beta_0$, the probability $P(\tilde{\beta}Z \neq \beta_0 Z) > 0$. Under such conditions, the Cox model is identifiable with case I interval censored data.

The parameter β is compact by assumption (A1). The parameter set for Λ is compact with respect to the weak topology. Theorem 5.14 in van der Vaart (2000) suggests that the distance between $(\hat{\beta}, \hat{\Lambda})$ and the set of maximizers of the Kullback-Leibler distance converges to zero. The consistency result follows from the identifiability result. Huang (1996) provides a different proof, which is lengthy but shares the same spirit as the above arguments.

Convergence Rate.   The rate of convergence is obtained using the general theorems for semiparametric models provided in van der Vaart (2000) and van der Vaart and Wellner (1996). The essence is that the convergence rate is directly linked with the "size" of the functional set of likelihood function, which can be measured with entropy (defined below). For commonly used functions, such as bounded monotone functions (e.g., cumulative baseline hazard functions) and splines (e.g., smooth covariate effects), the sharp bounds of the entropy are available.

DEFINITION: Bracketing number. Let $(\mathbb{F}, ||\cdot||)$ be a subset of a normed space of real function f on some set. Given two functions f_1 and f_2, the bracket $[f_1, f_2]$ is the set of all functions f with $f_1 \leq f \leq f_2$. An ϵ bracket is a bracket $[f_1, f_2]$ with $||f_1 - f_2|| \leq \epsilon$. The bracketing number $N_{[]}(\epsilon, \mathbb{F}, ||\cdot||)$ is the minimum number of ϵ brackets needed to cover $\mathbb{F}$. The entropy with bracketing is the logarithm of the bracketing number.

Lemma 25.84 of van der Vaart (2000) shows that there exists a constant K such that for every $\epsilon > 0$,

$$\log N_{[]}(\epsilon, \Phi, L_2(P)) \leq K\left(\frac{1}{\epsilon}\right).$$

Since the log-likelihood function l is Hellinger differentiable and considering the compactness assumptions, we have

$$\log N_{[]}(\epsilon, \{l(\beta, \Lambda) : \beta \in \Theta, \Lambda \in \Phi\}, L_2(P)) \leq K_1\left(\frac{1}{\epsilon}\right),$$

for a positive constant K_1.

Apply Theorem 3.2.5 of van der Vaart and Wellner (1996). Considering the consistency result, we have

$$E^* \sup_{d((\tilde{\beta}, \tilde{\Lambda}), (\beta, \Lambda)) < \eta} |n^{1/2}(P_n - E)(l(\tilde{\beta}, \tilde{\Lambda}) - l(\beta, \Lambda))| = O_p(1)\eta^{1/2}\left(1 + \frac{\eta^{1/2}}{\eta^2 n^{1/2}}K_2\right),$$

for a positive constant K_2, where E^* is the outer expectation. So conditions of Theorem 3.2.1 of van der Vaart and Wellner (1996) are satisfied. The above equation and the consistency result imply that

$$d((\hat{\beta}, \hat{\Lambda}), (\beta, \Lambda)) = O_p(n^{-1/3}).$$

Asymptotic Normality. Theorem 3.4 of Huang (1996) provides a generically applicable approach to establish the $n^{1/2}$ consistency and asymptotic normality of the parametric parameter estimate. A slightly different version is presented as Theorem 1 in Ma and Huang (2005). With the present Cox model, as $P_n l(\beta, \Lambda)$ is maximized at $(\hat{\beta}, \hat{\Lambda})$, we have

$$P_n l(\hat{\beta}, \hat{\Lambda}) \geq P_n l(\tilde{\beta}, \tilde{\Lambda}),$$

for any $(\tilde{\beta}, \tilde{\Lambda})$. In addition, we have the following results

1. (Consistency and convergence rate). $|\hat{\beta} - \beta_0| = O_p(n^{-1/3})$ and $||\hat{\Lambda} - \Lambda_0||_2 = O_p(n^{-1/3})$.

2. (Positive information). The Fisher Information matrix is positive definite and component-wise bounded.

3. (Stochastic equicontinuity). Denote $S_1(\beta, \Lambda) = E\dot{l}_\beta(\beta, \Lambda)$ and $S_2(\beta, \Lambda)[a] = E\dot{l}_f(a)$. Their empirical versions are $\hat{S}_1(\beta, \Lambda) = P_n\dot{l}_\beta(\beta, \Lambda)$ and $\hat{S}_2(\beta, \Lambda)[a] = P_n\dot{l}_f(a)$, respectively. In addition, define

$$\dot{S}_{11}(\beta, \Lambda) = -E(\dot{l}_\beta(\beta, \Lambda)\dot{l}'_\beta(\beta, \Lambda))$$
$$\dot{S}_{12}(\beta, \Lambda)[a] = \dot{S}'_{21}(\beta, \Lambda)[a] = -E(\dot{l}_\beta(\beta, \Lambda)\dot{l}_f(\beta, \Lambda)[a])$$
$$\dot{S}_{22}(\beta, \Lambda)[a_1][a_2] = -E(\dot{l}_f(\beta, \Lambda)[a_1]\dot{l}_f(\beta, \Lambda)[a_2]).$$

For any $\delta_n \to 0$ and constant $K_3 > 0$, within the neighborhood $\{|\tilde{\beta} - \beta_0| < \delta_n, ||\tilde{\Lambda} - \Lambda_0||_2 < K_3 n^{-1/3}\}$,

$$\sup|n^{1/2}(\hat{S}_1 - S_1)(\tilde{\beta}, \tilde{\Lambda}) - n^{1/2}(\hat{S}_1 - S_1)(\beta_0, \Lambda_0)| = o_p(1)$$

and

$$\sup|n^{1/2}(\hat{S}_2 - S_2)(\tilde{\beta}, \tilde{\Lambda})[a_*] - n^{1/2}(\hat{S}_2 - S_2)(\beta_0, \Lambda_0)[a_*]| = o_p(1).$$

Here, $\hat{S}_1$ is the empirical estimation of S_1 and defined as $\hat{S}_1(\beta, \Lambda) = P_n \dot{l}_\beta(\beta, \Lambda)$. $\hat{S}_2$ can be defined in a similar manner. The above equations can be proved by applying Theorem 3.2.5 of van der Vaart and Wellner (1996) and the entropy result.

4. (Smoothness of the model.) For $(\tilde{\beta}, \tilde{\Lambda})$ within the neighborhood $\{|\tilde{\beta} - \beta_0| < \delta_n, ||\tilde{\Lambda} - \Lambda_0||_2 < K_3 n^{-1/3}\}$, the expectations of $\dot{l}_\beta$ and $\dot{l}_f$ are Hellinger differentiable. That is,

$$|\dot{S}_1(\tilde{\beta}, \tilde{\Lambda}) - \dot{S}_1(\beta_0, \Lambda_0) - \dot{S}_{11}(\beta_0, \Lambda_0)[\tilde{\beta} - \beta_0] - \dot{S}_{12}(\beta_0, \Lambda_0)[\tilde{\Lambda} - \Lambda_0]|$$
$$= o(|\tilde{\beta} - \beta_0|) + O(||\tilde{\Lambda} - \Lambda_0||_2^2)$$

and

$$|\dot{S}_2(\tilde{\beta}, \tilde{\Lambda})[a_*] - \dot{S}_2(\beta_0, \Lambda_0)[a_*] - \dot{S}_{21}(\beta_0, \Lambda_0)[a_*][\tilde{\beta} - \beta_0]$$
$$- \dot{S}_{22}(\beta_0, \Lambda_0)[a_*][\tilde{\Lambda} - \Lambda_0]| = o(|\tilde{\beta} - \beta_0|) + O(||\tilde{\Lambda} - \Lambda_0||_2^2).$$

Conditions in Theorem 3.4 of Huang (1996) are satisfied, and hence, the $n^{1/2}$ consistency and asymptotic normality follow.

3.9   Exercises

1. Consider a small data set with multiple types of censoring: 2, 3, 4+, 5-, 6, 7+, [8, 9], [6, 11], where +(-) means right (left) censoring, and [a,b] represents an interval censored observation. Suppose that the distribution of the underlying survival time is exponential with a constant hazard λ. Write down the likelihood function, and find the MLE of λ and compute its variance.

2. In Section 3.2 (Parametric Modeling), the breast cancer data set is analyzed under the exponential model. Consider the nested structure exponential $\subset$ gamma $\subset$ generalized gamma. Fit this data set under the gamma and generalized gamma models. Compute the MLEs. Conduct hypothesis testing and examine the sufficiency of the exponential model.

3. Consider the RFM mice data. For the CE and GE groups separately, compute the NPMLE of the survival function. Examine the survival functions, and see if there is any obvious difference.

4. Consider the RFM mice data. Conduct hypothesis testing, and examine if there is any significant difference between the survival functions of the CE and GE groups. Apply rank-based, survival-based, and Wilcoxon-type approaches. Compare the significance level of different approaches.

5. Consider the breast cancer data. Apply the Turnbull approach, and compute the NPMLE of the survival function. Examine if the estimate satisfies the self-consistency equation. Compare the result with that using the R function EMICM.

6. Consider the semiparametric additive risk model. In this chapter, two estimation approaches and their asymptotic properties have been described. Now reestablish properties of the MLE following the same strategy as for the Cox model. First, derive an iterative convex minorant algorithm for computing the MLE. Then, derive the Fisher information matrix using the projection approach. Last, check conditions for the convergence rate and asymptotic normality of $\hat{\beta}$ as described in the Appendix (Section 3.8).

7. In this chapter, we described the multiple imputation approach for semiparametric models. This approach is also directly applicable to parametric models, although such an approach is expected to be computationally much more expensive than the maximum likelihood approach. Consider again the breast cancer data. Assume an exponential model. Apply the Poor Man's and asymptotic normal multiple imputation approaches, and estimate the parameter λ and its variance.

8. Consider the proportional odds model under case II interval censoring. Derive the Fisher information matrix for the MLE.

9. With the Cox model under case I interval censoring, a least squares estimation approach is described. Now, consider the additive risk model. Derive the form of the least squares estimate. Derive an iterative convex minorant computational algorithm.

Chapter 4

Special Modeling Methodology

The first three chapters cover some common statistical methods for survival analysis. Chapter 3 is particularly written for interval censored data. Right censoring is still the most prevalent data incompleteness encountered in clinical and epidemiological research. In the remainder of the book we will focus mainly on right censored data.

Thanks to various advancements in statistical research fields including nonparametric smoothing, multivariate modeling, and Bayesian analysis, we are now capable of handling more complicated problems in survival analysis. For example, for the Cox PH model, it is usually assumed that covariates affect failure rates in a log-linear form. Such a simple parametric structure may be too limited for some applications. In Section 4.1, we relax such a simple parameterization condition and introduce nonparametric regression methods to describe functional effects of covariates. A second topic of this chapter is multivariate survival analysis. The research on multivariate survival data has been fruitful, yielding results for dependence investigation, marginal modeling of clustered data, random effects models, competing risks, recurrent events and others. We present a few selective results in Section 4.2. The next section of this chapter discusses a special type of data where some individuals may never develop the failure outcome, and the data admit a "cured" or "immune" subgroup. Statistical modeling methods for such data are introduced in Section 4.3. Finally, we present some fundamental results for Bayesian analysis of survival data in Section 4.4, which provides a useful alternative to frequentist likelihood based estimation and inference.

We introduce the aforementioned topics in separate sections. However, it should be noted that many studies address problems that are a combination of two or more complicated features. For example, nonparametric smoothing and multivariate failure time have been discussed in Cai et al. (2007a,b); Bayesian analysis and shared frailty model have been discussed in Clayton (1991); and Bayesian analysis and cure rate model have been investigated in Chen et al. (1999). There are still many other similar examples in the literature that we are not able to summarize here. Therefore sometimes the solution to one complicated problem requires the development of two or more research areas. Understanding and being able to implement the individual tasks is a logical first step.

4.1 Nonparametric Regression

In a Cox PH regression model, the covariate effects are expressed as a functional form similar to linear regression, with the slope parameter multiplied on measured covariates. There is no easy way to check such a linearity assumption imposed in this model. Although multiple graphical and testing approaches have been developed, implementation of these procedures can be time-consuming, and the interpretation is rather subjective in most situations. Without the assumption of linearity, the parameterization in the Cox model loses its meaning, and it is not immediately clear what quantity we are trying to estimate in this model. It is thus sometimes necessary to consider extending the model forms beyond the simple Cox regression model with parametric covariate effects.

Modern nonparametric regression techniques have advanced greatly on various theoretical and applied aspects. Two commonly adopted model fitting methods, namely, the spline method and local polynomial regression method, have been adequately developed to benefit a wide range of applications such as finance, economics, medicine, and genetics. It is not difficult to adapt some of these state-of-the-art procedures in the analysis of survival data.

In this section, we start with the simplest case in the Cox PH regression model with a single covariate whose effect is modeled as a nonparametric function. After that, two specific extensions, namely the partly linear model and varying-coefficients model, are introduced.

4.1.1 Functional Covariate Effects

We now consider extending the Cox proportional hazards model (Equation 2.29), which was introduced in Chapter 2. For the sake of simplicity, consider a univariate predictor X. Under the Cox model, the hazard function is

$$h(t; X) = h_0(t) \exp(\psi(X)), \tag{4.1}$$

where $h_0(t)$ is the baseline hazard function and $\psi(\cdot)$ is an *unknown* function representing the covariate effect. Model (Equation 2.29) may be regarded as a special case of (4.1) when a parametric form is assumed such that $\psi(x) = \beta x$. The estimation of function ψ can be completed when an estimator for β is obtained. The linear function is considered rather restrictive, and model (4.1) may be more flexible since it can accommodate all kinds of nonlinear effects on the log-hazard scale. When the function $\psi(\cdot)$ is not parameterized, we may use nonparametric model-fitting methods or the so-called *smoothing* methods to obtain a functional estimator $\hat{\psi}$. The underlying rationale is that most practical survival mechanisms are "smooth." Two mainstream smoothing approaches have been widely practiced, namely, local polynomial regression and spline regression. Both types of analysis methods assume that the unknown function $\psi(\cdot)$ has a certain smoothness property. For example, a certain order of derivative exists (and integratable in some cases). Note that all the parametric models investigated in the previous chapters satisfy the smoothness conditions. Obviously such a condition is much weaker than the parametric structural requirement.

First, consider local polynomial regression. Without loss of generality, we discuss local linear regression only since it enjoys nice theoretical properties and may be relatively easier to implement than higher-order polynomials. Suppose that the first order derivative of $\psi(\cdot)$ at point x exists. Then by Taylor's expansion,

$$\begin{aligned} \psi(X) &\approx \psi(x) + \psi'(x)(X - x), \\ &\equiv a + b(X - x), \end{aligned}$$

for X in a small neighborhood of x. Let ς be the bandwidth parameter that controls the size of the local neighborhood and K be a kernel function that smoothly weighs down the contribution of remote data points. We then construct the partial local log-likelihood function as

$$l_x(a, b) = \sum_{i=1}^{m} \log \left[\frac{\exp(a + b(X_i - x))}{\sum_{j \in R(t_i)} \exp(a + b(X_j - x))} \right] K_\varsigma(X_i - x), \tag{4.2}$$

where $K_\varsigma(\cdot) = K(\cdot/\varsigma)/\varsigma$ is a rescaled kernel function. Maximizing formulation (4.2) leads to the estimate for local intercept and local slope $(\hat{a}, \hat{b})$. We then obtain $\hat{\psi}(x) = \hat{a}$ and $\hat{\psi}'(x) = \hat{b}$. However, it should be noted that the functional value $\psi(x)$ is not directly estimable; Formulation (4.2) does not involve the parameter $a = \psi(x)$ at all since it cancels out. This is not surprising since from the proportional hazards model (4.1), it is already clear that $\psi(x)$ is only identifiable up to a positive constant. The identifiability of $\psi(x)$ is ensured by imposing the condition that $\psi(0) = 0$. Then the function $\psi(x) = \int_0^x \psi'(u)\,du$ may be estimated by

$$\hat{\psi}(x) = \int_0^x \hat{\psi}'(u)\,du.$$

For practical implementation, we may approximate the integral by the trapezoidal rule (Tibshirani and Hastie, 1987; Fan et al., 1997). As x varies across the range of a set, we obtain an estimated curve on that set.

One crucial operation detail is the choice of the bandwidth parameter ς. ς is not a distribution parameter and does not need to be estimated. Instead, it resembles a numeric option that must be specified in the calculation procedure. If ς is too small, the resulting estimates may exhibit a large variance and be highly irregular (low bias, high variation). On the other hand, if ς is too large, the estimates may not be close to the true functions (high bias, low variation). Similar problems have been commonly observed in smoothed estimation. Therefore, an optimal bandwidth must be sought to balance variance and bias. In some studies, researchers rely on personal expertise or empirical guidelines to set the bandwidth parameter. For a few specific problems, such *ad hoc* approaches may have reasonable performance; however, they are not expected to be applicable in general. Data analysts are recommended to use objective data-driven bandwidth selection procedures such as K-fold cross-validation or generalized cross-validation when empirical bandwidth selection rules cannot be applied.

The choice of the kernel function used in the local likelihood function is relatively less important. We only require the kernel to be symmetric and unimodal. Most commonly adopted kernels include the Gaussian kernel (which is the density function of the standard normal distribution) and the beta family kernel given by

$$K(u; \gamma) = \frac{(1 - u^2)^\gamma I(|u| \le 1)}{B(0.5, 1 + \gamma)},$$

where $B(\cdot, \cdot)$ is the beta function. Setting $\gamma = 0, 1, 2$, and 3 correspond to the uniform, Epanechnikov, biweight, and triweight kernel functions, respectively.

Many authors have applied the local polynomial regression modeling principle to study the nonparametric proportional hazards model, particularly including Tibshirani and Hastie (1987), Gentleman and Crowley (1991), Fan et al. (1997), and others. The asymptotic properties such as consistency and asymptotic normality of the estimators $\hat{\psi}(x)$ have been rigorously established in Fan et al. (1997). Despite all sorts of theoretical advantages of local modeling, to our best knowledge, it has not yet been implemented for survival data analysis in common statistical packages such as R in a friendly manner.

Below we introduce the spline-based approach to fit model (4.1). Such an approach is now available in many statistical packages. Given model (4.1), we may write ψ as a generic functional argument and construct the usual partial likelihood function for the unknown function ψ as

$$L(\psi) = \prod_{i=1}^{m} \frac{\exp(\psi_i)}{\sum_{j \in R(t_i)} \exp(\psi_j)}, \tag{4.3}$$

where $\psi_i = \psi(x_i)$. Let $l(\psi) = \log L(\psi)$ be the log partial likelihood function. Denote Q to be a properly defined functional space for ψ. Then we search for $\psi \in Q$ to maximize

$$l(\psi) - \mathcal{P}(\psi; \lambda), \tag{4.4}$$

where $\mathcal{P}$ is a penalty function and $\lambda \geq 0$ is a smoothing parameter. The motivation for this maximization procedure is to explicitly balance fidelity to data and smoothness of the function. The log partial likelihood function is a natural measure of fidelity of data, while $\mathcal{P}$ is a roughness penalty and a decreasing function of the function smoothness. The most common choice for the penalty function is

$$\mathcal{P}(\psi; \lambda) = \lambda \int_{\mathcal{X}} \psi^{(2)}(x)^2 \, dx, \tag{4.5}$$

where $\psi^{(2)}$ is the second-order derivative of ψ. The corresponding Q is the space of functions with square integrable second-order derivatives on $\mathcal{X}$, the domain of the predictor x. In this case, Q is the so-called second-order Sobolev space. Under such a specification, the first term in (4.4) measures the closeness of fit to data, while the second term penalizes the curvature of fitted function.

Following some profound mathematical theories (e.g., Buja et al., 1989), it can be shown that the solution for maximizing (4.4) must be a cubic spline function with knots at unique predictor values $\{x_i\}$. This approach is thus referred to as a spline approach.

In mathematics, a spline is a sufficiently smooth piecewise-polynomial function. In statistics, the most commonly used splines are cubic spline and cubic B-spline, among others. A cubic spline is a piecewise cubic polynomial function on any subinterval defined by adjacent knots, has first- and second-order continuous derivatives, and has a third-order derivative that is a step function with jumps at the knots. Suppose that there are K distinct values of $\{x_i\}$. A cubic spline function can be represented as a linear basis expansion of the truncated power basis functions $1, x, x^2, x^3, (x - x_1)^3, \cdots, (x - x_K)^3$. The problem of estimating an infinite-dimensional function ψ is reduced to a finite dimensional problem of estimating coefficients corresponding to the chosen basis functions. Other sets of basis functions may also be employed to this end, such as B-spline basis, reproducing kernel Hilbert space basis, and wavelet basis. See deBoor (1978), Eubank (1999), and Wahba (1990) for their definitions.

Suppose now that $x_1 < x_2 < \cdots < x_K$ are the ordered distinct values of the predictors in the sample. By choosing a set of suitable basis functions such as the truncated power basis functions, we may rewrite the complicated integrated penalty function as $\int_{\mathcal{X}} \psi^{(2)}(x)^2 \, dx = \boldsymbol{\psi}^T \mathbf{P} \boldsymbol{\psi}$, where $\boldsymbol{\psi} = (\psi_1, \cdots, \psi_K)^T$ and $\mathbf{P}$ is a symmetric penalty matrix constructed from data. The numerical optimization of (4.4) can then be carried out effectively with this matrix representation.

The spline-based method for the Cox model has appeared earlier in O'Sullivan (1988), Hastie and Tibshirani (1990a), and Kooperberg et al. (1995a,b), and is now applied in medical research for a wide range of applications. It is implemented in the R function coxph with a nested function pspline. The following code is an example of fitting a nonparametric function of a predictor x for censored survival time,

```
fm = coxph(Surv(time, status) ~ psline(x, df = 4))
plot(set$x, predict(fm))
```

where the option df = 4 specifies the user-selected degrees of freedom to be four. The function plot is used to display the estimated curve over the range of x.

Similar to the bandwidth choice in the local polynomial fitting method, we also need to find a proper choice of the smoothing parameter λ when using the spline-based approach. Under the extreme case when $\lambda = 0$, ψ can be any function that interpolates data (*undersmoothing*). Such an estimate has zero bias for estimation but may have large variation. In contrast, when $\lambda = \infty$, we end up with a linear function since no second derivative can be allowed (*oversmoothing*). Here the variation is low but the bias may be high. Often it is satisfactory to choose a smoothing parameter subjectively, but sometimes it is useful to have a method for automatic selection. One possibility is a global cross-validation approach, leaving one datum point at a time and applying the entire estimation procedure repeatedly. This is too computationally expensive and may not be feasible.

The selection of the order of splines, the number of knots, and their positions are relatively less important. First, there is seldom any strong reason to consider polynomial splines other than cubic splines unless one needs higher-order smoothness. Nevertheless, in Chapter 3, we briefly discussed the problem of higher-order smoothness and choosing the order data dependently from a theoretical point of view. Second, the knots may be simply the set of distinct predictor values $\{x_i\}$ when the set is not very large. For a large data set with many distinct predictor values, one may apply a thinning strategy by selecting a moderate number of equally spaced grid points covering the domain of the predictor. The resulting fits are usually visibly satisfactory. Theoretically, Xiang and Wahba (1998) suggest that under mild regularity conditions, if the order of knots grows at least at the order of $n^{1/5}$, full efficiency can be achieved asymptotically.

Remark. In parametric inference, we provide confidence intervals for parameters after computing the point estimates. The construction of confidence intervals is based on the exact or asymptotic distributions of the point estimates. In the above nonparametric regression problem, we may also use the asymptotic distribution property of the estimated function and construct a $100(1 - \alpha)\%$ asymptotic confidence interval as

$$\hat{\psi}(x) \pm z_{\alpha/2} \sqrt{\operatorname{var}(\hat{\psi}(x))},$$

where $z_{\alpha/2}$ is the upper $\alpha/2$ quantile of the standard normal distribution and $\operatorname{var}(\hat{\psi}(x))$ is the asymptotic variance of $\hat{\psi}(x)$. It should be noted that such an interval is only a point-wise confidence interval in that it only ensures that $\psi(x)$ is covered by the interval with $1 - \alpha$ confidence for a single point x. Nonparametric inference for the function ψ requires a more sophisticated development of simultaneous confidence band over the entire of range of x. Bootstrap is a sensible method (Hall, 1993; Cheng et al., 2006). The Bayesian approach, which is briefly introduced in Section 4.4, is sometimes employed toward this end because of its satisfactory numerical performance in real studies (Nychka, 1988; Wang and Wahba, 1995).

When there is more than one covariate, one may be tempted to consider a similar form of Cox PH model by extending model (4.1) to

$$h(t; \mathbf{X}) = h_0(t) \exp\{\psi(\mathbf{X})\}, \tag{4.6}$$

where $\psi(\cdot)$ is an unknown multidimensional function for the p-dimensional covariates $\mathbf{X} = (X_1, \cdots, X_p)^T$. However, estimating a multidimensional function is never an easy task for statisticians, especially when the dimension of $\mathbf{X}$ is greater than two. Usually, the information available from sample does not allow us to fit an arbitrary multidimensional function without further assumptions. In practice, many authors have suggested imposing further structure or constraints to the unknown function $\psi(\cdot)$, including, for example, the use of additive modeling (Hastie and Tibshirani, 1990b), low-order interaction function modeling (Kooperberg et al., 1995a,b), and others.

Specifically, an additive modeling approach decomposes $\psi(\mathbf{X})$ into a sum of functions $\psi_1(X_1) + \cdots + \psi_p(X_p)$, each being a function of a one-dimensional covariate. Model (4.6) is thus simplified to

$$h(t; \mathbf{X}) = h_0(t) \exp\{\psi_1(X_1) + \cdots + \psi_p(X_p)\}, \tag{4.7}$$

where the unknown quantities are p univariate functions. Other components being fixed, the estimation of each of the p univariate functions is relatively easy and can be carried out with the computation methods introduced earlier in this section. We may then employ a backfitting algorithm to fit all unknown functions in (4.7). The advantage of this modeling approach is that one can use the well-established low-dimensional smoothing methods as building blocks to fit the whole model.

Additive modeling methods for penalized spline fitting are now available in R. One can use a program code like the following to estimate additive functions for individual covariates.

```
fm5 = coxph(Surv(time, status) ~ pspline(x1, df = 4)
+ pspline(x2, df = 4), data = set1)
```

The `plot` function can then be used twice to show the estimated functions for the two predictors `x1` and `x2` by extracting the appropriate numeric results stored in the fitted object `fm5`.

4.1.2   Partly Linear Model

In developing nonparametric methods for analyzing censored lifetime data, high-dimensional covariates may cause the *curse of dimensionality* problem. That is, even with the additive model (4.7), when the number of predictors p is large, we may need to estimate many unknown functions. The estimation of unknown function is not always as easy as the estimation of a Euclidean parameter.

One method for attenuating this difficulty is to model the covariate effects under a partly linear framework, which postulates a combination of parametric and nonparametric parts in the hazard model. It allows one to explore nonlinearity of certain covariates when the covariate effects are unknown and avoids the curse of dimensionality inherent in the saturated nonparametric regression models.

A partly linear model for Cox PH regression assumes the following form for the hazard function

$$h(t; \mathbf{X}, Z) = h_0(t) \exp\{\mathbf{X}^T \boldsymbol{\beta} + \psi(Z)\}, \tag{4.8}$$

where both $\mathbf{X}$ and Z are predictors for survival. In some applications, such as Cai et al. (2007a), Z is the main exposure variable of interest whose effect on the logarithm of the hazard may be nonlinear while $\mathbf{X}$ is a vector of covariates with linear effects. Comparing model (4.8) to the ordinary Cox PH model (2.29), we can see that the partly linear model allows us to model the nonlinear contribution of Z and thus can capture the actual covariate effects more accurately. Comparing model (4.8) to the saturated additive model (4.7), we can see that we no longer need to estimate unknown covariate effect functions for individual variables in $\mathbf{X}$ and thus can reduce computation cost. The partly linear model presents a compromise between the standard Cox PH model and additive nonparametric model in that it achieves model parsimony while retaining estimation accuracy.

Both the spline method (Huang, 1999a) and local polynomial regression method (Cai et al., 2007a) can be applied to fit model (4.8). The computation is less intensive than that for an additive model. One may mimic the following code in R to fit a partly linear model:

```
fm6 = coxph(Surv(time, status) ~ x1 + pspline(x2))
```

where x1 is a predictor expected to have a linear effect, and x2 is a predictor with nonlinear functional effect.

Example (Accident). We consider the accident study for university students in Germany, which was described in Section 2.3. Consider fitting a similar model as in Section 2.3.3, except that we now consider a nonparametric function for the Age variable. In R, estimation and inference can be achieved using the following code

```
> ac = coxph(Surv(time,cen) ~ factor(Gender) + factor(unitype)
 + factor(faculty) + pspline(Age, df = 4) , data = acci)
> ss = predict(ac , type = 'terms')
> plot(acci$Age , ss[,4])
```

The pspline estimates are contained in the fitted object acci.coxph, and we can use the predict function to extract the corresponding nonparametric estimates with an option type = 'terms'. The object ss is a matrix with four columns whose fourth column is the estimated function for the age variable. We then plot the nonparametric estimates in Figure 4.1.

We experiment with different choices of degrees of freedom for the pspline function. When df is too small, the fitted curve is just a straight line and may give an inaccurate extrapolation for large values of age. When df is too large, the function may fit the data very well but produce a somewhat wiggly curve. Similar phenomena have been commonly observed in non/semiparametric studies. The default choice df = 4 seems to provide a good balance between model fitting and smoothness (the solid line in Figure 4.1).

4.1.3 Varying-Coefficient Model

An important extension of the ordinary linear regression model with constant coefficients is the varying-coefficient model. The varying-coefficient model addresses an issue frequently encountered by investigators in practical studies. For example, the effect of an exposure variable on the hazard function may change with the level of a confounding covariate. This is traditionally modeled by including an interaction term in the model. Such an approach is a simplification of the true underlying association since a cross product of the exposure and the confounding variable only allows the effect of the exposure to change linearly with the confounding variable. In many studies, however, investigators may express the belief that the rate of change is not linear and seek to examine how each level of the exposure interacts with the confounding variables. Parametric models for the varying-coefficient functions are most efficient if the underlying functions are correctly specified. However, misspecification may cause serious bias, and the model constraints may distort the trend in local areas. Nonparametric modeling is appealing in these situations.

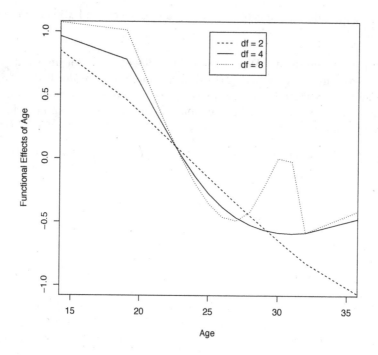

FIGURE 4.1: Estimated functional effects of age with different degrees of freedom for the accident study.

A varying-coefficient model for Cox PH regression assumes the following form for the hazard function

$$h(t; \mathbf{X}, U) = h_0(t) \exp\{\mathbf{X}^T \boldsymbol{\beta}(U) + \beta_0(U)\}, \qquad (4.9)$$

where $\boldsymbol{\beta}(\cdot) = (\beta_1(\cdot), \cdots, \beta_p(\cdot))^T$ and $\beta_0(\cdot)$ are unknown coefficient functions for a continuous variable U, characterizing the extent to which the association varies with the level of U. U acts like an effect modifier in biostatistical research studies (Thompson, 1991). Practically, U can be a confounder (Cai et al., 2007b) or an exposure variable (Fan et al., 2006). The particular forms of the nonlinear functional coefficients are not parameterized. Much effort has been spent on finding efficient estimates of the coefficient functions.

The estimation of coefficient functions can be carried out under a local polynomial regression framework. A local partial likelihood can be constructed in a similar way as (4.2), and functional estimates can be obtained by maximizing the local partial likelihood function. Note that the term $\beta_0(U)$ in (4.9) may be incorporated into the covariates by introducing a column variable of ones. There is a technical reason that many theoreticians opt to not do so.

In fact, the local intercept for $\beta_0(\cdot)$ will cancel out in the local partial like-lihood, similar to the phenomenon evidenced earlier for a single covariate in Section 4.1.1, leading to a different estimation rule and consistency justification for $\beta_0(\cdot)$ from other $\beta(\cdot)$. More computation details can be found in Fan et al. (2006).

When the variable U is time, rather than a covariate, model (4.9) becomes a time-dependent coefficient Cox model, which has been examined by a number of studies, including Zucker and Karr (1990), Murphy and Sen (1991), Marzec and Marzec (1997), Tian et al. (2005), and others. In this case, the model is no longer a proportional hazards model, since the hazards ratio also depends on time. In contrast, model (4.9) is still a proportional hazards model when U is not time. In general, although model (4.9) looks similar to a time-dependent coefficient Cox model, it is more complicated to establish asymptotic properties for the coefficient estimators Fan et al. (2006) under this model.

Currently, research on model (4.9) is confined to local polynomial regression methods, and little advancement has been made so far in the field of spline-based regression methods. As a consequence, there is still no user-friendly software available to fit model (4.9). The authors of this book have written some `MATLAB` programs to fit similar models and can help readers of interest to implement (4.9) upon request.

4.2 Multivariate Survival Data

Multivariate survival data have also been referred to as correlated survival data. The correlation among multiple survival times may be due to temporal or spatial dependence. In some follow-up studies, for one individual patient, multiple events due to multiple types of diseases may occur, or the same type of event may occur multiple times. Under the second scenario, the order of these repeated multiple events is also informative and can be used in analysis. In studies involving subjects from common clusters, their survival times for a failure event can also be correlated because of the spatial dependence. For example, a cluster can be a pair of twin brothers/sisters, a household with more than one member, or a community under the same environmental hazard exposures.

Multivariate survival times may arise from many contexts. The analysis methods are thus quite diversified. We target to offer a streamline below according to statistical modeling differences. In Section 4.2.1, we provide some necessary preparation on how to specify appropriate multivariate distributions with probability tools. Then, our attention will shift to the Cox regression model again, with multivariate failure time outcomes. A very general modeling strategy for correlated survival data is introduced in Section 4.2.2, followed by special treatments on competing risks in Section 4.2.3 and recurrent events in Section 4.2.4.

4.2.1 Distribution Specification

Most multivariate analysis focuses on multivariate normal distribution. However, just as univariate normal distribution does not fit survival data well, multivariate normal distribution is also not a good fit for multivariate survival time. It is necessary to first go through some distribution results for multidimensional data other than normal variates.

Consider M distinct failure times $T_1, T_2, \cdots, T_M$ associated with a typical unit. Here the unit can be an individual subject where the times are for M types of failure events such as the occurrence of different diseases. The unit can also be a cluster such as a household or community where the times are the failure times for the same event of M members within the cluster. Of interest is to model the joint survival function defined as

$$S(t_1, t_2, \cdots, t_M) = P(T_1 > t_1, T_2 > t_2, \cdots, T_M > t_M), \qquad (4.10)$$

which can be more useful than the corresponding joint density function $f(t_1, \cdots, t_M)$. When the unit is a cluster of M members with identical failure rates, the definition (4.10) implies that the M survival times are exchangeable. However, when the failure times are for M different types of events of the same subject, the failure rate for each T_m may be different from the others, and thus the arguments of S are unexchangeable. Thereafter we assume an unexchangeable structure since exchangeable distribution functions are much easier to deal with.

We may in a similar manner define the joint distribution function as

$$F(t_1, t_2, \cdots, t_M) = P(T_1 \leq t_1, T_2 \leq t_2, \cdots, T_M \leq t_M), \qquad (4.11)$$

and it is easy to see that $S(t_1, t_2, \cdots, t_M) + F(t_1, t_2, \cdots, t_M) \leq 1$. Here the equality only holds for $M = 1$. The law of complement becomes more complicated for multivariate distributions. The joint density function is then given by the following multidimensional differential

$$f(t_1, t_2, \cdots, t_M) = \frac{(-1)^M \partial^M S(u_1, u_2, \cdots, u_M)}{\partial u_1 \partial u_2 \cdots \partial u_M} \Big|_{u_1 = t_1, \cdots, u_M = t_M}. \qquad (4.12)$$

We can deduce any marginal survival function from the joint survival function. For any $1 \leq m < M$, the marginal survival function of the first m failure times is

$$S(t_1, \cdots, t_m) = P(T_1 > t_1, \cdots, T_m > t_m) = S(t_1, \cdots, t_m, 0, \cdots, 0) \quad (4.13)$$

which may be obtained from $S(t_1, t_2, \cdots, t_M)$ by setting the last $M - m$ arguments to zero. We can derive other marginal survival functions in a similar manner. The univariate marginal survival function for T_m is defined as

$S_m(t_m) = P(T_m > t_m)$, which may be derived from $S(t_1, \cdots, t_M)$ by setting all except the mth arguments as zero.

Hazard function is always a useful tool to study lifetime data, especially for regression analysis. We define the multivariate component-specific hazard function as

$$h_m(t) = \lim_{\Delta t \to 0} \frac{P(T_m \in [t, t + \Delta t)|T_1 \geq t, \cdots, T_M \geq t)}{\Delta t}, \qquad m = 1, \cdots, M$$

and the multivariate conditional hazard function as

$$h_{m,k}(t) = \lim_{\Delta t \to 0} \frac{P(T_m \in [t, t + \Delta t)|T_1 = t_1, \cdots, T_k = t_k, T_{k+1} \geq t, \cdots, T_M \geq t)}{\Delta t},$$

where $m > k$ and $t_j < t$ for $j = 1, \cdots, k$. These univariate functions are easy to deal with and often turn out to be quite useful in practical applications (Hougaard, 1987; Shaked and Shanthikumar, 1997). In particular, the component-specific hazard function has been used in competing risks modeling where there are M different failure types under consideration, and each T_m is the failure time for an individual to arrive at the mth type or cause of failure. See Section 4.2.3 for more details.

Nonparametric estimation of the joint survival function is difficult with censored lifetime data. Only for the simple bivariate case ($M = 2$), solutions extending the Kaplan-Meier's estimator exist and have been widely used. See Section 5.3 for more details. When $M > 2$, nonparametric tools are less frequently employed, and people usually resort to parametric approaches.

In order to apply standard likelihood procedures for parametric models, we need to parameterize the survival distribution. This requires us to generalize univariate densities described in Table 2.1 to their multivariate versions. Below we introduce a popular generalization approach.

Following some probability theories, we can represent the joint survival function as

$$S(t_1, t_2, \cdots, t_M) = C(S_1(t_1), \cdots, S_M(t_M)), \tag{4.14}$$

where $C : [0, 1]^M \mapsto [0, 1]$ is an M-dimensional distribution function on $[0, 1]^M$ called *copula*. Each marginal of $C(u_1, \cdots, u_M)$, which is obtained by setting the rest of the arguments to be 1, is a uniform distribution on $[0, 1]$. Equation (4.14) suggests a way of linking the multivariate survival function (left-hand side) to the M univariate survival functions (right-hand side). In practice, investigators may be familiar with the univariate behavior of each T_m, and may use a proper distribution such as Weibull or exponential to specify S_m. After the M marginals are well specified, we may then use (4.14) to specify the joint distribution by choosing a copula C.

There are many choices of copulas that may lead to interesting joint distributions. The most famous class of copulas is called the Archimedean copula and defined by

$$C(u_1, \cdots, u_M) = \phi(\phi^{-1}(u_1) + \cdots + \phi^{-1}(u_M)), \tag{4.15}$$

where ϕ is a monotone function called *generator*, and ϕ^{-1} is the inverse function of ϕ. The generator may involve some unknown parameter(s) θ. The actual copula construction is dependent on the selection of generator function. For example, if choosing $\phi(u) = (1+t)^{-1/\theta}$, then we attain the Clayton copula (Clayton, 1978), which leads to a multivariate survival function as

$$S(t_1, t_2, \cdots, t_M) = \left(\sum_{m=1}^{M} S_m(t_m)^{-\theta} - (M-1) \right)^{-1/\theta}. \tag{4.16}$$

For a complete list of copulas and their generators from the Archimedean family, we refer to Nelsen (1999). Most familiar copulas belong to this large family, except for the Gaussian copula given by

$$C(u_1, \cdots, u_M) = \Phi_\Sigma \big(\Phi^{-1}(u_1), \cdots, \Phi^{-1}(u_M) \big), \tag{4.17}$$

where Φ^{-1} is the quantile function of the standard normal distribution, and Φ_Σ is an $M-$dimensional multivariate normal distribution with mean $\mathbf{0}$ and covariance matrix Σ. Applications of these copulas have been welcomed in longitudinal data analysis (Song et al., 2009), dose response experiments (Li and Wong, 2011), quantitative finance (Meucci, 2011), climate research (Li et al., 2008), in addition to survival analysis.

We note that the two aforementioned copulas are associated with parameters θ and Σ, respectively. These parameters quantify the dependence among the multiple failure times. For example, when $\theta \to 0$, Clayton copula becomes $S(t_1, t_2, \cdots, t_M) = \prod_{m=1}^{M} S_m(t_m)$, indicating that the M failure times are independent. In general a larger θ indicates a stronger dependence between failure times under the Clayton model. Similarly, when Σ becomes an identity matrix, the corresponding Gaussian copula also produces M independent components. In this case, the off-diagonal elements in Σ characterize the correlation between pairs of survival times on a transformed scale.

The dependence parameters, θ or Σ in the above examples, involved in copulas may be difficult to interpret. In multivariate survival analysis, we have a set of measures that can provide explicit interpretations. It may be helpful to consider some of these commonly used measures. In particular dependence parameters are usually considered for the bivariate case. Here we consider a couple of important measures.

A traditional association measure is the Pearson correlation coefficient which is defined as

$$\text{corr}(T_1, T_2) = \frac{\text{cov}(T_1, T_2)}{\sqrt{\text{var}(T_1)\text{var}(T_2)}}. \tag{4.18}$$

Many textbooks introduce the Pearson correlation and especially emphasize its applications in linear regression analysis. Unfortunately, this simple correlation measure is not appropriate for survival data for the same reason why linear regression is not appropriate in the current context. First, the Pearson correlation can only measure linear dependence and does not account for nonlinear dependence. Second, survival data can be censored, and it can be difficult to evaluate this measure efficiently with an incomplete sample.

Kendall's tau (Kendall, 1938) is another popular measure which is defined by

$$\tau = P\{(T_{1a} - T_{2a})(T_{1b} - T_{2b}) > 0\}, \tag{4.19}$$

where a and b index two different units. For example, consider T_1 and T_2 to be the times to develop one of two types of cancers. The event described in the definition of tau is that subject a develops the first type of cancer earlier (later) than the second type of cancer, while subject b also develops the first type of cancer earlier (later) than the second type of cancer. Therefore, as a probability, if tau is large, we are able to infer that the order of the two events may be likely to be the same for two randomly selected individuals. Observing that one subject has a shorter (longer) T_1 than T_2 may imply that another subject in the same population also has a shorter (longer) T_1 than T_2.

Tau is also called a measure of *concordance*. It is invariant under monotone transformations of T_m ($m = 1, 2$), and is useful to measure their dependence even if T_1 and T_2 are nonlinearly dependent. The estimation of tau may be carried out nonparametrically for a given sample $\{(T_{1i}, T_{2i}) : i = 1, \cdots, n\}$. When data are complete, an unbiased estimator is

$$\hat{\tau} = 2\{n(n-1)\}^{-1} \sum_{i=1}^{n-1} \sum_{j=i+1}^{n} I\{(T_{1i} - T_{2i})(T_{1j} - T_{2j}) > 0\}. \tag{4.20}$$

When data are incomplete such as censored survival data, the estimation must be carried out in a more sophisticated fashion. We defer such discussions to Chapter 5 (see Sections 5.2 and 5.3).

Another association measure worth mentioning is Spearman's rho which is given by

$$\rho = \text{corr}(S_1(T_1), S_2(T_2)). \tag{4.21}$$

The transformation of T_m by its survival function S_m may be regarded as a way to standardize the data so that ρ can be compared across studies. However, Spearman's rho is still a correlation-based measure and thus may suffer similar criticisms as the Pearson correlation.

The distribution specification and parameterization in the above discussions are applied in the stage of exploratory analysis for survival data. After describing the dependence features of distributions with these tools, we usually still need to build proper regression models to study how covariates affect multivariate failure times.

4.2.2 Regression Analysis of Cluster Data

Now we consider regression models for multivariate or correlated failure time data. Though we have introduced multiple multivariate distribution functions in the preceding section, they are more appropriate for exploratory analysis, and none of the multiargument functions is recommended for regression modeling. We still borrow univariate model forms such as the semiparametric Cox PH model (2.29) or some parametric AFT models and incorporate necessary multivariate dependence structure in the estimation and inference procedure. To facilitate presentation, we restrict our attention to the correlation structure due to clustering in this section. Other types of correlation such as multiple failure times for the same individuals can be similarly addressed with the methods described here.

Let T_{ij} denote the survival time for the jth subject in the ith cluster, $i = 1, 2, \cdots, n$ and $j = 1, 2, \cdots, n_i$. Thus, n_i is the number of subjects in the ith cluster, and the total number of observations is $N = \sum_{i=1}^{n} n_i$. Survival times corresponding to the same clusters tend to be correlated, whereas survival times in different clusters tend to be independent. For event time T_{ij}, the corresponding covariates are denoted by $\mathbf{X}_{ij}$. We consider a cluster-specific Cox regression model given by

$$h_i(t; \mathbf{X}) = h_{0i}(t) \exp(\mathbf{X}^T \boldsymbol{\beta}), \qquad i = 1, \cdots, n \qquad (4.22)$$

where $h_i(t)$ is the hazard function function of T for the ith cluster. The unknown regression coefficient $\boldsymbol{\beta}$ remains the same across all n different clusters, while the baseline hazard functions h_{0i} may be different for different i.

Remarks. In the literature on lifetime data analysis, Equation (4.22) has been used to represent the stratified Cox regression model as well. Such a model is useful when the PH assumption does not hold for the whole sample but is suitable for each individual strata of the sample. Figure 4.2 gives a hypothetical illustration where the hazards ratios between the treatment group and control group remain constant for both strata (top two panels). However, after pooling the two strata together, the resulting hazard is no longer proportional between the two groups (bottom panel). We want to stress that despite the similarity of model forms, the models for multivariate survival times considered in this section are different from those for stratified analysis. For example, for the stratified Cox regression, the event times are all independent, and the usual estimation procedure based on the joint product of partial likelihood functions for all strata may carry forward with little modification. On the other hand, the likelihood-based inference for clustered data is more evolved. Another practical difference is that in stratified analysis, the number of strata is usually quite small, whereas in a cluster analysis, n can be very large.

Since there can be a large number of clusters, it is very hard to work on all these unknown baseline hazard functions $h_{0i}(t)$. A common practice is to pose further assumptions on $h_{0i}(t)$, for example one possibility is to assume that they are all proportional to a common function $h_0(t)$. Then model (4.22) reduces to the following form:

$$h_i(t; \mathbf{X}, w_i) = h_0(t) w_i \exp(\mathbf{X}^T \boldsymbol{\beta}), \qquad i = 1, \cdots, n \qquad (4.23)$$

where w_i is treated as an unobserved random variable called *frailty* (Vaupel et al., 1979). Subjects in the same clusters have the same w_is, and model (4.23) has been referred to as a *shared frailty model* (Clayton and Cuzick, 1985; Oaks, 1989; Aalen, 1994). Assume that w_is $(i = 1, \cdots, n)$ are i.i.d. from a common parametric distribution $g(w)$ instead of treating them as n fixed parameters. Model (4.23) is thus considered as a random-effects model. Not having w_is as fixed-effects parameters may largely reduce the number of unknown parameters, and it is also relatively easier to justify the asymptotic distribution theory. Normally, in order to estimate w_is consistently, we will have to require that $n_i \to \infty$, which is often not true since each cluster (such as a household) only has a small number of within-cluster subjects. Under the

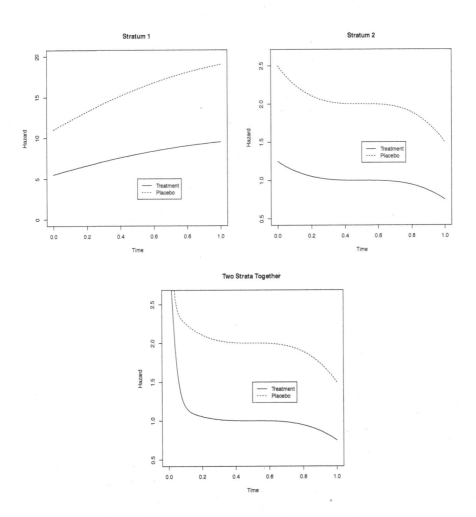

FIGURE 4.2: Hazards for two strata and the combined data.

random-effects model, we only need to estimate a few distribution parameters. Gamma distribution and log-normal distribution are two popular parametric choices for the frailty distribution $g(w)$. We usually assume the mean of w_i to be one, and that leaves only one parameter to be estimated for the frailty distribution.

The basic assumption of the shared frailty model is that the failure times of all subjects in the same cluster are conditionally independent given the common frailty. This shared w_i is the cause of dependence between failure times within the clusters. Consider the multivariate conditional survival functions for the ith cluster (with n_i subjects) given by

$$S(t_{i1}, \cdots, t_{in_i} | w_i, \mathbf{X}_i) = \prod_{j=1}^{n_i} S(t_{ij} | w_i, \mathbf{X}_{ij})$$

$$= \exp \left\{ -w_i \sum_{j=1}^{n_i} H_0(t_{ij}) e^{\mathbf{X}_{ij}^T \boldsymbol{\beta}} \right\},$$

where the first equality follows from the conditional independence assumption and $H_0 = \int h_0$. Averaging with respect to w_i leads to the unconditional marginal survival function

$$S(t_{i1}, \cdots, t_{in_i} | \mathbf{X}_i) = E_{w_i} \{ S(t_{i1}, \cdots, t_{in_i} | w_i, \mathbf{X}_i) \}$$

$$= E_{w_i} \exp \left\{ -w_i \sum_{j=1}^{n_i} H_0(t_{ij}) e^{\mathbf{X}_{ij}^T \boldsymbol{\beta}} \right\}$$

$$= \mathbb{L} \left\{ \sum_{j=1}^{n_i} H_0(t_{ij}) e^{\mathbf{X}_{ij}^T \boldsymbol{\beta}} \right\},$$

where $\mathbb{L}$ denotes the Laplace transformation of the frailty variable. The joint survival function for all clusters is thus a product

$$\prod_{i=1}^{n} \mathbb{L} \left\{ \sum_{j=1}^{n_i} H_0(t_{ij}) e^{\mathbf{X}_{ij}^T \boldsymbol{\beta}} \right\}.$$

One may also derive easily the univariate unconditional survival function as

$$S(t_{ij} | \mathbf{X}_{ij}) = E_{w_i} \{ S(t_{ij} | w_i, \mathbf{X}_{ij}) \}$$

$$= E_{w_i} \exp \left\{ -w_i H_0(t_{ij}) e^{\mathbf{X}_{ij}^T \boldsymbol{\beta}} \right\}$$

$$= \mathbb{L} \left\{ H_0(t_{ij}) e^{\mathbf{X}_{ij}^T \boldsymbol{\beta}} \right\},$$

and find the following connection between the joint and marginal distribution

$$S(t_{i1}, \cdots, t_{in_i} | \mathbf{X}) = \mathbb{L} \left[\mathbb{L}^{-1} \{ S(t_{i1} | \mathbf{X}_{i1}) \} + \cdots + \mathbb{L}^{-1} \{ S(t_{in_i} | \mathbf{X}_{in_i}) \} \right],$$

where $\mathbb{L}^{-1}$ is the inverse function of $\mathbb{L}$. This indicates that the frailty model is a special case of the Archimedian copula given in (4.15).

For a cluster of size $n_i = 2$, sometimes it is of interest to evaluate Kendall's tau (4.19) provided by the following formula

$$
\begin{aligned}
& E\{\text{sign}((T_{i1} - T_{j1})(T_{i2} - T_{j2}))\} \\
= \; & E\left[E\{\text{sign}(T_{i1} - T_{j1})|w_i, w_j\}E\{\text{sign}(T_{i2} - T_{j2})|w_i, w_j\}\right] \\
= \; & E\left(\frac{w_i - w_j}{w_i + w_j}\right)^2,
\end{aligned}
$$

where the first equality follows from the fact $P(T_{i1} < T_{j1}|w_i, w_j) = w_i/(w_i + w_j)$ under the assumed model. We note that tau does not have an analytic form for most frailty distributions, but estimation of this dependence measure can be easily carried out with a Monte Carlo method for a given distribution of w.

In addition to the cluster data setting, frailty has also been applied to univariate survival time to incorporate unobservable heterogeneity. Instead of assuming a homogeneous hazard function for the entire population, some researchers argue that individuals in an observed sample may have different frailties that modify their hazard function and that the most frail dies earlier than the least frail. When all covariates associated with the individuals are observed in a study, we may adjust the effects easily with regression models. However, this is impossible for most epidemiological and biological studies (Keiding et al., 1997; Komarek et al., 2007). We leave out further details since implementation of the univariate frailty analysis is quite similar to that of the shared frailty model.

In addition to the Cox PH model, frailty has been introduced in additive models (Silva and Amaral-Turkman, 2004; Gupta and Gupta, 2009), proportional odds models (Murphy et al., 1997; Lam et al., 2002; Lam and Lee, 2004), and accelerated failure time models (Anderson and Louis, 1995; Keiding et al., 1997; Klein et al., 1999; Pan, 2001; Chang, 2004; Lambert et al., 2004; Komarek et al., 2007). These references cover more computational strategies for frailty models than what we introduce in this section. Furthermore, two monographs Duchateau and Janssen (2008) and Wienke (2011) have been written on this subject, and interested readers may find more in-depth discussion on the modeling methods for frailty.

We now consider how to fit the marginal model (4.22). As for the standard Cox PH model, we usually regard the unknown baseline hazard functions as nuisance parameters and focus mainly on the regression coefficients $\boldsymbol{\beta}$.

In the cluster data setting, T_{ij} and $T_{ij'}$ $(j \neq j')$ are correlated, while T_{ij} and $T_{i'j'}$ $(i \neq i')$ are independent. A working independence assumption is sometimes made for all T_{ij} even if they are from the same cluster. This leads to a marginal model and allows the construction of the ordinary partial likelihood for $\boldsymbol{\beta}$, removing the nuisance baseline hazards from the likelihood function,

$$L(\boldsymbol{\beta}) = \prod_{i=1}^{n} \prod_{j=1}^{n_i} \Big[\frac{\exp(\mathbf{x}_{ij}^T \boldsymbol{\beta})}{\sum_{l \in R(y_{ij})} \exp(\mathbf{x}_{il}^T \boldsymbol{\beta})} \Big]^{\delta_{ij}}, \qquad (4.24)$$

where $(y_{ij} = \min(T_{ij}, C_{ij}), \delta_{ij} = I(T_{ij} \leq C_{ij}))$ is the sample observation, C_{ij} is the censoring time, and $R(y_{ij})$ is the at-risk set for the jth subject in the ith cluster. The likelihood function (4.24) is called a pseudo likelihood function since it is obtained under the independence assumption which is unrealistic for the data.

This likelihood function is exactly the same as the one we use for the stratified analysis with independent data. Therefore, we may end up with the same estimates $\hat{\boldsymbol{\beta}}$ from maximizing (4.24) were we treating different clusters as different strata. Indeed we can justify that the estimates $\hat{\boldsymbol{\beta}}$ from the marginal model are consistent for the true parameters $\boldsymbol{\beta}$ even with the unrealistic working independence assumption. However, the estimates for $\text{var}(\hat{\boldsymbol{\beta}})$ are inconsistent if we ignore the cluster dependence. One has to choose a so-called robust estimate (Reid and Crepeau, 1985; Lin and Wei, 1989) for the variance to yield correct inference. The variance estimator has a sandwich form, and some details for its construction will be covered in the technical notes of this chapter.

In R, we can perform stratified analysis for independent data by calling a nested function `strata` in `coxph` and use a code line like the following

```
coxph(Surv(time, status) ~ x1 + strata(id))
```

where `id` is a column which specifies the different strata, and `time`, `status`, `x1` represent the observed time, event indicator, and covariate as usual. The baseline hazard functions are estimated separately for each stratum.

A standard way to perform clustered analysis under the marginal model (4.23) is to call a nested function `cluster` in `coxph` and use the following code line

```
coxph(Surv(time, status) ~ x1 + strata(id) + cluster(id))
```

where `id` is the variable indicating different clusters. Note that dropping the term `strata(id)` in the above R code leads to fitting an ordinary Cox PH model with h_{0i} being the same across clusters. The risk sets in the likelihood function (4.24) are modified accordingly. From the default output, a robust standard error can be attained, and correct inference can then be drawn. Similarly named functions for stratified or cluster analyses are also available in STATA and SAS.

Sometimes it is not plausible to assume common coefficient values across the clusters, or the sets of covariates may differ across clusters. It may be appealing to relax model (4.22) to be

$$h_i(t; \mathbf{X}) = h_{0i}(t) \exp(\mathbf{X}^T \boldsymbol{\beta}_i), \qquad i = 1, \cdots, n, \qquad (4.25)$$

and so regression analysis has to be done in a coordinate-wise manner for each cluster. We usually do not recommend this approach unless the sample size for the smallest cluster is sufficiently large. To fit model (4.26) we only need to slightly modify the previous notations by introducing a concatenated covariate vector $\mathbf{X}^* = (\mathbf{X}^T I(k=1), \cdots, \mathbf{X}^T I(k=n))^T$ and parameter vector $\boldsymbol{\beta}^* = (\boldsymbol{\beta}_1^T, \cdots, \boldsymbol{\beta}_n^T)^T$ for the kth subject and applying the same programs to fit a marginal model

$$h_i(t; \mathbf{X}) = h_{0i}(t) \exp(\mathbf{X}^{*T} \boldsymbol{\beta}^*), \qquad i = 1, \cdots, n. \tag{4.26}$$

Finally, when it is necessary to assume that the coefficients are the same across clusters, we may use the weighted average of $\hat{\boldsymbol{\beta}}_i$s to construct the final estimator $\hat{\boldsymbol{\beta}}$. The weight is constructed from the covariance of the estimators (see for example, Spiekerman and Lin, 1998), incorporating the relative variability of the cluster-specific estimates. The theoretical justification of this approach can be found in Wei et al. (1989) and Spiekerman and Lin (1998), under the assumption that the marginal model is correctly specified and that the censoring is independent.

Hypothesis testing for the regression coefficients may follow the Wald test procedure. The likelihood ratio test is usually not valid since the asymptotic distribution of the test statistic may not be an exact chi-square distribution. A modified distribution needs to be used in this test (Cai, 1999).

Using marginal models may lose some information since we ignore the correlation of multivariate data. Some simulation studies have also found a slight underestimation of the model variability (Gao and Zhou, 1997) with this approach. Now we take into account the data dependence structure and consider estimation of the shared frailty model (4.23). Under the assumed model, we may construct the complete likelihood function as

$$\prod_{i=1}^{n} \prod_{j=1}^{n_i} \left[\{h(y_{ij}|\mathbf{x}_{ij}, w_i)\}^{\delta_{ij}} S(y_{ij}|\mathbf{x}_{ij}, w_i) \right] \prod_{i=1}^{n} g(w_i), \tag{4.27}$$

where g is the density function of the frailty. Since w_is are not observable, we have to integrate the above likelihood function with respect to all w_is. Numerical optimization methods such as the EM algorithm then need to be implemented to find the maximizer of the observed likelihood function.

When the gamma distribution is chosen for the frailty, integration of (4.27) may enjoy a nice closed form. Since we require the mean of the frailty to be one, the two parameters in the gamma density (Table 2.1) must be equal, that is, $\lambda = k$. We thus obtain a one-parameter gamma density, in this case

$$g(w) = \frac{\lambda^{\lambda} w^{\lambda-1} \exp(-\lambda w)}{\Gamma(\lambda)}.$$

Integrating (4.27) with respect to $w_1, w_2, \cdots, w_n$, we obtain the likelihood function as

$$L(\boldsymbol{\beta}, \lambda, H) = \frac{1}{\Gamma(\lambda)} \exp\left\{\sum_{i=1}^{n}\sum_{j=1}^{n_i} \delta_{ij}\mathbf{x}_{ij}^T\boldsymbol{\beta}\right\}$$

$$\times \prod_{i=1}^{n} \frac{\lambda^\lambda \prod_{j=1}^{n_j}\{h_0(y_{ij})\}^{\delta_{ij}}}{\{\sum_{j=1}^{n_j} H_0(y_{ij})\exp(\mathbf{x}_{ij}^T\boldsymbol{\beta}) + \lambda\}^{\sum_{j=1}^{n_i}\delta_{ij}+\lambda}}. \quad (4.28)$$

One can then use a profile likelihood method to numerically estimate the parametric and nonparametric components in (4.28) in an iterative manner.

Currently, R can accommodate frailty modeling by adopting a penalized likelihood approach. We may follow a code line like

```
coxph(Surv(time, status) ~ x1 + frailty.gamma(id), data = set1)
```

by using the nested function `frailty.gamma` to fit model (4.23) with the gamma frailty. Another option in R is `frailty.gaussian` which allows users to specify a log-normal distribution for the frailty terms. Computation turns out to be much faster than the EM algorithm-based solutions. The similarity between frailty model and penalized Cox model has been recognized by many authors (Klein, 1992; Nielsen et al., 1992; MacGilchrist, 1993) and consequently led to the computation advancement in R. Another computationally appealing approach for frailty models is via Bayesian methods. We will briefly discuss this idea in Section 4.4.

Asymptotic theories for the estimators under shared frailty models are covered in Murphy (1994, 1995), Parner (1998), and Fine et al. (2003), among others. The consistency and asymptotic normality of the model estimates have been established under some technical conditions. However, these results are only justified when the true model is indeed as assumed. Because frailty models place stronger structural requirements for the distribution of the lifetime, model diagnostics and assumption checking must be performed as one necessary step in data analysis. Graphical and numerical methods have been proposed in the literature to examine the adequacy of frailty models. See Shih and Louis (1995), Glidden (1999), Manatunga and Oaks (1999), Fine and Jiang (2000), Glidden (2007), and Economou and Caroni (2005) for more details.

Remark. After introducing the two mainstream clustered data analysis approaches in this section, we point out the following major differences between marginal and frailty models. First, the interpretation of regression coefficients $\boldsymbol{\beta}$ differs in the two types of models. The hazards ratio in the marginal model reflects a comparison of failure risks between subjects randomly sampled from the population regardless of which clusters they are from. On the other hand, the relative risk interpretation for a shared frailty model refers to a comparison for subjects drawn from the same cluster in the population. The estimated effects of covariates can be quite different from the two models since they are estimating different parameters of the population. The second difference is the way a prediction based on the model is made. The marginal model can predict for a single subject while the frailty model can predict the joint survival of all subjects from the same cluster. The latter analysis may be useful in a sibship study where investigators may make prediction of survival based on the current status of other members in the cluster. Knowing these differences may guide practitioners to select a proper model for analysis.

4.2.3 Competing Risks

In a typical follow-up study, patients are often at risk for more than one failure outcome. For example, a group of patients under observation may die of cancer, cardiovascular disease, or kidney disease. If one of the survival event occurs for an individual, death due to other diseases cannot be exactly observed, and so their failure times are censored. This situation is usually referred to as *competing risks* in survival analysis. We consider special modeling methods for such data in this section.

Consider two mainstream regression approaches for competing risks data. The first approach is based on modeling *cause specific hazards*. Suppose that when failure occurs, it must be one of the J distinct types or causes denoted by $\mathcal{J} \in \{1, \cdots, J\}$. The overall hazard function $h(t)$ may still follow the definition in (1.1). In addition, it is important to distinguish hazards due to different causes. We define the cause-specific hazard function as

$$h_j(t) = \lim_{\Delta t \to 0} \frac{P(T \in [t, t + \Delta t), \mathcal{J} = j | T \geq t)}{\Delta t}, \tag{4.29}$$

for $j = 1, \cdots, J$ and $t > 0$. This is a multivariate component-specific hazard function introduced earlier in Section 4.2.1. The overall hazard is related to the cause-specific hazards by noting the equality

$$h(t) = \sum_{j=1}^{J} h_j(t), \tag{4.30}$$

assuming that only one of the failure types can occur at time t.

Using Cox PH regression, we may model the covariate effects for the jth cause-specific hazard function at time t as

$$h_j(t; \mathbf{X}) = h_{0,j}(t) \exp(\mathbf{X}^T \boldsymbol{\beta}_j), \qquad j = 1, \cdots, J \tag{4.31}$$

where $h_{0,j}(t)$ is the baseline cause-specific hazard for cause j, and $\boldsymbol{\beta}_j$ represents the covariate effects on the log hazard of failure time for cause j. The analysis is standard and can be performed with ordinary packages with a slight modification to structure data in a right form. In the data set, individuals whose failures are due to causes $j' \neq j$ are all considered as censored for the jth cause. The implication is that when we deal with the jth failure types, we completely ignore likely failure due to other causes. A subject failing due to causes $j' \neq j$ at time t is still on his/her way to develop the jth type of failure, but that future moment of failure cannot be observed anymore.

For competing risks data, we usually keep a column in the data set to represent the type of failure. In R, we can obtain the model fitting results by using

```
coxph(Surv(time, status == j) ~ x1 + x2, data = set1)
```

where **status** is coded to be a J-level categorical variable with different values of j representing different causes. The logic statement **status == j** labels the uncensored observation for type j failure in the original data set. We repeat the previous analysis for the J categories to achieve the estimated cause-specific coefficients and baseline hazards for all causes. There are several other specification methods to achieve the same output in R. One may follow examples in the supporting documents of R. For other software packages such as **STATA** and **SAS**, similar functions exist for fitting model (4.31).

Most "classic" textbooks on survival analysis only cover this approach. Nonetheless, it has a major limitation for practical interpretation. We consider a second approach that addresses this limitation and may be more appealing to some data analysts.

To understand the limitation of the first approach and facilitate introduction of the second approach, we need a few more notations. Given the definition of cause-specific hazard function (4.29), we may define the cumulative cause-specific hazard by

$$H_j(t) = \int_0^t h_j(u)\,du, \tag{4.32}$$

and cause-specific survival function by

$$S_j(t) = \exp\{-H_j(t)\}. \tag{4.33}$$

We note that (4.33) should not be interpreted as a marginal survival function. The marginal survival function is not a quantity that we can estimate for reasons given in the remarks of this section. The corresponding jth subdensity function is

$$f_j(t) = \lim_{\Delta t \to 0} \frac{P(T \in [t, t + \Delta t), \mathcal{J} = j)}{\Delta t} \tag{4.34}$$

which is also not a proper density function.

The overall survival function is defined as

$$S(t) = \exp\left\{-\sum_{j=1}^{J} H_j(t)\right\}. \tag{4.35}$$

This is a properly defined survival function which describes the probability of not having failed from any of the J causes by time t. Correspondingly, the overall density of time to failure is $f(t) = \sum_{j=1}^{J} f_j(t)$.

Normally, the complement of the survival function is called an incidence function. For the jth failure type, the cause-specific cumulative incidence function is

$$F_j(t) = \int_0^t f_j(u)\,du, \tag{4.36}$$

which is equal to the probability $P(T \le t, \mathcal{J} = j)$. The incidence function (4.36) is not a distribution function and is usually termed as a *subdistribution function* (Gray, 1988).

In the absence of competing risks, the covariate effects in hazard regression models can be easily translated into effects on survival functions. For example, when X is a one-dimensional binary indicator for treatment assignment (for example, $X = 1$ for the treatment group and $X = 0$ for the control group) in a clinical trial, the survival functions for the two groups can be related by

$$S(t|X=1) = S(t|X=0)^{\exp(\beta)}. \tag{4.37}$$

Such a relationship can be further transferred into incidence functions, and hence β has a meaningful interpretation regarding the change in incidence function values associated with the change of covariate values. However, in the presence of competing risks, this simple relationship cannot be extended to cause-specific cumulative incidence functions, and the coefficients β_ks can be hard to interpret for their effects on the actual probabilities of failure.

Therefore, the second approach for competing risks modeling is to directly model the following subdistribution hazard function

$$\hbar_j(t) = \lim_{\Delta t \to 0} \frac{P(T \in [t, t+\Delta t), \mathcal{J} = j | (T \ge t) \,\text{or}\, (T < t; \mathcal{J} \ne j))}{\Delta t}, \tag{4.38}$$

and it is related to the quantities introduced earlier through

$$\hbar_j(t) = -\frac{d\log(1 - F_j(t))}{dt} \tag{4.39}$$

$$= \frac{f_j(t)}{1 - F_j(t)}. \tag{4.40}$$

Note that function (4.38) is not a cause-specific hazard and has a similar but distinct probabilistic interpretation from the familiar hazard function.

Following Fine and Gray (1999), we characterize the covariate dependence for (4.38) through a Cox-type regression model

$$\hbar_j(t; \mathbf{X}) = \hbar_{0,j}(t) \exp(\mathbf{X}^T \mathbf{b}_j), \tag{4.41}$$

where $\hbar_{0,j}$ is a baseline hazard of the subdistribution. Estimation of this model can follow a similar partial likelihood estimation method with a modification on the risk set $R(t_i)$ in (2.30). In this case, the risk set contains not only those who are still alive before time t_i but also subjects who fail from another cause $j' \neq j$ before time t_i. Weights (see for example, Fine and Gray, 1999) may be included to adjust the distance between those earlier other-cause failure times and t_i.

In model (4.41), coefficients $\mathbf{b}_j$ may be interpreted as the logarithm of the subdistribution hazard ratio. Most important, model (4.41) leads to a simple interpretation in terms of cause-specific cumulative incidence function. For example, when X is a one-dimensional binary indicator for treatment in a clinical trial, the jth cause-specific cumulative incidence functions for the two treatment groups can be related by

$$1 - F_j(t|X = 1) = [1 - F_j(t|X = 0)]^{\exp(b_j)}. \qquad (4.42)$$

Such a simple connection is not available for the first approach.

This second regression approach for competing risks is available in R using library `cmprsk`. There are some self-written `ado` code for `STATA` and macro code for `SAS` as well. One may follow examples published by the authors of these programs to implement model (4.41).

The two methods introduced in this section have been both extensively applied to competing risks data. The first method is not wrong since it simply models cause-specific hazard functions. However, interpretation of most statistical results should be based on a probability-scale statement. Since for competing risks data we care about individual cause-specific incidence rates more than overall incidence rates, the second method presents obvious advantages. Many other approaches for competing risks have been developed for special data sets (see for example, Tai et al., 2002; Cheng et al., 2009; Cheng et al., 2010; Xu et al., 2010; and Lee and Fine, 2011; among others).

Remark. Competing risks data are multivariate survival data by definition. However, we should not attempt to estimate the multivariate joint distribution for the J types of failure times. Let $\tilde{T}_j$ denote the latent time to failure of cause j, and we observe $T = \min\{\tilde{T}_j\}$ for each subject. One may want to consider the joint survival function similar to (4.10) as

$$\bar{S}(t_1, \cdots, t_M) = P(\tilde{T}_1 > t_1, \cdots, \tilde{T}_M > t_M), \qquad (4.43)$$

and all kinds of marginal distribution functions derived from (4.43). The fundamental problem with this approach is that the joint survival function (4.43) is not identifiable, nor are the marginal functions from the observed sample. In fact, two models with completely different dependence structures, hence different joint distributions, can lead to the same cause-specific hazard and hence the same partial likelihood for a given sample. The dependence parameters specified in the joint distributions, if any, can be completely meaningless. This is mainly because competing risks data only contain a single failure time for each subject and therefore cannot be treated as a real multivariate data.

It is thus not sensible to consider modeling joint and marginal distributions of these latent failure times.

4.2.4 Ordered Multiple Failure Times Data

For many cancers such as melanoma, breast cancer, bladder tumor, and gastrointestinal cancer, repeated occurrences of detectable tumors for a given subject are common. For a single subject, we may have more than one observation of failure event. These multiple events are naturally ordered, that is, the first tumor must appear before the second one. We call such data *recurrent events* data. The recurring events may be of the same type in an ordered sequence or of different types. We consider regression methods for such ordered multiple failure times in this section.

A rather convenient approach to avoid problems with correlated observations among recurrent events is to keep only the time to the first event for each individual. Regression analysis for the remaining data can be easily done, and interpretation is straightforward. However, this naive approach suffers criticisms that information may be wasted when pretending that recurrent events (after the first one) do not exist.

We introduce three more appropriate approaches for modeling failure times for ordered outcomes. All three approaches are based on marginal models given in Section 4.2.2, leaving the nature of dependence among related failure times completely unspecified. Marginal models are considered appropriate for multivariate ordered or unordered survival data by many authors, when the association parameters for correlated failure times in regression analysis are not of major concern. Theoretical development has also justified the asymptotic consistency properties of estimated regression parameters resulting from marginal models.

The first approach is the simplest and ignores the order of multiple events. We simply use the Cox PH model (2.29) to study the dependence of failure times on covariates. This approach is usually attributed to Andersen and Gill (1982) whose article justified the asymptotic properties of Cox regression using counting process techniques. Note that when constructing the partial likelihood function (4.2), we must arrange the time measurements on a proper scale for this method. Specifically, the length of the first failure time is the difference between the time of the first failure event and the entry time; $\cdots$; the length of the jth failure time is the difference between the time of the jth failure event and the time of the $j-1$th failure event $\cdots$. That is, after one failure occurs, we treat the subject as if he/she has never experienced such an outcome and reset the clock to zero. The data are then regarded as a sample of independent observations.

This approach is easy to implement. After preparing data into the right form, we simply fit a Cox PH model as we do in Chapter 2. One additional requirement is to modify the standard error of the estimated coefficients by using the robust estimator which was introduced in Section 4.2.2. Then we can carry out correct statistical inference.

The first approach is useful especially when we are interested in the overall rate for recurrence of the same nature. However, this approach does not differentiate the order of the recurrent events and may lose some information. In particular, sometimes the researchers' interest may be on the gap time between adjacent failure events. The failure probability for later failures may be thought as different from earlier failures. This is obviously true when the recurrent events are, for example, for different types of deaths. The second approach, attributed to Prentice et al. (1981), addresses this issue by treating the jth failure time as a type j random variable that is correlated with type $j' \neq j$. For each type we specify a possibly different baseline hazard function and therefore attain a Cox model exactly the same as (4.22). In other words, the cluster is defined according to the order of failure events. However, the time needs to be processed similarly as we do with the first approach. That is, the time length is calculated as the difference between the recurring failure and the immediately preceding failure. Computationally, the risk sets for the $(j + 1)$th recurrences are restricted to individuals who have experienced the first j recurrences only. We may then use the same computational methods as for model (4.22) in Section 4.2.2.

The third approach is similar to the second approach. We still consider fitting model (4.22) but use a different time scale for the failure time data. Specifically, the failure time length is just the observed study time, that is, the difference between the recurring failure and the entry time. The rest of the computation is the same as with the second method. This approach, attributed to Wei et al. (1989), is still a marginal approach based on (4.22) as we ignore the dependence structure of related failure times. Compared to the previous two approaches, this approach is the most sensible when the recurrent events are due to different causes.

To use R, STATA, or SAS to implement the three approaches, we only need to prepare our data in the right form and then use code similar to that in Section 4.2.2.

When one is unsatisfied with the marginal model approach because of overlooking the underlying dependence, the frailty model approach may be entertained. The implementation for ordered multivariate failure times is no different from that for unordered ones after data are structured either in Anderson and Gill's form or Wei, Lin, and Weissfeld's form. We may follow the procedures in Section 4.2.2 to obtain desired results.

4.3 Cure Rate Model

In the previous chapters, particularly in the analysis with parametric survival models, it has been directly or indirectly assumed that $S(+\infty) = 0$. Survival functions satisfying such a condition are referred to as *proper*. Under proper survival functions, eventually all subjects in the population would experience the event of interest. Such an assumption may be too strong in practice. Consider, for example, a breast cancer recurrence study. For breast cancer patients who have been successfully treated, we are interested in the time to recurrence. For some patients, recurrence may occur within a finite observation period. For others, breast cancer may never happen again in their lifetime. That is, those patients are *cured*, or equivalently, the event times are equal to infinity.

In this section, we consider survival analysis where the cohort contains a cured subgroup. In the literature, cured subjects have also been referred to as *immune* or *extremely long-term survivors*. When the cured subgroup exists, the overall survival function satisfies $S(+\infty) > 0$, that is, it is *improper*.

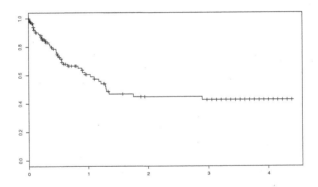

FIGURE 4.3: Survival function for a sample with a cured subgroup.

In Figure 4.3, we display the estimated survival function (KM estimate) for a simulated data set with a cured subgroup. We can see that the last event happens before time $= 3$, and there is no more event in the time interval $(3, 4.5)$. In practice, *a long plateau to the right end in the survival curve, as the one observed in Figure 4.3, may suggest the existence of a cured subgroup*. On the other hand, it is worth noting that an observed plateau to the right end is not sufficient to justify the existence of a cured subgroup. In Figure 4.3, it is possible that the waiting time is not long enough, and that events may happen again after time $= 4.5$.

Denote T as the event time of interest. Most of the statistical literature on cure rate modeling is based on one of the following two models. The first is the two-component *mixture model*, where the survival function is

$$S(t) = p + (1 - p)S_0(t), \tag{4.44}$$

with $S_0(t)$ being a proper survival function. Under the mixture model, the cohort is composed of two subgroups. The first group, with proportion p, has infinitely long survival time. The notion of survival time being ∞ is mathematically convenient, and its practical implication is that individuals in this group tend to have very long survival time. In contrast, subjects in the second group have finite survival times. Model (4.44) is a special case of the more general discrete m-component mixture model for survival time where an individual randomly selected from the population has a survival function given by

$$S(t) = p_1 S_1(t) + \cdots p_m S_m(t), \tag{4.45}$$

where $p_j > 0$ and $\sum_{j=1}^{m} p_j = 1$. Each mixing component $S_j(t)$ may contribute some distribution features (such as proportional hazards or bathtub hazard) toward the overall $S(t)$, and this mixture approach may be useful for very complicated survival data. In cure rate models $m = 2$, and in most other real applications we only consider $m \le 3$.

The second model to incorporate a cure probability is the *bounded cumulative hazard (BCH) model*, where

$$S(t) = \exp\left(-\int_0^t \lambda(s)\,d\,s\right), \quad \text{with}$$

$$p = \exp\left(-\int_0^\infty \lambda(t)dt\right) < 1.$$

Under the BCH model, all the subjects share the same hazard (survival) function. The hazard of event becomes ignorably small for large t. Every subject has a chance p of surviving to infinity.

For parametric modeling, the mixture model approach may be more convenient, as a large family of commonly adopted proper parametric survival distributions can be used for model construction. From a nonparametric point of view, the two models are equivalent. For semiparametric modeling, in this section, we focus on the mixture model approach, but note that the BCH model is also feasible, for example, under a change-point modeling framework.

In what follows, we first consider parametric modeling. For the susceptible subjects, the same models as described in Table 2.1 can be adopted. Thus, the key is to properly accommodate the *mixture* of cured and susceptible subjects. We then investigate nonparametric modeling, where the focus is on the estimation of the cured probability. Right censored and case I interval censored data are separately investigated. Last, we consider the scenario where the susceptible subjects satisfy a proper semiparametric survival function. In this section, we focus on likelihood based estimation and inference. In a series of recent publications, Bayesian approaches have been extensively adopted for cure models. For reference, see for example, Ibrahim et al. (2001b).

4.3.1 Parametric Modeling

Consider the mixture modeling approach. Assume, for example, that the susceptible subjects have an exponential survival function with $S_0(t) = \exp(-\lambda t)$ for $t \geq 0$. Other parametric survival functions can be analyzed in a similar manner. Consider right censored survival data, where the censoring is independent of the event process. Denote T and C as the event and censoring times, respectively. Then one observation consists of $(Y = \min(T, C), \delta = I(T \leq C))$. Consider the scenario with a cured subgroup, then the overall distribution function is

$$F(t) = pF_0(t) = p(1 - \exp(-\lambda t)),$$

where the additional parameter p denotes the susceptible proportion. As the case with $p = 0$ can be easily ruled out, assume that $p \in (0, 1]$. With the boundary case $p = 1$, this model simplifies to the standard exponential model.

Assume n i.i.d. observations $\{(y_i, \delta_i), i = 1, \ldots, n\}$. Then the log-likelihood function is

$$l(\lambda, p) = n^{-1} \sum_{i=1}^{n} \delta_i \log(p\lambda \exp(-\lambda y_i)) + (1 - \delta_i) \log(1 - p + p \exp(-\lambda y_i)).$$

Consider the MLE, which is defined as the maximizer of $l(\lambda, p)$. For computation, as the log-likelihood function has a simple form under parametric modeling, various Newton methods can be employed. In what follows, we briefly discuss asymptotic properties of the MLE, which serve as the basis for inference. We refer to Chapters 5 and 7 in Maller and Zhou (1996) for more detailed development.

Denote the true value of (λ, p) as (λ_0, p_0). First consider the *interior* case, where $0 < p_0 < 1$. As with an ordinary exponential model, it is assumed that $0 < \lambda_0 < \infty$ and the censoring distribution does not degenerate at time zero. Then it can be proved that *the MLE $(\hat{\lambda}, \hat{p})$ of (λ_0, p_0) exists and is locally uniquely defined with probability approaching one. In addition, it is $\sqrt{n}$ consistent and asymptotically normally distributed.* This result is stated as Theorem 7.3 in Maller and Zhou (1996). Here, the asymptotic distribution is the same as that with ordinary parametric models. Various inference and hypothesis testing can be conducted in the same manner as described in Section 2.3.4.

Next, we consider the more challenging case with $p_0 = 1$. That is, the true distribution function is that of the ordinary exponential distribution. Again, consider $(\hat{\lambda}, \hat{p})$, maximizer of the log-likelihood function. Further assume that the censoring time C is not degenerate at 0 and satisfies

$$E(\exp(\lambda_0 C)) < \infty.$$

Then as $n \to \infty$, with probability approaching one, $(\hat{\lambda}, \hat{p})$ is uniquely determined in a neighborhood of (λ_0, p_0). In addition, $(\hat{\lambda}, \hat{p})$ is $\sqrt{n}$ consistent for (λ_0, p_0). Define the matrix

$$D = \begin{bmatrix} \lambda_0^{-2} E(1 - e^{-\lambda_0 C}) & E(C) \\ E(C) & E(e^{\lambda_0 C} - 1) \end{bmatrix},$$

which is finite under the above assumptions. Let $(X, Z)^T$ be a vector of length-2 which has the bivariate normal distribution with mean vector 0 and covariance matrix D^{-1}. In addition, assume that

$$E(\exp((\lambda_0 + \eta)C)) < \infty$$

for some $\eta > 0$. Then for each real x and all $z < 0$, the joint asymptotic distribution of $\sqrt{n}(\hat{\lambda} - \lambda_0)$ and $\sqrt{n}(\hat{p} - 1)$ is given by

$$\lim_{n \to \infty} P\{\sqrt{n}(\hat{\lambda} - \lambda_0) \leq x, \sqrt{n}(\hat{p} - 1) \leq z\} = P(X \leq x, Z \leq z).$$

In settings where the censoring proportion is heavy, there may be a high correlation between the MLE $\hat{\lambda}$ and $\hat{p}$ and also an imprecise estimate of p. We have to check the data to ensure that the follow-up of individuals is long enough and very few failures at large values of t occur.

Consider the deviance defined as

$$d_n = 2(\max_{\lambda, p} l(\lambda, p) - \max_{\lambda} l(\lambda, 1)),$$

which is two times the difference of maximized log-likelihood functions without and with the constraint of $p = 1$. Then it can be proved that

$$\lim_{n \to \infty} P(d_n \leq x) = \frac{1}{2} + \frac{1}{2} P\{\chi_1^2 \leq x\},$$

for $x \geq 0$.

With the interior case, the asymptotic properties are similar to those without the cured subgroup. Other parametric survival models have properties similar to those for the exponential model. However, with the boundary case, as can be seen from above, the assumptions and hence asymptotic results depend on the parametric distribution. Thus, even though other parametric models can be studied using similar techniques, the asymptotic properties need to be established on a case-by-case basis. We refer to Maller and Zhou (1996) for more detailed discussions.

4.3.2 Nonparametric Modeling

In nonparametric modeling, the main goal is to estimate the survival function. For right censored and interval censored data, the nonparametric estimates of survival functions were presented in Sections 2.1 and 3.3, respectively. We note that, *at a fixed time point, the validity of those estimates does not depend on whether the survival function is proper or not.* That is, those estimates can be directly adopted when there is a cured subgroup. Under nonparametric modeling, when it is suspected that there may exist a cured subgroup, a quantity of special interest is the proportion of cure, which is the right end of the survival curve. Because of their significant differences, in what follows, we investigate right censored and case I interval censored data separately.

Right Censored Data. Denote T and C as the event and censoring times, respectively. Under right censoring, we observe $(Y = \min(T, C), \delta = I(T \leq C))$. Assume a sample of n i.i.d. observations. Consider the distribution function $F = pF_0$, where F_0 is a proper distribution function. From the definition, $p = F(+\infty)$. So, a natural estimate of p is

$$\hat{p} = \hat{F}(y_{(n)}),$$

where $\hat{F}(t) = 1 - \hat{S}(t)$ and $\hat{S}(t)$ is the Kaplan-Meier estimate (2.2) described in Section 2.1, and $y_{(n)} = \max(y_1, \ldots, y_n)$. Denote G as the distribution function of the censoring time. Further define that

$$\tau_F = \inf\{t \geq 0 : F(t) = 1\},$$
$$\tau_G = \inf\{t \geq 0 : G(t) = 1\},$$
$$\tau_{F_0} = \inf\{t \geq 0 : F_0(t) = 1\}$$

with the convention that $\tau_F = \infty$ if $F(t) < 1$ for all $t \geq 0$. In addition, define

$$\tau_H = \min(\tau_F, \tau_G).$$

Assume that the censoring is independent of event process, $0 < p \leq 1$, and F is continuous at τ_H in case $\tau_H < \infty$. Then it can be proved that

$$\hat{p} \to_P p, \text{ as } n \to \infty \text{ if and only if } \tau_{F_0} \leq \tau_G.$$

That is, when the event distribution is dominated by that of the censoring distribution (in another word, the waiting period is "long enough"), the right end of the KM estimate is consistent for the cured proportion.

Next we consider the estimate of F_0, the distribution function of the susceptible. From the definition, $F_0(t) = F(t)/p$. Thus, a natural estimate of $F_0(t)$ is the nondecreasing, right continuous statistic given by

$$\hat{F}_0(t) = \hat{F}(t)/\hat{p}, \quad t \geq 0,$$

with $\hat{F}(t)$ and $\hat{p}$ being defined above. This estimate has the following consistency property. Assume independent censoring, $0 < p \le 1$, and F is continuous at τ_H if $\tau_H < \infty$. Then

$$\sup_{t \ge 0} |\hat{F}_0(t) - F_0(t)| \to_P 0,$$

as $n \to \infty$ if and only if $\tau_{F_0} \le \tau_G$. That is, when the waiting is long enough, a simple rescaling leads to a consistent estimate of the distribution function for the susceptible.

For inference, the consistency result is insufficient. We further need the asymptotic distribution result. Assume independent censoring and $0 < p \le 1$. In addition, assume that

$$\int_{[0,\tau_{F_0}]} \frac{dF_0(t)}{1 - G(t-)} < \infty,$$

and that F is continuous at τ_{F_0} in case $\tau_{F_0} < \infty$. Then as $n \to \infty$,

$$\sqrt{n}(\hat{p} - p) \to_D N(0, (1 - p)^2 v_0)$$

for some $v_0 < \infty$. This result has different implications when the cured subgroup is present ($p < 1$) or not ($p = 1$). When $p < 1$, the above result suggests that the estimate $\hat{p}$ is $\sqrt{n}$ consistent and the asymptotic distribution is normal. However, when $p = 1$ (that is, there is no cured subject), $\hat{p}$ has a convergence rate faster than $\sqrt{n}$. It is theoretically difficult to scale $\hat{p} - p$ to yield a non-degenerate proper limiting distribution when $p = 1$ (Maller and Zhou, 1996). For this boundary case, researchers usually have to rely on simulations to gain some idea of the asymptotic distribution of $\hat{p}$.

When a cured subgroup exists ($p < 1$), for variance, Maller and Zhou (1992) prove that

$$v_0 = \int_0^\infty \frac{dF(s)}{(1 - F(s))(1 - F(s-))(1 - G(s-))}.$$

Use notations described in Section 2.5, where $N(s)$ and $A(s)$ denote the event and at-risk processes at time s. Consider the estimate

$$\hat{v}(t) = \int_{[0,t]} \frac{nI(\Delta N(s) < A(s))}{(A(s) - \Delta N(s))} \frac{dN(s)}{A(s)}.$$

Following Gill (1980), it can be proved that as $n \to \infty$,

$$\hat{v}(y_{(n)}) \to_P v_0.$$

Inference on p when $0 < p < 1$ can be based on the result

$$\frac{\sqrt{n}(\hat{p} - p)}{(1 - \hat{p})\sqrt{\hat{v}(y_{(n)})}} \to_D N(0, 1).$$

Maller and Zhou (1996) present a heuristic simulation-based approach for generating the asymptotic distribution of $\hat{p}$ when $p = 1$. The researchers suggest that the analytic form of the asymptotic distribution is still not known, and it depends on the unknown distributions F and G. Additional asymptotic research on this topic may be needed to solve this theoretical conundrum. However, when practitioners are facing sufficient information that p may be close to one, they may avoid attempting to estimate the overparameterized cure rate model and simply focus on $S_0(t)$.

Case I Interval Censored Data. Denote T and C as the event and censoring times, respectively. Under case I interval censoring, one observation consists of $(C, I(T \leq C))$. As shown in Chapter 3, the nonparametric estimate of the survival function under case I interval censoring differs significantly from that under right censoring. Under nonparametric modeling, the cure rate estimation under case I interval censoring has been investigated in Sen and Tan (2008).

Denote the n i.i.d. observations as $\{(C_i, \delta_i), i = 1, \ldots, n\}$. Following the same rationale as with right censored data, it may seem that a natural estimate is $\hat{p} = \hat{F}(C_{(n)})$, where $C_{(n)} = max(C_1, \ldots, C_n)$. However, Sen and Tan (2008) establish that $\hat{p}$ *is unique if and only if* $\hat{F}(C_{(n)}) = 1$. In addition, from Chapter 3, we have

$$\hat{F}(C_{(n)}) = \max_{i \leq n} \frac{\sum_{j=i}^{n} \delta_{(j)}}{n - i + 1},$$

where $\delta_{(j)}$ is the event indicator associated with $C_{(j)}$. Thus, $\hat{F}(C_{(n)}) = 1$ if and only if $\delta_{(n)} = 1$.

For $0 < p < 1$ and any $0 < \epsilon < p$, it can be proved that

$$P(|\hat{F}(C_{(n)}) - p| > \epsilon) \geq P(\hat{F}(C_{(n)}) = 1) \to pF(\tau_G).$$

That is, $\hat{F}(C_{(n)})$ is *not* a consistent estimate of F, which significantly contradicts the case with right censored data. To construct a consistent estimate, consider that

$$\hat{F}(C_{(n)}) = \max_{i \leq n} \frac{\sum_{j=i}^{n} \delta_{(j)}}{n - i + 1} = \max_{x \leq C_{(n)}} \frac{\sum_{j=1}^{n} \delta_j I(C_j \geq x)}{\sum_{j=1}^{n} I(C_j \geq x)}.$$

That is, $\hat{F}(C_{(n)})$ is the maximum of the *tail-averages* of the concomitants, $\delta_{(i)}$, $1 \leq i \leq n$. Such an observation motivates the following ratio empirical process

$$\tilde{p}_1(x) = \frac{\sum_{j=1}^{n} \delta_j I(C_j \geq x)}{\sum_{j=1}^{n} I(C_j \geq x)} \to p_1(x) = p \frac{\int_x^{\infty} F dG}{\int_x^{\infty} dG}$$

almost surely for each $x \geq 0$ as $n \to \infty$. Moreover, note that

$$p_1(x) \uparrow p, \quad \text{as } x \uparrow \infty.$$

In addition,

$$\tilde{p}_2(x) = \max_{y \leq x} \tilde{p}_1(y) \to p \max_{y \leq x} \frac{\int_y^\infty F dG}{\int_y^\infty dG} \uparrow p \text{ as } x \uparrow \infty.$$

Motivated by the above observations, consider the following two estimates:

ESTIMATE I: $\hat{p}_1 = \tilde{p}_1(x_n)$, that is, tail-average at a suitable sequence $x_n \uparrow \infty$ of "cut-off" points.

ESTIMATE II: $\hat{p}_2 = \tilde{p}_2(x_n) = \max_{y \leq x_n} \tilde{p}_1(y)$, that is, partial maximum of the tail average.

The above estimates have simple forms. As long as the "tuning" parameter sequence x_n is properly specified, they can be easily computed. Asymptotically, as shown in Sen and Tan (2008), the requirement on x_n is fairly weak. Below we describe how to select x_n for a given sample of size n. Consider that

$$
\begin{aligned}
\hat{p}_1 - p &= \frac{\sum_{j=1}^n \delta_j I(C_j \geq x_n) - p(\int_{x_n}^\infty F dG/\bar{G}(x_n)) I(C_j \geq x_n)}{n\bar{G}(x_n)} \\
&\quad \times \frac{\bar{G}(x_n)}{n^{-1} \sum_{j=1}^n I(C_j \geq x_n)} - p \int_{x_n}^\infty (1 - F) dG/\bar{G}(x_n) \\
&= A(x_n)C(x_n) - B(x_n),
\end{aligned}
$$

where $\bar{G}(x) = \int_x^\infty dG = 1 - G(x)$. Now

$$
\begin{aligned}
n\bar{G}(x_n) var(A(x_n)) &= p \int_{x_n}^\infty F dG/\bar{G}(x_n) - \left[p \left(\int_{x_n}^\infty f dG/\bar{G}(x_n) \right) \right]^2 \\
&\to p(1 - p)
\end{aligned}
$$

as $x_n \to \infty$. In addition, $C(x_n) = O_p(1)$ and $B(x_n) = o(1)$ as $x_n \to \infty$. Hence,

$$\hat{p}_1 - p = O_p((n\bar{G}(x_n))^{-1/2}) + o(1)$$

as $x_n \to \infty$.

Thus, we have the following trade-off. As $n \to \infty$, we must have $x_n \uparrow \infty$, so that the bias $-B(x_n) \to 0$ and also $\bar{G}(x_n) \to 0$. On the other hand, x_n needs to grow slowly enough so that $n\bar{G}(x_n) \to \infty$. In view of the above results, optimal order of $x_n \to \infty$ can be determined by minimizing, with respect to x, the function

$$M(x) = \frac{p(1 - p)}{n\bar{G}(x)} + p^2 \left(\int_x^\infty (1 - F) dG/\bar{G}(x) \right)^2.$$

With a fixed sample of size n, x_n can be chosen data dependently as the minimizer of

$$\hat{M}(x) = \widehat{var}(A(x)) + \hat{B}^2(x)$$

with respect to x, where $\widehat{var}(A(x))$ and $\hat{B}^2(x)$ are the estimates of $var(A(x))$ and $B(x)$, respectively.

For detailed proofs of the above procedure, we refer to Sen and Tan (2008). Numerical studies with sample size as small as 100 show reasonable performance of the proposed estimates.

4.3.3 Semiparametric Modeling

As described in earlier chapters, there are a large number of semiparametric survival models. In principle, all those models can be adopted for the susceptible subjects when there is a cured subgroup. In this section, we demonstrate semiparametric modeling with a cured subgroup using two specific examples. The first example is the Cox model with right censored data, and the second is the additive risk model with case I interval censored data.

Right Censored Data with the Cox Model. This model has been investigated in Fang et al. (2005). Denote Z as the vector of covariates. Consider the mixture model, under which there are two subgroups. Subjects in the first subgroup are "cured" or "immune" and have event times $T = \infty$. Subjects in the second subgroup satisfy the Cox model with $\lambda(t|Z = z) = \lambda_0(t) \exp(z^T\beta)$. Denote $\Lambda_0(t)$ as the cumulative baseline hazard function. We introduce the *unobservable* cure indicator: $U = 0$ if $T = \infty$ (the first subgroup). Assume that, conditional on Z, the censoring time C is independent of T. For the cure indicator, assume the logistic regression model, where

$$P(z) = P(U = 1|Z = z) = \frac{1}{1 + \exp(-x^T\gamma)}$$

with $x = (1, z^T)^T$. Here the logistic model can be replaced by other parametric generalized linear models. In principle, it is also possible to adopt a semiparametric model for $P(z)$.

Denote C as the censoring time. Assume n i.i.d. copies of $(Y = \min(T, C), \delta = I(T \leq C), Z)$. Then the likelihood function is

$$
\begin{aligned}
L(\beta, \gamma, \Lambda_0) &= \prod_{i=1}^{n} \left(\frac{\exp(X_i^T\gamma) \exp(Z_i^T\beta) \exp(-\Lambda_0(Y_i) \exp(Z_i^T\beta))}{1 + \exp(X_i^T\gamma)} d\Lambda_0(Y_i) \right)^{\delta_i} \\
&\times \left(\frac{1 + \exp(X_i^T\gamma) \exp(-\Lambda_0(Y_i) \exp(Z_i^T\beta))}{1 + \exp(X_i^T\gamma)} \right)^{1-\delta_i}
\end{aligned}
$$

where $d\Lambda_0(Y_i) = \Lambda_0(Y_i) - \Lambda_0(Y_i-)$.

Assume that $\beta \in B_1$, a compact set of $\mathbb{R}^p$, and that $\gamma \in B_2$, a compact set of $\mathbb{R}^{p+1}$. First, it can be proved that *there exists $(\hat{\beta}, \hat{\gamma}, \hat{\Lambda})$ that maximizes the likelihood function $L(\beta, \gamma, \Lambda)$ over the parameter space $\{\beta \in B_1, \gamma \in B_2, \Lambda$ is a right continuous, nonnegative, nondecreasing function$\}$.* That is, with a sample of size n, the MLE exists. The proof is built on the compactness conditions and smoothness of the likelihood function.

Without the cured subgroup, the profile likelihood approach described in Section 2.5 can effectively "remove" the nonparametric parameter Λ from the estimation process. However, when the cured subgroup exists, such an approach is not applicable. We have to simultaneously maximize over two unknown parametric parameters and one nonparametric parameter. Here computation can be realized using an EM algorithm as suggested in Fang et al. (2005). In addition, an iterative algorithm can also be developed.

The asymptotic properties of MLE can be briefly summarized as follows. Denote $(\beta_0, \gamma_0, \Lambda_0)$ as the true value of parameters. Let τ be a time point satisfying $E(\delta I(Y \geq \tau)) > 0$. Then as $n \to \infty$,

$$\sup_{t \in [0,\tau]} |\hat{\Lambda}(t) - \Lambda_0(t)|, \quad ||\hat{\beta} - \beta_0||_2, \quad ||\hat{\gamma} - \gamma_0||_2 \to_{a.s.} 0,$$

where $|| \cdot ||_2$ is the ℓ_2 norm.

As demonstrated in Chapter 3, in the process of establishing the asymptotic distribution for the parametric parameter estimates, a necessary condition is the nonsigularity of Fisher information. In Chapter 3, we demonstrate information calculation with several semiparametric survival models. With the present model, (β, γ) can be viewed as one mega parametric parameter. Information calculation can be carried out using the projection approach described in Chapter 3 and Groeneboom and Wellner (1992). For more details, we refer to Fang et al. (2005). Denote I_β and I_γ as the Fisher information matrices for β and γ respectively. It can be proved that

$$\sqrt{n}(\hat{\beta} - \beta_0) \to_d N(0, I_\beta^{-1}), \quad \sqrt{n}(\hat{\gamma} - \gamma_0) \to_d N(0, I_\gamma^{-1})$$

as $n \to \infty$. In addition, for any fixed $t \in (0, \tau)$,

$$\sqrt{n}(\hat{\Lambda}(t) - \Lambda_0(t)) \to_d N(0, v^2(t)),$$

where $0 < v(t) < \infty$ and its exact form is defined in Fang et al. (2005). For inference with β and γ, multiple approaches are applicable, particularly including the plug-in approach and weighted bootstrap.

With right censored data and Cox model, the presence of a cured subgroup makes the objective function considerably more complicated. In particular, the nonparametric baseline hazard cannot be "profiled out" and needs to be estimated along with the parametric regression parameters. However, the main asymptotic properties of the estimates are similar to those when there is no cured subgroup.

Case I Interval Censored Data with Additive Risk Model. This setting has been studied in Ma (2011). For simplicity of notation, assume only two covariates Z_1 and Z_2. Denote $Z = (Z_1, Z_2)'$. Extension to the case with more covariates is almost trivial. For case I interval censored data, one observation consists of $(C, \delta = I(T \leq C), Z_1, Z_2)$. To account for the cured subgroup, we introduce the unobservable cure indicator U as described above. Again assume the logistic regression model for U where

$$P(U = 1|Z = z) = \frac{1}{1 + \exp(-x^T\gamma)}$$

with $x = (1, z^T)^T$. For subjects with $U = 1$, we model the survival hazard of event time T using the partly linear additive risk model:

$$\lambda(T|Z) = \lambda(T) + f(Z), \text{ where } f(Z) = \beta Z_1 + h(Z_2). \tag{4.46}$$

Here, $\lambda(T|Z)$ is the conditional hazard function, $\lambda(T)$ is the unknown baseline hazard function, β is the parametric regression coefficient, and h is the smooth, nonparametric covariate effect. The cumulative hazard function is:

$$\Lambda(T|Z) = \Lambda(T) + f(Z)T.$$

The partly linear additive risk model is an extension of the standard additive risk model, allowing for semiparametric covariate effects. It is noted that the above model accommodates the standard additive risk model with parametric covariate effects and time-dependent covariates.

Under the assumption that the event time and censoring are conditionally independent, for a single observation, the log-likelihood (up to a constant) is equal to

$$\begin{aligned}
l(\gamma, f, \Lambda) &= \delta \log(\phi(X^T\gamma)) + \delta \log[1 - \exp(-\Lambda(C) - f(Z)C)] \\
&\quad + (1 - \delta) \log\{1 - \phi(X^T\gamma)[1 - \exp(-\Lambda(C) - f(Z)C)]\}
\end{aligned}$$

where ϕ is the logit function. We use the notation ϕ so that the results described below can be directly extended to other parametric generalized linear models for U.

There are two unknown nonparametric parameters h and Λ. It is assumed that the nonparametric covariate effect h is a smooth function. Particularly, we assume that h is a spline function and adopt a penalized estimation approach. Assume that n i.i.d. observations $(C_1, \delta_1, Z_{11}, Z_{21}), \ldots, (C_n, \delta_n, Z_{1n}, Z_{2n})$ are available. Consider the penalized maximum likelihood estimate (PMLE)

$$(\hat{\gamma}, \hat{\beta}, \hat{h}, \hat{\Lambda}) = argmax_{\gamma,\beta,h,\Lambda} P_n l - \lambda_n^2 J^2(h), \tag{4.47}$$

where λ_n is the data-dependent tuning parameter and can be chosen via cross-validation, $J^2(h)$ is the penalty on smoothness defined as $J^2(h) = \int_{Z_2} (h^{(s_0)}(Z_2))^2 dZ_2$, $h^{(s_0)}$ is the s_0^{th} derivative of h, and P_n is the empirical measure. In practical data analysis, it is commonly assumed that $s_0 = 2$. It is operationally acceptable to adopt other s_0 values. In addition, it is also possible to adaptively select the order of smoothness, as demonstrated in Chapter 3.

Let $C_{(1)}, \ldots, C_{(n)}$ be the ordered $C_1, \ldots, C_n$. Let $\delta_{(i)}, Z_{(1i)}, Z_{(2i)}$ be the event indicator and covariates corresponding to $C_{(i)}$. Since only the values of Λ at $C_{(i)}$ matter in the log-likelihood function, we set the PMLE $\hat{\Lambda}$ as the right-continuous nondecreasing step function with jumps only at $C_{(i)}$. Assume that $\delta_{(1)} = 1$ and $\delta_{(n)} = 0$. The PMLE defined in (4.47) has the following finite-sample properties:

$$\frac{\partial P_n l}{\partial \alpha}\bigg|_{\alpha=\hat{\alpha},\beta=\hat{\beta},h=\hat{h},\Lambda=\hat{\Lambda}} = 0, \quad \frac{\partial P_n l}{\partial \beta}\bigg|_{\alpha=\hat{\alpha},\beta=\hat{\beta},h=\hat{h},\Lambda=\hat{\Lambda}} = 0, \qquad (4.48)$$

$$\sum_{j \geq i} \left(\frac{\delta_{(j)}}{1 - \exp(-\hat{\Lambda}_{(j)} - (\hat{\beta}Z_{(1j)} + \hat{h}(Z_{2j}))C_{(j)})} \right.$$
$$- \frac{(1 - \delta_{(j)})\phi}{1 - \phi[1 - \exp(-\hat{\Lambda}_{(j)} - (\hat{\beta}Z_{(1j)} + \hat{h}(Z_{2j}))C_{(j)})]} \Bigg)$$
$$\times \exp(-\hat{\Lambda}_{(j)} - (\hat{\beta}Z_{(1j)} + \hat{h}(Z_{2j}))C_{(j)}) \leq 0, \qquad (4.49)$$

$$\sum_{i=1}^{n} \left(\frac{\delta_{(i)}}{1 - \exp(-\hat{\Lambda}_{(i)} - (\hat{\beta}Z_{(1i)} + \hat{h}(Z_{2i}))C_{(i)})} \right.$$
$$- \frac{(1 - \delta_{(i)})\phi}{1 - \phi[1 - \exp(-\hat{\Lambda}_{(i)} - (\hat{\beta}Z_{(1i)} + \hat{h}(Z_{2i}))C_{(i)})]} \Bigg)$$
$$\times \exp(-\hat{\Lambda}_{(i)} - (\hat{\beta}Z_{(1i)} + \hat{h}(Z_{2i}))C_{(i)})\hat{\Lambda}_{(i)} = 0, \qquad (4.50)$$

for $i = 1, \ldots, n$. Equation (4.48) holds following the definition of PMLE and smoothness of the likelihood function. Equations (4.49) and (4.50) can be proved by following Proposition 1.1 of Groeneboom and Wellner (1992).

It can be further shown that $\hat{h}$ defined in (4.47) is a spline function. Specifically, suppose that $\tilde{h}$ maximizes the penalized log-likelihood function. Then there exists a spline function $\hat{h}$, such that $\hat{h}(Z_{2i}) = \tilde{h}(Z_{2i})$ for $i = 1, \ldots, n$ and $J(\hat{h}) \leq J(\tilde{h})$ (Wahba, 1990).

A few regularity conditions are needed to establish the asymptotic results. The details are provided in Ma (2011). A particularly interesting condition is that $\lambda_n = O_p(n^{-1/3})$, which is slower than that commonly assumed for spline functions. Denote the unknown true value of $(\gamma, \beta, h, \Lambda)$ as $(\gamma_0, \beta_0, h_0, \Lambda_0)$. Define the distance between $(\alpha, \beta, h, \Lambda)$ and $(\alpha_0, \beta_0, h_0, \Lambda_0)$ as

$$d((\alpha, \beta, h, \Lambda), (\alpha_0, \beta_0, h_0, \Lambda_0)) = ||\alpha - \alpha_0|| + ||\beta - \beta_0|| + ||h - h_0||_2 + ||\Lambda - \Lambda_0||_2,$$

where $||h - h_0||_2^2 = \int_{Z_2} (h(Z_2) - h_0(Z_2))^2 dZ_2$ and $||\Lambda(c) - \Lambda_0(c)||_2^2 = \int_{l_C}^{u_C} (\Lambda(c) - \Lambda_0(c))^2 dc$. The consistency and convergence result can be summarized as follows

$$d((\hat{\gamma}, \hat{\beta}, \hat{h}, \hat{\Lambda}), (\gamma_0, \beta_0, h_0, \Lambda_0)) = O_p(n^{-1/3}) \text{ and } J(\hat{h}) = O_p(1).$$

Interesting implications of the above result include the $n^{1/3}$ convergence rate of $\hat{\Lambda}$, which is sharp. In addition, the convergence rate of $\hat{h}$ is "suboptimal" (in the sense that it is slower than that with ordinary spline penalized regression) and cannot be further improved. $\hat{h}$ has a "proper" degree of smoothness, being not too smooth or too wiggly.

To establish the asymptotic distribution of $(\hat{\gamma}, \hat{\beta})$, a necessary condition is the nonsingularity of the information matrix. A nonorthogonal projection approach can be applied here. We refer to Ma (2011) for more details. With a nonsingular information matrix, the asymptotic distribution result can be summarized as

$$\sqrt{n}(\hat{\alpha} - \alpha_0, \hat{\beta} - \beta_0) \rightarrow_d N(0, I^{-1})$$

where I is the information matrix. That is, the estimates of parametric parameters are still asymptotically normal and fully efficient.

For inference, we resort to the weighted bootstrap approach. In particular, denote $w_1, w_2, \ldots, w_n$ as n i.i.d. positive random weights generated from a known distribution with $E(w) = 1$ and $var(w) = 1$. Denote $(\hat{\gamma}^*, \hat{\beta}^*, \hat{h}^*\hat{\Lambda}^*)$ as the weighted PMLE

$$(\hat{\gamma}^*, \hat{\beta}^*, \hat{h}^*, \hat{\Lambda}^*) = argmax_{\alpha,\beta,h,\Lambda} P_n\{w \times l(\gamma, \beta, h, \Lambda)\} - \lambda_n^2 J^2(h).$$

Then conditional on the observed data, $(\hat{\gamma}^* - \hat{\gamma}, \hat{\beta}^* - \hat{\beta})$ has the same asymptotic variance as $(\hat{\gamma} - \gamma_0, \hat{\beta} - \beta_0)$.

Example (Calcification). Consider the Calcification study (Yu et al., 2001), which investigated the calcification associated with hydrogel intraocular lenses—an infrequently reported complication of cataract treatment. The quantity of interest is the effect of clinical risk factors on the time to calcification of lenses after implantation. The patients were examined by an ophthalmologist to determine the status of calcification at a random time ranging from 0 to 36 months after implantation of the intraocular lenses. Case I interval censored data arises since only the examination time and calcification status at examination are available. The longest follow-up was three years, but no new case was observed after two years. As has been noted by Yu et al. (2001), there is not enough evidence to conclude that the unaffected intraocular lenses will remain calcification free after two years. However, it is highly likely that some patients are subject to much less risk of calcification, that is, those patients may consist of a cured subgroup. The severity of calcification was graded on a discrete scale ranging from 0 to 4. Those with severity ≤ 1 were classified as "not calcified." The clinical risk factors of interest include gender, incision length, and age at implantation. The data set contains 379 records. We exclude the one record with missing measurement, resulting in $n = 378$. For more discussions of the experimental setup, see Yu et al. (2001) and Lam and Xue (2005).

Let $Z_1 = incision\ length$, $Z_2 = gender$ and $Z_3 = age\ at\ implantation/10$.

Denote $Z = (Z_1, Z_2, Z_3)$. Assume the logistic model for cure

$$P(not \ cure) = \frac{\exp(\gamma_1 + \gamma_2 Z_3)}{1 + \exp(\gamma_1 + \gamma_2 Z_3)}. \tag{4.51}$$

For subjects susceptible to calcification, the conditional hazard satisfies the linear additive risk model

$$\lambda(T|Z) = \lambda(T) - (\beta_1 Z_1 + \beta_2 Z_2 + \beta_3 Z_3) \tag{4.52}$$

or the partly linear additive risk model

$$\lambda(T|Z) = \lambda(T) - (\beta_1 Z_1 + \beta_2 Z_2 + h(Z_3)). \tag{4.53}$$

The cure model (4.51) and form of the covariate effects have also been partly motivated by Lam and Xue (2005). We employ the penalized maximum likelihood approach and select the tuning parameter λ_n using cross-validation. Inference is based on the weighted bootstrap with 500 realizations of random $\exp(1)$ weights. For model (4.52), set $\lambda_n = 0$, and the PMLE simplifies to the MLE.

For model (4.52), the MLEs are

$$\hat{\gamma}_1 = -0.310(1.567); \ \hat{\gamma}_2 = 1.225(0.332);$$

$$\hat{\beta}_1 = 0.022(0.036); \ \hat{\beta}_2 = -0.047(0.044); \ \hat{\beta}_3 = -0.051(0.009),$$

where values in "()" are the corresponding bootstrap standard deviation estimates. Based on model (4.52), age has a significant effect on the cure probability: the cure rate decreases as age increases. Out of the three covariates considered for survival, only the age effect is significant: older people have higher survival risks, which is intuitively reasonable.

We also consider model (4.53), the model that allows for a nonparametric age effect in the additive risk model. The PMLEs are

$$\hat{\gamma}_1 = 1.686(1.462); \ \hat{\gamma}_2 = 0.795(0.807); \ \hat{\beta}_1 = 0.028(0.058); \ \hat{\beta}_2 = -0.042(0.100).$$

We can see that estimates from model (4.53) are considerably different from their counterparts from model (4.52). The age effect on the cure rate is no longer significant. Incision length and gender still have no significant effect on survival. The age effect on survival has a bell shape (Figure 4.4). We also show the estimated cumulative baseline hazard function with its lowess smoother in Figure 4.4. The cumulative baseline function is almost linear, which suggests a constant hazard function.

In models (4.51) and (4.52), we have assumed specific forms of the covariate effects. Our choices have been partly motivated by Lam and Xue (2005). Such choices may be subject to more rigorous model checking. Since tools for formal model checking are still not available, interpretation of the data analysis results should be with extreme caution.

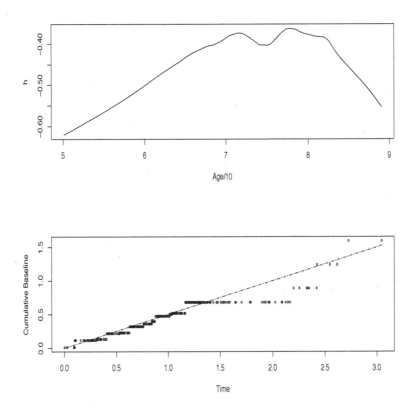

FIGURE 4.4: Calcification data under the partly linear model. Estimates of h_0 and Λ_0 (with lowess smoothers).

Remarks. As with ordinary survival analysis, when there is a cured subgroup, semiparametric models can be more flexible than parametric and nonparametric models. Under the mixture model framework, there are multiple ways of introducing semiparametric modeling. The first, as described above, is to assume a semiparametric model for the survival of susceptible subjects. In addition, it is also possible to introduce semiparametric modeling when describing the cure indicator. The third possibility is to adopt semiparametric modeling for both the cure indicator and survival of susceptible. Compared to the first approach, the latter two have been less investigated.

The estimation and inference aspects of cure rate models have been investigated in quite a few publications. Most of these studies, including some of those describe above, *assume* the existence of a cured subgroup. Thus, the parameters of interest are *interior points* of the corresponding parameter spaces.

Even with simple parametric models, the boundary case can be nontrivial. With semiparametric models, asymptotic properties with the boundary case have not been well established. In addition, the study on model diagnostics with cure rate models remains scarce.

4.4   Bayesian Analysis

Bayesian methods have been always popular in modern statistical analysis and gained success in various applications, owing to their easy availability to practitioners. The analysis of survival data can also follow the well-established Bayesian paradigm.

The advantage of the Bayes approach is primarily two-fold. First, it allows small sample analysis by incorporating prior information from historical data or subject matter experts' opinions. There has been a systematic development in Bayesian statistics on how to objectively quantify information available from similar problems. These procedures compensate the small sample inefficiency usually encountered in frequentist analysis. Second, Bayes approach avoids the derivation of complicated asymptotic distributions. Distribution-based inference is settled upon sampling the posterior distribution of parameters, attainable from well-established numeric procedures.

There are possibly other benefits of Bayes methods, such as an objective model selection with Bayesian information criterion (Chapter 2), and their contribution for missing value problems. In this section, we consider a few important results for survival analysis. In Section 4.4.1, we provide a simple account on how Bayes methods work in general. Then in Section 4.4.2 we focus on the Cox model again, with a Bayes approach for model fitting. Additional applications of Bayesian analysis for survival data are briefly summarized in Section 4.4.3.

4.4.1   Basic Paradigm of Bayesian Inference

In Bayesian analysis, all parameters are treated as random, living in a suitably defined parameter space Θ with a probability distribution. The basic idea of Bayesian inference is to obtain the distribution of parameter given information from the prior experience and current sample. Monte Carlo-based sampling methods can then generate a large sample from the distribution and facilitate inference.

Suppose that we have a probability model for the observed data $\mathcal{D}$, given a vector of unknown parameters $\boldsymbol{\theta}$. A likelihood function can be constructed as $L(\boldsymbol{\theta}|\mathcal{D})$, which is mathematically equivalent to the distribution density function of data given the parameter. Also assume a probability distribution $\pi(\boldsymbol{\theta})$ for $\boldsymbol{\theta}$ that summarizes any information we have not related to that provided by data $\mathcal{D}$, called the *prior* distribution. Inference concerning $\boldsymbol{\theta}$ is then based on its *posterior* distribution obtained by using Bayes' theorem. The posterior distribution of $\boldsymbol{\theta}$ is given by

$$\pi(\boldsymbol{\theta}|\mathcal{D}) = \frac{L(\boldsymbol{\theta}|\mathcal{D})\pi(\boldsymbol{\theta})}{\int_{\Theta} L(\boldsymbol{\theta}|\mathcal{D})\pi(\boldsymbol{\theta})\,d\boldsymbol{\theta}}. \tag{4.54}$$

Since the denominator of (4.54) is just a normalizing constant independent of $\boldsymbol{\theta}$, sometimes it suffices to specify the analytic structure in the numerator, and we may write

$$\pi(\boldsymbol{\theta}|\mathcal{D}) \propto L(\boldsymbol{\theta}|\mathcal{D})\pi(\boldsymbol{\theta}). \tag{4.55}$$

From (4.55) we can see that the posterior density invites a contribution from the observed data through $L(\boldsymbol{\theta}|\mathcal{D})$, and a contribution from prior information quantified through $\pi(\boldsymbol{\theta})$.

Our goal is to find $\pi(\boldsymbol{\theta}|\mathcal{D})$. With an available posterior density we may carry out all kinds of inferences. For instance, in order to obtain a point estimate for $\boldsymbol{\theta}$, one can evaluate the mean, median, or mode of such a posterior distribution. When the distribution is symmetric and unimodal, the three agree with each other. In most situations, the mode is the easiest to compute and equivalent to the frequentist maximum likelihood estimate when the prior $\pi(\boldsymbol{\theta})$ is noninformative. In addition to the point estimate, the posterior distribution can also allow the construction of intervals similar to the frequentist confidence intervals. Specifically, in order to obtain an interval estimate, one can report the highest density *credible region* from the posterior distribution defined as

$$C = \{\boldsymbol{\theta} \in \Theta : \pi(\boldsymbol{\theta}|\mathcal{D}) \geq c_\alpha\} \tag{4.56}$$

where c_α is chosen so that $\int_C \pi(\boldsymbol{\theta}|\mathcal{D})\,d\boldsymbol{\theta} \geq 1 - \alpha$. Such a credible interval indicates that the chance that $\boldsymbol{\theta}$ lies inside C given the observed sample $\mathcal{D}$ is at least $1 - \alpha$. The interpretation differs from that of a frequentist confidence interval based on the sample variability.

Other inference results can be obtained similarly from the posterior distribution when it is available. However, only in very few classic cases, we can derive the analytic expression of the posterior distribution (4.54) by using conjugate properties of the distributions, such as a normal likelihood coupled with a normal prior for the mean or a binomial likelihood coupled with a beta prior for the success probability. In most modern applications such as the Cox PH regression, $\pi(\boldsymbol{\theta}|\mathcal{D})$ does not have a closed form because the integration in the denominator of (4.54) does not have a closed form. Therefore, we have to consider numerical techniques such as the Markov chain Monte Carlo (MCMC) to sample observations from $\pi(\boldsymbol{\theta}|\mathcal{D})$. Inferences can then be drawn based on an arbitrarily large number of posterior samples.

The *Gibbs sampler* may be one of the best known MCMC sampling algorithms in the Bayesian literature. It samples from $\pi(\boldsymbol{\theta}|\mathcal{D})$ without knowledge of the normalizing constant. We give a brief introduction of this method here. Let $\boldsymbol{\theta} = (\theta_1, \cdots, \theta_p)^T$ be a p-dimensional vector of parameters. Given a set of freely chosen starting values $\boldsymbol{\theta}^{(0)} = (\theta_1^{(0)}, \cdots, \theta_p^{(0)})^T$, the algorithm proceeds as follows:

Step 1 At current iteration $k \geq 1$, generate $\boldsymbol{\theta}^{(k)} = (\theta_1^{(k)}, \cdots, \theta_p^{(k)})^T$ as follows:

- Draw $\theta_1^{(k)}$ from $\pi(\theta_1^{(k)}|\theta_2^{(k-1)}, \theta_3^{(k-1)} \cdots, \theta_p^{(k-1)}, \mathcal{D})$;
- Draw $\theta_2^{(k)}$ from $\pi(\theta_2^{(k)}|\theta_1^{(k)}, \theta_3^{(k-1)} \cdots, \theta_p^{(k-1)}, \mathcal{D})$;

  $\cdots\cdots$

- Draw $\theta_p^{(k)}$ from $\pi(\theta_p^{(k)}|\theta_1^{(k)}, \theta_3^{(k)} \cdots, \theta_{p-1}^{(k)}, \mathcal{D})$.

Step 2 Set $k = k + 1$, and go to Step 1.

It can be shown that under mild regularity conditions, the vector sequence $\{\boldsymbol{\theta}^{(k)}, k = 1, 2, \cdots\}$ has a stationary distribution exactly equal to our target $\pi(\boldsymbol{\theta}|\mathcal{D})$ (Gelfand and Smith, 1990; Schervish and Carlin, 1992; Roberts and Polson, 1994). In other words, when k is sufficiently large, say greater than k_0, $\{\boldsymbol{\theta}^{(k)}, k = k_0 + 1, \cdots, k_0 + n\}$ is a sample of size n from the true posterior, from which inference can be conducted.

The conditional distributions involved in Step 1 are called *full conditionals*. They are univariate distributions and thus easier to deal with than the original joint distribution for multivariate $\boldsymbol{\theta}$. Sometimes all full conditionals are familiar univariate distributions such as normal distribution or gamma distribution, and we can directly draw samples from such distributions at each step of the Gibbs sampler. Otherwise we may have to consider indirect sampling approaches given below.

Many indirect sampling approaches are based on the idea of rejection sampling. The most famous approach is the Metropolis-Hastings algorithm. Choose a proposal density $q(\boldsymbol{\theta}^*|\boldsymbol{\theta})$ that is a valid density function for every possible value of the conditional variable $\boldsymbol{\theta}$. Given a set of freely chosen starting values $\boldsymbol{\theta}^{(0)}$, the algorithm proceeds as follows:

Step 1 At current iteration $k \geq 1$, draw a candidate point $\boldsymbol{\theta}^*$ from $q(\boldsymbol{\theta}^*|\boldsymbol{\theta})$ and U from a uniform distribution over the unit interval $U[0, 1]$.

Step 2 If $U < \min\{\frac{\pi(\boldsymbol{\theta}^*|\mathcal{D})q(\boldsymbol{\theta}^*|\boldsymbol{\theta}^{(k-1)})}{\pi(\boldsymbol{\theta}^{(k-1)}|\mathcal{D})q(\boldsymbol{\theta}^{(k-1)}|\boldsymbol{\theta}^*)}, 1\}$, set $\boldsymbol{\theta}^{(k)} = \boldsymbol{\theta}^*$; otherwise set $\boldsymbol{\theta}^{(k)} = \boldsymbol{\theta}^{(k-1)}$.

Step 3 Set $k = k + 1$, and go to Step 1.

The proposal density is usually chosen to be a function easy to work with. Computation involved in Step 2 in each iteration also does not require us to know the exact normalizing constant in the posterior. It is thus desirable to incorporate the Metropolis-Hastings algorithm into the Gibbs sampler when direct sampling from the full conditionals is difficult. The majority of Bayesian MCMC computing is accomplished using their combination.

In addition to the Metropolis-Hastings algorithm, there are a few other popular indirect sampling methods that can be used within Gibbs samplers, including for example the adaptive rejection algorithm. This algorithm can be used for any log-concave conditionals. A function f is log-concave if the second-order derivative of $\log f$ is nonpositive everywhere. Checking this property is easy, and conditionals of this type can be relatively easy to generate without using the Metropolis-Hastings algorithm.

Bayesian inference with the implementation of MCMC can be achieved using WinBUGS (Lunn et al., 2000). It provides a flexible system for Gibbs samplers for all kinds of complicated problems and can be adapted in R, STATA, and SAS.

4.4.2 Bayes Approach for Cox Model

Bayes methods can be applied to many problems in survival analysis. We focus on the Cox PH regression model in this section. Even for this model, there are multiple Bayesian solutions addressing different aspects of the model. What we present in this section is perhaps one of the most common approaches which may be useful to illustrate the general practice of Bayesian survival analysis.

In Chapter 2 we introduce the Cox PH model (2.29) to study how covariates $\mathbf{X}$ may affect failure rate. Inference for the covariate effects $\boldsymbol{\beta}$ is based on the partial likelihood (2.30). In practice, if we are only interested in these coefficient parameters, we may simply treat the partial likelihood as the likelihood function for the purpose of computing the posterior density. A justification for doing this is given in Kalbfleisch (1978a). It then remains to select a prior for $\boldsymbol{\beta}$ for this rather simple parametric problem.

In fact, the Cox PH model is a semiparametric model with unknown $\boldsymbol{\beta}$ and unknown baseline hazard function h_0, and joint consideration of these two unknown components is necessary. We now provide a complete solution for both $\boldsymbol{\beta}$ and h_0 in this section. This requires us to assign a prior for an infinite dimensional function. Often we consider a more convenient approach by assigning a prior for the cumulative baseline hazard $H_0 = \int h_0$. From a Bayesian's point of view, the function H_0 is considered as a random function. In practice, we associate a distribution for a random variable, a multivariate distribution for a random vector, and a process for a random function. The most commonly used prior for H_0 is the gamma process prior.

We need some notations to facilitate presentation. Let $\Gamma(k, \lambda)$ be the gamma density function given in Table 2.1 with a shape parameter $k > 0$ and a scale parameter $\lambda > 0$. Let $k(t)$ be an increasing left continuous function for $t \geq 0$ such that $k(0) = 0$. Consider a stochastic process $G(t)$ with the following properties:

- $G(0) = 0$;

- $G(t)$ has independent increments in disjoint intervals;

- for $t > s$, $G(t) - G(s)$ follows a gamma distribution $\Gamma[\lambda\{k(t) - k(s)\}, \lambda]$.

The process $\{G(t) : t \geq 0\}$ is called a gamma process and is denoted by $G(t) \sim \mathcal{GP}(\lambda k(t), \lambda)$.

In the Cox model, we assume that the cumulative hazard function follows a gamma process prior

$$H_0 \sim \mathcal{GP}(\lambda_0 H^*, \lambda_0), \tag{4.57}$$

where H^* is an increasing function with $H^*(0) = 0$. Often we assume H^* to be a known parametric function with a hyperparameter η_0. For example, if we consider that the baseline hazard is a hazard for Weibull distribution (see Table 2.1), then $H^*(t) = t^{\gamma_0}/\sigma_0$, and $\eta_0 = (\gamma_0, \sigma_0)^T$ is the hyperparameter.

We now begin to construct the full likelihood function for survival data. To this end, we partition the time axis into J disjoint intervals, $0 = s_0 < s_1 < \cdots < s_J$, with $s_J > \max\{y_i : i = 1, \cdots, n\}$, and denote $I_j = (s_{j-1}, s_j]$. Let h_j be the increment in the cumulative baseline hazard in the jth interval

$$h_j = H_0(s_j) - H_0(s_{j-1}), \qquad j = 1, 2, \cdots, J. \tag{4.58}$$

The gamma process prior for $H_0(t) \sim \mathcal{GP}(\lambda_0 H^*, \lambda)$ implies that the h_js are independent and $h_j \sim \Gamma(k_{0,j} - k_{0,j-1}, \lambda_0)$, where $k_{0,j} = \lambda_0 H^*(s_j)$.

We may regard $\mathbf{h} = (h_1, \cdots, h_J)^T$ as a vector of parameters. Let R_j be the set of subjects who are at risk, and D_j be the set of failure subjects for the jth interval I_j. We may obtain the following grouped data likelihood for $(\boldsymbol{\beta}, \mathbf{h})$ as

$$L(\boldsymbol{\beta}, \mathbf{h}|\mathcal{D}) \propto \prod_{j=1}^{J} L_j,$$

where

$$L_j = \exp\left\{-h_j \sum_{k \in R_j - D_j} \exp(\mathbf{x}_k^T \boldsymbol{\beta})\right\} \prod_{l \in D_j} [1 - \exp\{-h_j \exp(\mathbf{x}_l^T \boldsymbol{\beta})\}].$$

This kind of grouped data likelihood is a common representation for survival data and may be used when $\mathbf{h}$ follows other priors as well.

A typical prior for $\boldsymbol{\beta}$ is multivariate normal $N(\mu_0, \Sigma_0)$ with hyperparameters μ_0 and Σ_0. Given the above specification, we may arrive at the posterior

$$\pi(\boldsymbol{\beta}, \mathbf{h}|\mathcal{D}) \propto \prod_{j=1}^{J} [L_j h_j^{k_{0,j} - k_{0,j-1} - 1} \exp(-\lambda_0 h_j)]$$

$$\times \exp\left\{-\frac{1}{2}(\boldsymbol{\beta} - \mu_0)^T \Sigma_0^{-1}(\boldsymbol{\beta} - \mu_0)\right\}. \tag{4.59}$$

This multivariate distribution is hard to sample and we may implement the Gibbs sampler with the following full conditionals:

$$\pi(\beta_l | \boldsymbol{\beta}^{(-l)}, \mathbf{h}, \mathcal{D}) \quad \propto \quad \prod_{j=1}^{J} L_j \exp\left\{-\frac{1}{2}(\boldsymbol{\beta} - \mu_0)^T \Sigma_0^{-1}(\boldsymbol{\beta} - \mu_0)\right\},$$

$$\pi(h_j | \mathbf{h}^{(-j)}, \boldsymbol{\beta}, \mathcal{D}) \quad \propto \quad h_j^{k_{0,j} - k_{0,j-1} - 1}$$

$$\times \exp\left\{-(\lambda_0 + \sum_{k \in R_j - D_j} \exp(\mathbf{x}_k^T \boldsymbol{\beta}))h_j\right\}.$$

The Metropolis-Hasting algorithm can be used at each step of sampling, since the above two conditionals are not familiar univariate distributions and do not allow a direct sampling. However, it is unnecessary in this case. We may find that both conditionals are log-concave and thus allow the use of adaptive rejection algorithm. Furthermore, the full conditional of $\mathbf{h}$ given $(\boldsymbol{\beta}, \mathcal{D})$ can even be approximated by a gamma distribution and thus allows a much quicker direct sampling.

Prior elicitation is an important step for all Bayesian analysis. When similar analysis has been performed on past data and the pertinent results from previous studies are known to the current investigators, a relatively informative and objective approach is a *power prior* specification (Ibrahim and Chen, 2000; Ibrahim et al., 1999). For example, in the above Cox model, we may set the power prior for $(\boldsymbol{\beta}, \mathbf{h})$ to be

$$\pi(\boldsymbol{\beta}, \mathbf{h} | \mathcal{D}_0, a_0) \propto \{L(\boldsymbol{\beta}, \mathbf{h} | \mathcal{D}_0)\}^{a_0} \pi_0(\boldsymbol{\beta}, \mathbf{h}),$$

where $L(\boldsymbol{\beta}, \mathbf{h} | \mathcal{D}_0)$ is the likelihood of $(\boldsymbol{\beta}, \mathbf{h})$ based on a historical data set $\mathcal{D}_0$ with the same topology as $\mathcal{D}$ and $\pi_0(\boldsymbol{\beta}, \mathbf{h})$ is an initial prior. Under the present context of Cox regression analysis, $L(\boldsymbol{\beta}, \mathbf{h} | \mathcal{D}_0)$ can be the group-data likelihood (4.59) constructed from the previous analysis using the same model but different data (i.e., $\mathcal{D}_0$). The hyperparameter a_0 is a scalar that weighs the historical data relative to the likelihood of the current study.

Sometimes it is more convenient to assume conditional independence between the baseline hazard rate and regression coefficient (Ibrahim et al., 1999). This simply leads to

$$\pi(\boldsymbol{\beta}, \mathbf{h} | \mathcal{D}_0, a_0) = \pi(\boldsymbol{\beta} | \mathcal{D}_0, a_0)\pi(\mathbf{h} | \mathcal{D}_0, a_0).$$

For $\boldsymbol{\beta}$, we may assume a multivariate normal distribution $N(\mu_0, a_0^{-1}\Sigma_0)$, where μ_0 is the maximum partial likelihood estimate for $\boldsymbol{\beta}$ by fitting a Cox PH model to the historical data $\mathcal{D}_0$ and Σ_0 is the corresponding estimated covariance matrix. This is a practical and useful summary of data set $\mathcal{D}_0$ as indicated by many authors, including Cox (1972), Tsiatis (1981).

When the hazard function is considered nuisance, we may assume a non-informative prior for $\pi(\mathbf{h} | \mathcal{D}_0, a_0)$. Otherwise, we may assume a gamma prior as discussed earlier while the parameters in the gamma prior are determined from historical data $\mathcal{D}_0$ in a similar way.

There are other ways of using Bayesian approaches to fit Cox PH model, such as assuming a gamma process prior on the baseline hazard h_0 directly (see for example, Dykstra and Laud, 1981) or simply assuming a parametric structure for the baseline hazard and imposing prior for the a vector of parameters (see for example, Leonard, 1978 and Sinha, 1993). We cannot exhaustively summarize all of them in this chapter and refer readers to the cited literature. Two important alternative methods are described below.

Piecewise Constant Hazard Model. Consider the piecewise constant hazard model for $h_0(t)$, which is perhaps the simplest model. First partition time into J nonoverlapping intervals $(0, s_1], (s_1, s_2], \ldots, (s_{J-1}, s_J]$, where $s_J > \max\{y_j : j = 1, \ldots, n\}$. Assume that in the jth interval, $h_0(t) = h_j$ for $t \in (s_{j-1}, s_j]$. Denote $\mathbf{h} = (h_1, \ldots, h_J)^T$. For $i = 1, \ldots, n$ and $j = 1, \ldots, J$, define $\tilde{\delta}_{i,j} = I(y_i \in (s_{j-1}, s_j])$. Then, the likelihood function of $(\boldsymbol{\beta}, \mathbf{h})$ is

$$L(\boldsymbol{\beta}, \mathbf{h}|\mathcal{D}) = \prod_{i=1}^{n} \prod_{j=1}^{J} (h_j \exp(\mathbf{x}_i^T \boldsymbol{\beta}))^{\tilde{\delta}_{i,j}\delta_i}$$

$$\exp\left\{ -\tilde{\delta}_{i,j} \left[h_j(y_i - s_{j-1}) + \sum_{g=1}^{j-1} h_g(s_g - s_{g-1}) \right] \exp(\mathbf{x}_i^T \boldsymbol{\beta}) \right\}.$$

A common prior of the baseline hazard $\mathbf{h}$ is the independent gamma prior with $h_j \sim \Gamma(\alpha_{0j}, \lambda_{0j})$ for $j = 1, \ldots, J$. Here $(\alpha_{0j}, \lambda_{0j})$s are the prior parameters. We notice that this is similar to a discretized version of the gamma process prior. An alternative is to assume a correlated prior where $(\log(h_1), \ldots, \log(h_J)) \sim N_J(\phi_0, \Sigma_0)$ and (ϕ_0, Σ_0) are the prior parameters. After specifying the priors, the posterior distribution can be obtained. MCMC techniques are then needed for estimation and inference.

Smoothed Estimation of Baseline Hazard. With the simple piecewise constant hazard model, the baseline hazard has "jumps" and so is not continuous. The discontinuity problem can be solved with the gamma process approach. However, this approach needs to assume simple parametric $H^*(t)$. In Section 4.1, we describe nonparametric smooth estimation of the hazard function using spline approaches. It is also possible to work with the smooth condition for pertinent functions under the Bayesian framework.

Under the Cox model, we reparameterize $g(t) = \log(h_0(t))$. Then the likelihood contribution of the ith subject is

$$\exp\left\{ \delta_i(g(Y_i) + X_i^T \beta) - \exp(X_i^T \beta) \int_0^{Y_i} \exp(g(s)) ds \right\}.$$

Henschel et al. (2009) developed an approach that models $g(t)$ as a stepwise constant function as well as a cubic spline. The stepwise approach results in piecewise exponential survival distribution, whereas the cubic splines can make the estimated baseline hazard function smoother. Note that both ways can be formulated under the B-spline framework (deBoor, 1978).

Partition time into J nonoverlapping intervals $(0, s_1], (s_1, s_2], \ldots, (s_{J-1}, s_J]$, where $s_J > \max\{y_j : j = 1, \ldots, n\}$. Note that the partition sometimes can be subjective. A simple choice is to have equal lengths. It is suggested that the intervals should have "comparable information," for example, similar number of events. Henschel et al. (2009) suggest using the following, more objective way of determining the intervals. With right censored data, first compute the nonparametric Kaplan-Meier estimate of the survival function. Note that the KM estimates are only available at discrete time points where events happen. Instead of adopting constant survival between those points, interpolate linearly. Partition the range of the resulting survival curve into J equal parts and use the inverse function of the (linearly interpolated) survival function to compute the boundaries of the J time intervals. It is suggested that a large J value stabilize the mixing properties of the MCMC chain by resting in a state for long periods. Thus, Henschel et al. (2009) suggest "choosing J as large as possible by keeping an acceptable mixing of the chain."

The priors for components of β are chosen as independently normally distributed with zero means and large variances to accommodate a lack of precise information.

For the log baseline hazard function, we take a sieve approach, where

$$g(t) = \sum_{j=0}^{J} g_j b_j(t).$$

Here, $\{g_j : j = 0, \ldots, J\}$ are the unknown regression coefficients. $b_j(t)$ is the B-spline function defined in the jth interval. If $b_j(\cdot)$ has degree zero, we end up with a step function. If $b_j(\cdot)$ has degree three, we have a cubic spline function. The prior for g_j is taken as a first-order process, which accommodates prior information on smoothness. The first-order process is defined as

$$g_k = g_{k-1} + \epsilon_k,$$

where $\epsilon_k \sim N(0, \sigma_k^2)$, $g_0 \sim N(0, \sigma_0^2)$, and g_0 and ϵ_k $(k = 1, \ldots, J)$ are pairwise independent. The choice of the mean of the g_0 prior can be subjective. To compensate, a large value of σ_0 can be used. This defines a prior with little influence and allows the data to more objectively determine the value of the hazard function at time zero. The variances for later time points are chosen as $\sigma_k^2 = \Delta_k \sigma_1^2$ and Δ_k may be defined by the mean of the corresponding time interval length. The inverse of the covariance matrix $\Sigma = (E(g_k g_l))_{k,l=0,\ldots,J}$, Σ^{-1}, can be written as $\frac{1}{\sigma_0^2} Q_0 + \frac{1}{\sigma_1^2} Q_1$. Here matrix Q_0 has its $(0,0)$th element equal to 1 and the rest equal to 0. Q_1 is a simple structured band matrix with bandwidth one, reflecting the first-order process characteristic. The parameters $\frac{1}{\sigma_0^2}$ and $\frac{1}{\sigma_1^2}$ are treated as hyperparameters with flat gamma priors.

In the MCMC calculation, we will need to sample for the parameter vector, baseline hazard, and dispersion parameters iteratively. The sampling can be conducted as follows.

For the parameter vector, we first note that the Cox model can be interpreted as a generalized linear model (Aitkin and Clayton, 1980) by writing the likelihood function L in the form

$$L = [f(t)]^\delta [1 - S(t)]^{1-\delta} = [\mu^\delta \exp(-\mu)] \left[\frac{h_0(t)}{H_0(t)}\right]^\delta,$$

where $\mu = H_0(t) \exp(X^T \beta)$. When the cumulative baseline hazard $H_0(t)$ is fixed, the above is the likelihood function of a Poisson sample where the observation is the event indicator δ, the link function is the canonical "log" link, and the offset is $\log\{H_0(t)\}$. Gamerman (1997) proposes effectively sampling the parameter vector in generalized linear mixed models in a block updating step by combining the iterated weighted least squares estimation approach with a Metropolis-Hastings sampling. For the prior, a weak informative normal distribution $N(0, \frac{1}{\sigma^2}I)$ is chosen. The iteration starts with $\beta = \beta_0$ (a sensible initial value is component-wise zero) and $t = 1$. Then one needs to sample β^* from $N(m^{(t)}, C^{(t)})$ where

$$C^{(t)} = \left[\frac{1}{\sigma^2}I + X^T W(\beta^{(t-1)})X\right]^{-1}$$

$$m^{(t)} = C^{(t)} X^T W(\beta^{(t-1)}) \tilde{y}(\beta^{(t-1)}).$$

Here, W and $\tilde{y}$ are the weight matrix and pseudo-response in the iterated weighted least squares approach. Then we decide whether to accept β^* or keep the previous value based on the Metropolis-Hastings algorithm.

With the structure of log-baseline hazard function specified above, we need to sample the coefficients $\mathbf{g} = (g_1, \cdots, g_J)^T$ from a Gaussian Markov random field (GMRF). Following Knorr-Held and Rue (2002) and Rue (2001), the log-baseline hazard can be sampled in one step. The posterior of $\mathbf{g}$ is

$$\pi(\mathbf{g}|\beta, \sigma_0, \sigma_1) \propto \exp\left\{ -\frac{1}{2}\mathbf{g}^T \Sigma^{-1} \mathbf{g} + \sum_{i=1}^{n}[\delta_i \sum_{k=0}^{J} g_k b_k(t_i) - \exp(X_i^T \beta) \right.$$

$$\left. \int_0^{t_i} \exp\left[\sum_{k=0}^{J} g_k b_k(s)\right] ds]\right\}.$$

Knorr-Held and Rue (2002) propose approximating the exponent by a quadratic form, using the resulting Gaussian random field (multivariate normal distribution) to sample a proposal for $\mathbf{g}$, and accepting or rejecting the proposal based on a Metropolis-Hastings algorithm. The cumulative baseline hazard function, which is part of the exponent, can be calculated in a closed form if it is a step function (B-spline with a degree of zero). When the cumulative baseline hazard function is B-spline of degree three, it can be approximated by the trapezoidal rule. The trapezoidal rule results in complex terms which contain the exponent of linear terms of h. The calculation of a good quadratic approximation to this term allows one to derive the multivariate normal distribution.

For the dispersion parameters σ_0^2 and σ_1^2, a flat gamma prior with rate κ and shape ν is chosen. The full conditional distribution of σ_0^2 is gamma distributed and has rate $\kappa + \frac{h_0^2}{2}$ and shape $\nu + 1/2$, and the full conditional distribution of σ_1^2 has rate $\kappa + \frac{1}{2}\mathbf{g}^T Q_1 \mathbf{g}$ and shape $\nu + \frac{J}{2}$.

4.4.3 Other Applications in Survival Analysis

Because of their computational ease, Bayes approaches are usually applied in sophisticated analyses such as the advanced topics covered in this chapter. Specifically, they may be used for nonparametric regression coupled with the the spline approach. In nonparametric regression literature, authors have used truncated power basis or B-spline basis for the approximation of functions to be estimated and then applied Gibbs sampler to provide Bayes estimation of the nonparametric curves (Lang and Brezger, 2004; Nott and Li, 2010; Shi et al., 2012b). Although this approach is mainly seen in additive models, we see no difficulty of extending to partially linear models or varying-coefficient models. Bayes methods may potentially provide an easier way to solve the complicated estimation problems for nonparametric regression.

The Bayes approach is also a nice tool for multivariate survival data. We consider the frailty model as an illustrative example. The parameterization of dependence structure by using random effects has a strong appeal to Bayesian statisticians. It turns out that fitting the frailty model (4.23) with a Bayes approach is fairly convenient.

Denote $\mathbf{w} = (w_1, \cdots, w_n)$ to be the vector of i.i.d. frailty terms with one parameter gamma distribution $g(w; \lambda)$. We still assume a gamma process prior for the cumulative baseline hazard function. Using the same notations for group-data likelihood as given in Section 4.4.2, we may derive the likelihood function of model parameters given data and frailty as

$$\pi(\boldsymbol{\beta}, \mathbf{h}|\mathbf{w}, \mathcal{D}) \propto \prod_{j=1}^{J} [L_j^* h_j^{k_{0,j}-k_{0,j-1}-1} \exp(-\lambda_0 h_j)], \qquad (4.60)$$

where

$$
\begin{aligned}
L_j^* &= \exp\left\{-h_j \sum_{k \in R_j} \exp(\mathbf{x}_k^T \boldsymbol{\beta} + \log(w_k))\right\} \\
&\quad \times \prod_{l \in D_j} [1 - \exp\{-h_j \exp(\mathbf{x}_l^T \boldsymbol{\beta} + \log(w_l))\}].
\end{aligned}
$$

The likelihood of $(\boldsymbol{\beta}, \mathbf{h})$ based on the observed data $\mathcal{D}$ can be obtained by integrating out the w_is with respect to $g(w; \lambda)$. Such a likelihood function, with n integrals involved, is too complicated to work with, and thus it is impossible to evaluate the joint poseterior of $(\boldsymbol{\beta}, \mathbf{h})$ analytically. We have to use MCMC techniques such as Gibbs sampler to circumvent this problem.

Depending on the prior specification, we may carry out a Gibbs sampler approach to sample from the full conditional distributions. Specifically, $\pi(\boldsymbol{\beta}|\mathbf{w}, \mathbf{h}, \mathcal{D})$ can be approximated by the normal distribution when normal prior for $\boldsymbol{\beta}$ is used; $\pi(\mathbf{h}|\boldsymbol{\beta}, \mathbf{w}, \mathcal{D})$ is an independent-increment gamma process; $\pi(\mathbf{w}|\boldsymbol{\beta}, \mathbf{h}, \lambda, \mathcal{D})$ is an independent gamma distribution; and finally for the hyperparameter λ, we consider its inverse $\eta = \lambda^{-1}$, which is distributed independent of $\boldsymbol{\beta}$ and $\mathbf{h}$ as

$$\pi(\eta|\mathbf{w}) \propto \prod_{i=1}^{n} w_i^{\eta-1} \eta^{-n\eta} \exp\left\{-\eta \sum_{i=1}^{n} w_i\right\} [\Gamma(\eta)]^{-1}.$$

All these conditional densities are log-concave and can be easily sampled in various computing packages such as `WinBUGS`. See Clayton (1991) for more details on this problem.

Now consider the additive risk model (2.64), which, as described in Section 2.3.5, provides a useful alternative to the Cox model. Under this model, the conditional hazard function has the form

$$h(t|x) = h_0(t) + h_1(t|x), \quad t \geq 0.$$

Here, $h_0(t)$ is the unknown nonparametric parameter, and $h_1(t|x)$ is a parametric function. Consider the additive gamma-polygonal model, which is a special case of the above model. Under this model, $h_0(t)$ is assumed to be a nonnegative polygonal function. Denote $a_0 = 0 < a_1 < \ldots < a_g$ as a partition of the time interval. Assume that

$$h_0(t) = \begin{cases} \tau_{j-1} + \frac{(\tau_j - \tau_{j-1})(t - a_{j-1})}{a_j - a_{j-1}} & \text{if } a_{j-1} \leq t \leq a_j, \ j = 1, \ldots, g \\ \tau_g & \text{if } t \geq a_g \end{cases}$$

The parametric part $h_1(t|x)$ is the hazard function of a Gamma distribution with parameters α and β:

$$h_1(t|x) = \frac{t^{\alpha-1} \exp(-\beta t)}{\int s^{\alpha-1} \exp(-\beta s) ds},$$

for $t > 0$. The covarites are incorporated in the parameter α, as will become clear in the following development. Given the parameters $\tau = (\tau_0, \tau_1, \ldots, \tau_g)$, α and β, the survival function is

$$S(t|\tau, \alpha, \beta) = S_0(t|\tau) S_1(t|\alpha, \beta),$$

and the density function takes the form

$$\begin{aligned} f(t|\tau, \alpha, \beta) &= S(t|\tau, \alpha, \beta) h(t|\tau, \alpha, \beta) \\ &= S_0(t|\tau) S_1(t|\alpha, \beta)[h_0(t|\tau) + h_1(t|\alpha, \beta)]. \end{aligned}$$

To introduce the effect of covariates, we consider a hierarchical model, where

$$\alpha|\beta, x : \log\left(\frac{\alpha}{\beta}\right) \sim N(x^T b, \sigma_\alpha^2)$$

$$\beta : \log(\beta) \sim N(\mu_\beta, \sigma_\beta^2).$$

That is, given β and x, the logarithm of the mean, $\log(\alpha/\beta)$, is modeled as a normal distribution with mean $x^T b$, a linear combination of the covariates, and variance σ_α^2. The logarithm of β is also modeled as a normal distribution with mean μ_β and variance σ_β^2. The hyperparameters b, σ_α^2, and μ_β are unknown constants common to all individuals in the population. The above formulation is equivalent to consider that, given the hyperparameters, the mean and shape of the gamma distribution are independent and log-normal distributed.

Assume n i.i.d. observations $\{(Y_i, \delta_i, X_i) : i = 1, \ldots, n\}$. Then the likelihood function is proportional to

$$\prod_{i=1}^{n} S(Y_i|\tau, \alpha_i, \beta_i)[h(Y_i|\tau, \alpha_i, \beta_i)]^{\delta_i} f(\alpha_i, \beta_i|x_i, b, \sigma_\alpha^2, \mu_\beta, \sigma_\beta^2).$$

When there is a categorical covariate, there are two ways of introducing it in the model. The first is to create dummy variables and embed them in the vector x. The second is to partition the population in multiple strata based on this covariate. In the later case, one can assume different baseline hazard functions for different strata, however, under the condition that other parameters and hyperparameters are common to all strata.

In Bayesian analysis with the additive risk model, it seems natural to assume independence between τ, (b, σ_α^2), and $(\mu_\beta, \sigma_\beta^2)$. Consider the following priors for the hyperparameters

$$\mu_\beta|\sigma_\beta^2 : \mu_\beta \sim N(m_\beta, \sigma_\beta^2 v_\beta^2)$$

$$\sigma_\beta^2 : 1/\sigma_\beta^2 \sim \Gamma(a_\beta, b_\beta)$$

$$b|\sigma_\alpha^2 : b \sim N_p(m_\alpha, \sigma_\alpha^2 V_\alpha)$$

$$\sigma_\alpha^2 : 1/\sigma_\alpha^2 \sim \Gamma(a_\alpha, b_\alpha)$$

The prior for the vector τ can be specified as an autocorrelated first-order process, where

$$\tau_j = \tau_{j-1} \exp(\epsilon_j), \quad j = 1, \ldots, g,$$

where $(\epsilon_1, \ldots, \epsilon_g)$ are independent, normal, zero mean, and variance σ_ϵ^2 random variables, and

$$\tau_0 : \tau_0 \sim \Gamma(a_\tau, b_\tau)$$

$$\sigma_\epsilon^2 : 1/\sigma_\epsilon^2 \sim \Gamma(a_\epsilon, b_\epsilon).$$

Even though simple priors have been assumed, because of the complexity of the model, the posterior distribution does not have a simple closed form. For computation, we will have to resort to MCMC using Gibbs sampling. As each conditional distribution has a form much simpler than the joint distribution, sampling from the posterior distribution is feasible. In particular, the posterior for the hyperparameters $(\mu_\beta, \sigma_\beta^2)$ is proportional to

$$\left[\prod_{i=1}^{n} N(\log \beta_i | \mu_\beta, \sigma_\beta^2)\right] N(\mu_\beta | m_\beta, \sigma_\beta^2 v_\beta^2)\Gamma\left(\frac{1}{\sigma_\beta^2} | a_\beta, b_\beta\right).$$

This expression is the same as that appears in the usual conjugate prior analysis of normal data. Therefore, it is proportional to a normal-inverse gamma distribution

$$\mu_\beta | \sigma_\beta^2 : \mu_\beta \quad \sim \quad N\left(\frac{m_\beta + nv_\beta^2\overline{\log(\beta)}}{1 + nv_\beta^2}, \frac{\sigma_\beta^2 v_\beta^2}{1 + nv_\beta^2}\right),$$

$$\sigma_\beta^2 : \frac{1}{\sigma_\beta^2} \quad \sim \quad \Gamma\left(a_\beta + \frac{n}{2}, b_\beta + \frac{ns_{\log \beta}^2}{2} + \frac{n(\overline{\log(\beta)} - m_\beta)^2}{2(1 + nv_\beta^2)}\right)$$

where $\overline{\log(\beta)}$ and $s_{\log \beta}^2$ are the sample mean and variance of $\log \beta$, respectively.

In a similar way, the posterior distribution of (b, σ_α^2) is proportional to

$$\left[\prod_{i=1}^{n} N\left(\log \frac{\alpha_i}{\beta_i} | x_i^T b, \sigma_\alpha^2\right)\right] N(b | m_\alpha, \sigma_\alpha^2 V_\alpha)\Gamma\left(\frac{1}{\sigma_\alpha^2} | a_\alpha, b_\alpha\right),$$

the same expression as in the usual conjugate analysis of the normal linear homoscedastic model. It is then proportional to a multivariate normal-inverse gamma distribution

$$b | \sigma_\alpha^2 : b \quad \sim \quad N_p(\hat{b}, \sigma_\alpha^2(V_\alpha^{-1} + X^T X)^{-1})$$

$$\sigma_\alpha^2 : 1/\sigma_\alpha^2 \quad \sim \quad \Gamma\left(a_\alpha + \frac{n}{2}, b_\alpha + \frac{1}{2}\left[(\ell - X\hat{b})^T \ell + (m_\alpha - \hat{b})^T V_\alpha^{-1} m_\alpha\right]\right)$$

where ℓ is the vector composed of $\log \frac{\alpha}{\beta}$s, X is the covariate matrix, and $\hat{b} = (V_\alpha^{-1} + X^T X)^{-1}(V_\alpha^{-1} m_\alpha + X^T \ell)$.

Another parameter that has a conjugate prior is σ_ϵ^2. The corresponding posterior is proportional to

$$\left[\prod_{j=1}^{d} N(\log \tau_j | \log \tau_{j-1}, \sigma_\epsilon^2)\right] \Gamma(1/\sigma_\epsilon^2 | a_\epsilon, b_\epsilon),$$

and with some algebra,

$$\sigma_\epsilon^2 \sim \Gamma\left(1/\sigma_\epsilon^2 | a_\epsilon + \frac{g}{2}, b_\epsilon + \frac{1}{2}\sum_{j=1}^{g} \log \frac{\tau_j}{\tau_{j-1}}\right).$$

Other parameters do not have conjugate priors. For $i(= 1, \ldots, n)$, the conditional distribution for the pair (α_i, β_i) is proportional to

$$S(Y_i | \tau, \alpha_i, \beta_i)[h(Y_i | \tau, \alpha_i, \beta_i)]^{\delta_i} f(\alpha_i, \beta_i | x_i, b, \sigma_\alpha^2, \mu_\beta, \sigma_\beta^2),$$

which unfortunately does not have a closed form. However it is still possible to sample from it using a Metropolis algorithm with $f(\alpha_i, \beta_i | x_I, b, \sigma_\alpha^2, \mu_\beta, \sigma_\beta^2)$, the bivariate log-normal distribution, as an importance function.

The posterior conditional distribution of τ also does not have a closed form. It is proportional to

$$\left[\prod_{i=1}^{n} S_0(Y_i | \tau, \alpha_i, \beta_i)[h(Y_i | \tau, \alpha_i, \beta_i)]^{\delta_i} \right] \left[\prod_{j=1}^{d} N(\log \tau_j | \log \tau_{j-1}, \sigma_\epsilon^2) \right] \Gamma(\tau_0 | a_\tau, b_\tau).$$

To sample from it, we use the following Metropolis algorithm. We obtain each ϵ_i from a normal distribution with mean $\epsilon_i^* \sigma^2 / C$ and variance σ^2, where ϵ_i^* is the value obtained in the previous step, C is a Metropolis tuning parameter, and $\sigma^{-2} = \frac{1}{C} + \frac{1}{\sigma_\epsilon^2}$. Once the vector $(\epsilon_1, \ldots, \epsilon_g)$ has been simulated, we obtain τ_0 from a log-normal distribution with mean τ_0^* (the value of τ_0 at the previous step), and every other τ_j as $\tau_{j-1} \exp(\epsilon_j)$, $j = 1, \ldots, g$.

Bayesian approaches have also been applied to cure rate models (4.3). Useful references include Chen et al. (1999), Ibrahim et al. (2001a), Chen et al. (2002), and others.

4.5   Theoretic Notes

Functional Space. A space $\mathbb{X}$ is a metric space if it is equipped with a metric $d(\cdot, \cdot)$ which can measure the distance between two elements from the space. If we further introduce a norm $\| \cdot \|$ to this space, $\mathbb{X}$ becomes a normed space, and we may discuss the concept of convergence over this space. Finally, in order to study the angle between two elements, we introduce an inner product $\langle \cdot, \cdot \rangle$, and $\mathbb{X}$ further equipped with an inner product is called an inner product space. A complete inner product space is also called a Hilbert space.

If $\mathbb{S} = \{\phi_\alpha | \alpha \in A\}$ is a set of orthonormal basis for $\mathbb{X}$, then for any $x \in \mathbb{X}$, we have

$$x = \sum_{\alpha \in A} \langle x, \phi_a \rangle \phi_a, \tag{4.61}$$

where $\{\langle x, \phi_a \rangle | \alpha \in A\}$ is called the Fourier coefficient with respect to $\mathbb{S}$. Least square problems in regression can be regarded as a problem of finding the projection onto a subspace of $\mathcal{X}$. The idea can be further developed to justify the application of spline functions in nonparametric regression (Wahba, 1990).

Nonparametric Regression. Evaluation of how well an estimated function $\hat{\psi}(x)$ can approximate the true unknown function $\psi(x)$ depends on the following integrated squared error

$$\int_{\mathcal{X}} \left[\hat{\psi}(x) - \psi(x) \right]^2 dx$$

and its expected value, mean integrated squared error (MISE), where the expectation is taken with respect to the estimate $\hat{\psi}$.

At a single point x, we can usually decompose the mean squared error (MSE) as

$$
\begin{aligned}
E\left[\hat{\psi}(x) - \psi(x) \right]^2 &= E\left[\hat{\psi}(x) - E\hat{\psi}(x) + E\hat{\psi}(x) - \psi(x) \right]^2 \\
&= E\left[\hat{\psi}(x) - E\hat{\psi}(x) \right]^2 + \left[E\hat{\psi}(x) - \psi(x) \right]^2,
\end{aligned}
$$

where the first term is the variance of $\hat{\psi}(x)$ and the second term is the squared bias.

When $\hat{\psi}(x)$ is obtained from the local linear regression approach with a kernel function K, bandwidth ς, and sample size n, it can be shown that the variance is

$$\text{var}(\hat{\psi}(x)) = \frac{C_1}{n\varsigma} + O(n^{-1})$$

and the bias is

$$\text{bias}(\hat{\psi}(x)) = C_2\varsigma + O(\varsigma^2),$$

where C_1 and C_2 are finite constants. Clearly, one can see that the variance is a decreasing function of ς while the bias is an increasing function of ς. Also, the estimation bias is independent of the sample size n.

The MSE is then given by

$$\text{MSE}(\hat{\psi}(x)) = \frac{C_1}{n\varsigma} + C_2^2\varsigma^2 + O(n^{-1}) + O(\varsigma^3).$$

MISE has a similar expression, showing the trade-off between bias and variance. The usual technical conditions for nonparametric smoothing thus require $\varsigma \to 0$ and $n\varsigma \to \infty$. A theoretically optimal bandwidth that minimizes MISE may be obtained when C_1 and C_2 can be estimated well. Otherwise we can only know the order of the optimal ς.

Sandwich Estimator. The marginal models for multivariate failure time data ignore the dependence and still yield consistent regression coefficient estimators. However, the variance of estimates and covariance between the estimates cannot be based on the independence assumption.

Specific to the marginal model (4.22), the robust variance construction may be conducted following the derivation below. For the pseudo-likelihood $L(\boldsymbol{\beta})$ given (4.24), we further denote

$$\mathbf{U}(\boldsymbol{\beta}) = \frac{\partial \log L}{\partial \boldsymbol{\beta}}$$

as the score vector, and

$$\mathbf{I}(\boldsymbol{\beta}) = -\frac{\partial^2 \log L}{\partial \boldsymbol{\beta} \partial \boldsymbol{\beta}^T}$$

as the information matrix. A large sample argument gives the correct asymptotic covariance of $\hat{\boldsymbol{\beta}}$ that maximizes (4.24) to be

$$E\{\mathbf{I}^{-1}(\boldsymbol{\beta})\} E\{\mathbf{U}(\boldsymbol{\beta})\mathbf{U}(\boldsymbol{\beta})^T\} E\{\mathbf{I}^{-1}(\boldsymbol{\beta})\},$$

where the expectations are taken with respect to the true data distribution, which is completely different from the assumed independent structure (4.24). In practice, the expectations can be replaced by their empirical versions by taking sample averages.

For the univariate Cox model, $L(\boldsymbol{\beta})$ is indeed the correct model, and so the expectations coincide

$$E\{\mathbf{I}(\boldsymbol{\beta})\} = E\{\mathbf{U}(\boldsymbol{\beta})\mathbf{U}(\boldsymbol{\beta})^T\}$$

after simple algebra. The sandwich formula thus reduces to the simple covariance estimator $E\{\mathbf{I}^{-1}(\boldsymbol{\beta})$ given in Chapter 2.

4.6 Exercises

1. TRUE or FALSE?

   (a) When fitting model (4.1) with the local linear regression approach, selecting a large bandwidth may allow more data to be used in the likelihood function (4.2) and thus produce a more accurate estimate for $\psi(x)$.

   (b) In a clinical trial, the prior for the failure rate of the control group can be developed by using existing data.

2. For positive random variables T_1 and T_2, show that $E(T_1 T_2) = \int_0^\infty \int_0^\infty S(t_1, t_2)\, dt_1\, dt_2$, where $S(t_1, t_2)$ is the joint survival function of T_1 and T_2.

3. A bivariate joint survival function, whose support is $[0,1] \times [0,1]$, is specified by

$$P(T_1 > t_1, T_2 > t_2) = (1 - t_1)(1 - t_2)\{1 - \max(t_1, t_2)\}^\eta,$$

where $\eta > 0$.

(a) Obtain the marginal survival function for T_1 and determine its mean.

(b) Calculate $\mathrm{cov}(T_1, T_2)$.

(c) Write the likelihood of η given a random sample $\{(t_{1i}, t_{2i}) : i = 1, \cdots, n\}$ from the above joint distribution.

4. Suppose the random variables U, V_1, V_2 and M are all independent, $\eta \in (0,1)$ is a constant, U follows Uniform(0,1), V_i follows an exponential distribution with mean 1, $i = 1, 2$, and $M = 0$ or 1 with probability η and $1 - \eta$, respectively. Let $T_1 = U^\eta V$ and $T_2 = (1 - U)^\eta V$, where $V = V_1 + MV_2$.

(a) Fine the probability density function of V.

(b) Show that the joint survival function of (T_1, T_2) is

$$P(T_1 > t_1, T_2 > t_2) = \exp\{-(t_1^{1/\eta} + t_2^{1/\eta})^\eta\}.$$

(Hint: Consider first conditioning on V.)

5. Suppose that, in a two component system, each component fails when it receives shock. The shocks are generated by three independent Poisson processes. A shock generated from the first process with intensity $1/\eta$ causes only component 1 to fail. A shock generated by the second process with intensity $1/\eta$ causes only component 2 to fail. The third process with intensity $1/(2\eta)$ generates a shock to cause both components to fail. Let T_1 and T_2 be the life lengths of the first and the second component, respectively. Their joint survival function is given by

$$P(T_1 > t_1, T_2 > t_2) = \exp\left\{-\frac{2t_1 + 2t_2 + \max(t_1, t_2)}{2\eta}\right\}.$$

(a) Using this survival function, derive a probability density function for (T_1, T_2).

(b) Based on data $\{(t_{1i}, t_{2i}) : i = 1, \cdots, n\}$, determine the posterior distribution of η when the prior distribution is the inverted gamma distribution with a density function

$$\pi(\eta) = \frac{\beta^\alpha \exp(-\beta/\eta)}{\Gamma(\alpha)\eta^{\alpha+1}}$$

specialized to $\alpha = 2$ and $\beta = 1$.

(c) Under quadratic loss, obtain the Bayes estimator of expected life length when the two components are connected in a series. (The system fails if at least one component fails.)

6. For a shared frailty model, show $P(T_{i1} < T_{j1}|w_i, w_j) = w_i/(w_i + w_j)$.

7. In a clinical trial, usually treatment assignment X is a variable of primary interest. Each individual is randomly assigned to the treatment arm ($X = 1$) or the control arm ($X = 0$). In a Bayesian analysis, we usually consider a prior separately for the coefficient of this binary variable and assume the prior of this coefficient is independent to the prior of coefficients for all other covariates. Discuss whether it is appropriate to make such an assumption?

Chapter 5

Diagnostic Medicine for Survival Analysis

Screening and diagnosis are encountered frequently in medicine. Statistical methods are needed to rigorously evaluate the diagnostic accuracy, usually expressed as numerical probabilities. Sensible accuracy measures should reflect how often the test can correctly identify the disease status of a subject. Accuracy analysis may serve as an alternative approach to study the association between survival outcomes and predictors, taking a rather different angle from hazard regression. Statistical tools such as the ROC (receiver operating characteristics) analysis are introduced in this chapter.

5.1   Statistics in Diagnostic Medicine

We first present some common statistical methods in diagnostic medicine. Statistical methods have been commonly used to evaluate the accuracy of diagnostic procedures such as blood tests, urine tests, CT scans, MRIs, and X-rays. This part is not directly related to survival data analysis. However, contemporary statisticians working on survival analysis usually are also contributing significantly to the area of diagnostic medicine. First, an accurate diagnosis usually defines the starting point of the disease progression and is closely related to the calculation of the overall time-to-failure for a subject. Second, many classic methods for diagnostic data may be viewed as simpler versions of statistical methods for survival data. In particular, for continuous diagnostic tests to be introduced in Section 5.1.2, the parametric approaches mimic those used for survival data without censored observations.

In this section, consider the scenario where the true outcome for a patient is binary, for example, disease-present or disease-absent status. We use S to denote the true disease status, where $S = 1$ if the subject is diseased, and $S = 0$ if the subject is healthy. In the next section, we will extend the outcome to survival response.

5.1.1   Binary Diagnostic Test

5.1.1.1   Sensitivity and Specificity

The simplest diagnostic test may simply produce binary values and then the evaluation of accuracy can be done by matching the test value to the outcome value. Use Y to denote the test result, where $Y = 1$ if the test is positive (suggesting that the subject is diseased) and $Y = 0$ if the test is negative (suggesting that the subject is healthy). Then we may define the sensitivity (se) and specificity (sp) as follows:

$$\text{se} = P(Y = 1 | S = 1) \tag{5.1}$$

$$\text{sp} = P(Y = 0 | S = 0). \tag{5.2}$$

Note that the two probabilities are both conditional probabilities instead of joint probabilities of Y and S. The joint probability is the product of the conditional probability of Y given S and the marginal probability of S. For example, the joint probability $P(Y = 0, S = 0)$ may be high when $P(S = 0)$ is large. Therefore, a high agreement may be simply due to a high marginal probability. The conditionally defined probability may better reflect the intrinsic accuracy of the test, independent of how prevalent the disease is in the population.

These two probabilities are used to simultaneously judge the accuracy of a diagnostic test. A good test should have satisfactory values in both dimensions. For example, a naive test may always give $Y = 1$ for subjects in the population and therefore attain perfect sensitivity ($\mathsf{se} = 1$) all the time. However, such a test always gives false diagnosis for healthy subjects and thus has a degenerate specificity ($\mathsf{sp} = 0$). The message is that we should always report the two values in lieu of just looking at one of them.

In practice, we may estimate sensitivity and specificity based on a sample of subjects undergoing a diagnostic test. Suppose that there are n_1 subjects with true outcome $S = 1$ and n_0 subjects with true outcome $S = 0$. After we perform diagnostic tests on these subjects, we may tabulate their test results in a two-by-two table as Table 5.1. Among the n_1 diseased subjects, we observe X_1 positive diagnosis and $n_1 - X_1$ negative diagnosis; Among the n_0 healthy subjects, we observe X_0 negative diagnosis and $n_0 - X_0$ positive diagnosis.

TABLE 5.1: Diagnostic Results for a Binary Test

Test (Y)	True Status (S)		Total
	$S = 1$	$S = 0$	
Positive ($Y = 1$)	X_1	$n_0 - X_0$	$X_1 + n_0 - X_0$
Negative ($Y = 0$)	$n_1 - X_1$	X_0	$X_0 + n_1 - X_1$
Total	n_1	n_0	$n_1 + n_0$

S is the true disease status, Y is the test result, n_1 is the total number of diseased subjects, n_0 is the total number of healthy subjects, X_1 is the number of positive diagnosis among diseased subjects, and X_0 is the number of negative diagnosis among healthy subjects.

We may then estimate sensitivity and specificity by evaluating the sample fractions as

$$\widehat{\mathsf{se}} = \frac{X_1}{n_1} \tag{5.3}$$

$$\widehat{\mathsf{sp}} = \frac{X_0}{n_0}. \tag{5.4}$$

These estimators are easy to calculate and unbiased for the corresponding true parameters. This kind of evaluation is usually presented in elementary statistics modules to illustrate the estimation of conditional probabilities. For example, Agresti and Franklin (2009, p. 236) the authors analyzed a well-known data set (Haddow et al., 1994) consisting of a blood test on pregnant women for fetal Down syndrome, a genetic disorder causing mental impairment. The corresponding X_1, X_0, n_1, and n_0 are 48, 3921, 54, and 5228, respectively. The sensitivity and specificity for such a parental serum screening are therefore estimated to be $48/54 = 0.889$ and $3921/5228 = 0.75$, respectively.

After a patient undergoes a screening or diagnostic test, his or her test result may be viewed as a Bernoulli random variable, which is either equal to the true disease status or not. Consequently, given n_1, X_1 follows a binomial distribution with the success probability equal to the true **se**, while given n_0, X_0 follows a binomial distribution with the success probability equal to the true **sp**. Knowing these exact distribution results, we can conduct inference. Specifically, the variances for $\widehat{se}$ and $\widehat{sp}$ are

$$\sigma_{se}^2 = se(1-se)/n_1, \qquad \sigma_{sp}^2 = sp(1-sp)/n_0, \tag{5.5}$$

respectively. The variance estimates $\hat{\sigma}_{se}^2$ and $\hat{\sigma}_{sp}^2$ for these two estimators can be obtained using the plug-in method as $\hat{se}(1-\hat{se})/n_1$ and $\hat{sp}(1-\hat{sp})/n_2$, respectively.

Furthermore, as a straightforward application of central limit theorem, it can be shown that asymptotically,

$$\widehat{se} \sim N(se, \sigma_{se}^2)$$

and

$$\widehat{sp} \sim N(sp, \sigma_{sp}^2).$$

The asymptotic normality may be used to produce approximate confidence intervals. When n is large, the intervals based on the normal approximation are as good as those constructed from the exact binomial distribution.

There are other methods to construct confidence intervals for binomial parameter. See Newcombe (1998) for a comprehensive review. In particular, when **se** and **sp** are close to one, the upper limit of the interval may exceed one or the interval may be degenerate to be a single point. We may have to perform some necessary boundary adjustments by using methods described in Breslow (1983), Agresti and Coull (1998), Wilson (1927), Louis (1981), and Tai et al. (1997), among others.

5.1.1.2 Positive Predictive Value (PPV) and Negative Predictive Value (NPV)

The aforementioned sensitivity and specificity are useful measures for investigators who want to study the intrinsic accuracy of a diagnostic test. On the other hand, for a particular patient or a specific population, the likelihood of disease condition given the test results may be more relevant and can be reflected by another set of accuracy principles.

To this end, we usually consider positive predictive value (PPV) and negative predictive value (NPV), which are defined as

$$\text{PPV} = P(S = 1|Y = 1) \tag{5.6}$$
$$\text{NPV} = P(S = 0|Y = 0). \tag{5.7}$$

The definitions of these probabilities switch the positions of S and Y in the previous definitions for sensitivity and specificity. Such a seemingly small change results in a substantial conceptual difference. The sensitivity and specificity of a test remain the same for any population since they are considered as certain fixed features of the test. In contrast, PPV and NPV may vary across different populations with different disease prevalence. A calculated pair of PPV and NPV values thus indicate how often the test can correctly predict patient outcomes in the population under investigation.

To illustrate this point, let p be the disease prevalence for a population. That is, $p = P(S = 1)$ is the marginal probability that a randomly selected subject from the population has the disease condition. We may obtain the following algebraic relationship among the four accuracy measures by using the Bayesian formula:

$$\mathrm{PPV}(p) = \frac{p \times \mathsf{se}}{p \times \mathsf{se} + (1 - p) \times (1 - \mathsf{sp})}, \tag{5.8}$$

$$\mathrm{NPV}(p) = \frac{(1 - p) \times \mathsf{sp}}{p \times (1 - \mathsf{se}) + (1 - p) \times \mathsf{sp}}. \tag{5.9}$$

From these equations, we may note that for a fixed diagnostic test (hence fixed sensitivity and specificity), PPV and NPV are explicit functions of the prevalence p. For two tests with different sensitivities and specificities, PPV and NPV may not be strictly ordered for all p. One test may have a higher PPV (p) for a certain p, but not for other p values, and similarly for NPV (p). Such differences are very important when comparing the clinical utilities of diagnostic procedures, since their utilities depend on the disease prevalence of the population.

Consider the setting in Table 5.1, one can calculate the sample PPV and NPV by

$$\widehat{\mathrm{PPV}} = \frac{X_1}{X_1 + n_0 - X_0}, \tag{5.10}$$

$$\widehat{\mathrm{NPV}} = \frac{X_0}{X_0 + n_1 - X_1}. \tag{5.11}$$

These two empirical measures are routinely reported along with (5.3) and (5.4) in biomedical studies. However, we need to caution practitioners that these two values indicate the prediction capability of the test applied to the particular sample summarized in Table 5.1. Such a claimed accuracy may not be generalizable to a future population which may have a risk level different from this current sample. Often in a clinical study for the diagnostic accuracy of a test, we oversample diseased subjects and therefore inflate the sample prevalence which is n_1/n. PPV and NPV derived under such a setting may not be the same as those for a broader population where the disease prevalence p is much lower.

5.1.1.3   Inference for PPV and NPV

Using formulas (5.10) and (5.11) is a common procedure for the estimation of PPV and NPV. The implicit assumption is that we are studying a population whose risk level is close to the observed disease prevalence in the sample (n_1/n). Making inference for population parameters requires knowledge of the unknown prevalence. One may consider the following two *ad hoc* approaches:

(i) Calculate the proportion of disease-present patients in a study population, and use this proportion as an estimate for p.

(ii) Fix p at a prespecified value p^* according to past clinical experiences.

The first approach may be useful when the population under study is of interest. However, using an estimate of p when calculating the predictive values precludes studying their variability across populations. The second approach is rather subjective and only assesses the measures at a single value p^*. Neither approach gives simultaneous inference across prevalence values, as is needed to evaluate the clinical utility of the tests either in populations with different fixed prevalence values or in a single population where there may be prevalence heterogeneity according to unobserved risk factors.

Suppose that $\widehat{se}$ and $\widehat{sp}$ consistently estimate se and sp, respectively. These two estimators are independent and follow asymptotically normal distributions $N(se, \sigma_{se}^2)$ and $N(sp, \sigma_{sp}^2)$, respectively. Further assume that consistent variance estimators $\hat{\sigma}_{se}^2$ and $\hat{\sigma}_{sp}^2$ are available. At a fixed prevalence p, we estimate PPV and NPV by plugging in the estimates of se and sp in definitions (5.8) and (5.9):

$$\widehat{PPV}(p) = \frac{p \times \widehat{se}}{p \times \widehat{se} + (1 - p) \times (1 - \widehat{sp})}, \tag{5.12}$$

$$\widehat{NPV}(p) = \frac{(1 - p) \times \widehat{sp}}{p \times (1 - \widehat{se}) + (1 - p) \times \widehat{sp}}. \tag{5.13}$$

The estimators are consistent for PPV and NPV at any given p since the estimators for sensitivity and specificity are consistent.

Next, consider constructing confidence bands for PPV and NPV for all possible values of p. We begin by deriving confidence intervals for $\gamma_1 = \frac{1-sp}{se}$ and $\gamma_2 = \frac{1-se}{sp}$. It is later shown that these intervals can be used to construct simultaneous (in p) confidence bands for PPV(p) and NPV(p). In practice, γ_1^{-1} and γ_2 are usually referred to as the likelihood ratios for positive and negative test results, respectively.

Consider two methods for interval construction. The first is to apply Fieller's theorem (Zerbe, 1978). This theorem has been used in bioassay study to derive biologically effective dose or toxic dose. Let $Z_1 = \frac{(1-\hat{sp})-\gamma_1 \times \hat{se}}{\sqrt{\sigma_{sp}^2 + \gamma_1^2 \sigma_{se}^2}}$ which asymptotically follows a standard normal distribution. Since

$$1 - \alpha = P(Z_1^2 \leq z_{\alpha/2}^2) = P(A_1 \gamma_1^2 + B_1 \gamma_1 + C_1 \leq 0)$$

where $A_1 = \hat{se}^2 - z_{\alpha/2}^2 \sigma_{se}^2$, $B_1 = -2\hat{se}(1 - \hat{sp})$, and $C_1 = (1 - \hat{sp})^2 - z_{\alpha/2}^2 \sigma_{sp}^2$, a $100(1 - \alpha)\%$ confidence interval for γ_1 is

$$\frac{-B_1 - \sqrt{B_1^2 - 4A_1C_1}}{2A_1} \leq \gamma_1 \leq \frac{-B_1 + \sqrt{B_1^2 - 4A_1C_1}}{2A_1}, \tag{5.14}$$

provided that $A_1 > 0$ and $B_1^2 - 4A_1C_1 > 0$.

A similar argument yields a $100(1 - \alpha)\%$ confidence interval for γ_2 as

$$\frac{-B_2 - \sqrt{B_2^2 - 4A_2C_2}}{2A_2} \leq \gamma_2 \leq \frac{-B_2 + \sqrt{B_2^2 - 4A_2C_2}}{2A_2}, \tag{5.15}$$

provided that $A_2 > 0$ and $B_2^2 - 4A_2C_2 > 0$, where $A_2 = \hat{sp}^2 - z_{\alpha/2}^2 \sigma_{sp}^2$, $B_2 = -2\hat{sp}(1 - \hat{se})$, and $C_2 = (1 - \hat{se})^2 - z_{\alpha/2}^2 \sigma_{se}^2$.

The delta method may be used as a second approach to generate confidence intervals for γ_1 and γ_2. Assume that as the total sample size $n = n_0 + n_1 \to \infty$, $n_1/n_0 \to P/Q$ where $P + Q = 1$ and $P > 0, Q > 0$, and

$$\sqrt{n}\left(\begin{array}{c} \sqrt{P}(\hat{se} - se) \\ \sqrt{Q}(\hat{sp} - sp) \end{array}\right) \to N\left(0, \left(\begin{array}{cc} se(1 - se) & 0 \\ 0 & sp(1 - sp) \end{array}\right)\right).$$

The delta method gives that

$$\sqrt{n}\left(\frac{1 - \hat{sp}}{\hat{se}} - \frac{1 - sp}{se}\right) \to N(0, \sigma_{\gamma_1}^2),$$

$$\sqrt{n}\left(\frac{1 - \hat{se}}{\hat{sp}} - \frac{1 - se}{sp}\right) \to N(0, \sigma_{\gamma_2}^2),$$

where $\sigma_{\gamma_1}^2 = (1 - sp)^2(1 - se)/(P \times se^3) + sp(1 - sp)/(Q \times se^2)$ and $\sigma_{\gamma_2}^2 = (1 - se)^2(1 - sp)/(Q \times sp^3) + se(1 - se)/(P \times sp^2)$. These variances can be consistently estimated by substituting the estimates for se and sp. It follows that $100(1 - \alpha)\%$ confidence intervals for γ_1 and γ_2 are

$$\frac{1 - \hat{sp}}{\hat{se}} - z_{\alpha/2}\hat{\sigma}_{\gamma_1}/\sqrt{n} \leq \gamma_1 \leq \frac{1 - \hat{sp}}{\hat{se}} + z_{\alpha/2}\hat{\sigma}_{\gamma_1}/\sqrt{n}, \tag{5.16}$$

$$\frac{1 - \hat{se}}{\hat{sp}} - z_{\alpha/2}\hat{\sigma}_{\gamma_2}/\sqrt{n} \leq \gamma_2 \leq \frac{1 - \hat{se}}{\hat{sp}} + z_{\alpha/2}\hat{\sigma}_{\gamma_2}/\sqrt{n}. \tag{5.17}$$

Some algebra gives that

$$PPV(p) = \frac{p \times se}{p \times se + (1 - p)(1 - sp)} = \frac{1}{1 + \frac{(1-p)(1-sp)}{p \times se}}$$

and

$$NPV(p) = \frac{(1 - p) \times sp}{p \times (1 - se) + (1 - p) \times sp} = \frac{1}{1 + \frac{p(1-se)}{(1-p)sp}}$$

following (5.8) or (5.9).

Using either (5.14) or (5.16), we can construct a $100(1-\alpha)\%$ confidence interval for $PPV(p)$ by noticing the monotone functional relationship between $PPV(p)$ and γ_1. Suppose that the lower and upper limits for γ_1 are L_{γ_1} (as is the left-hand side of Equation 5.14 or 5.16) and U_{γ_1} (as is the right-hand side of Equation 5.14 or 5.16), respectively. Then,

$$
\begin{aligned}
1 - \alpha &= P(L_{\gamma_1} \leq \gamma_1 \leq U_{\gamma_1}) \\
&= P\left[\frac{1}{1 + \frac{1-p}{p} L_{\gamma_1}} \geq PPV(p) \geq \frac{1}{1 + \frac{1-p}{p} U_{\gamma_1}} \right]. \qquad (5.18)
\end{aligned}
$$

We only need to transform the left- and right-hand sides of (5.14) or (5.16) to obtain the confidence intervals for $PPV(p)$ at every fixed p. In fact, these confidence intervals form a simultaneous confidence band for PPV for all p, because the intervals are perfectly correlated as p varies, as shown below:

$$
\begin{aligned}
1 - \alpha &= P(L_{\gamma_1} \leq \gamma_1 \leq U_{\gamma_1}) \\
&= P\left(\frac{1}{1 + \frac{1-p}{p} L_{\gamma_1}} \geq PPV(p) \geq \frac{1}{1 + \frac{1-p}{p} U_{\gamma_1}}, \text{ for any } p \in [0,1] \right) \\
&= P\left[\cap_{p \in [0,1]} \left\{ \frac{1}{1 + \frac{1-p}{p} L_{\gamma_1}} \geq PPV(p) \geq \frac{1}{1 + \frac{1-p}{p} U_{\gamma_1}} \right\} \right].
\end{aligned}
$$

Similarly, a simultaneous confidence band for NPV can be formed based on the following confidence intervals:

$$
\begin{aligned}
1 - \alpha &= P(L_{\gamma_2} \leq \gamma_2 \leq U_{\gamma_2}) \\
&= P\left(\frac{1}{1 + \frac{p}{1-p} L_{\gamma_2}} \geq NPV(p) \geq \frac{1}{1 + \frac{p}{1-p} U_{\gamma_2}} \right), \qquad (5.19)
\end{aligned}
$$

where L_{γ_2} and U_{γ_2} are the lower and upper limits for γ_2 from (5.15) or (5.17), respectively.

Example (IVDR BSI). Safe and reliable vascular access is an essential requirement for intravascular devices (IVDs) in modern medical practice. The most common life-threatening complication of vascular access is bloodstream infection (BSI), caused by colonization of the implanted IVD or contamination of the catheter hub or infusate administered through the device. Accurate and early diagnosis is necessary for appropriate management of intravascular device-related bloodstream infection (IVDR BSI). A variety of diagnostic methods have been developed for this purpose. Safdar et al. (2005) carried out a meta-analysis to compare the accuracy of these diagnostic methods. Sensitivities and specificities for the tests can be estimated more efficiently after combining multiple studies and were obtained in Li et al. (2007). We here consider making inference for PPV and NPV.

We select two diagnostic tests, namely, semiquantitative catheter segment culture and differential time to positivity, and plot their PPV and NPV against p in Figures 5.1 and 5.2, respectively. PPV is an increasing concave function of the prevalence (Figure 5.1), while NPV is a decreasing concave function of the prevalence (Figure 5.2). The two confidence bands from Fieller's method and delta method are almost indistinguishable. This is due to the fact that these methods are asymptotically equivalent, and the sample sizes in the meta-analysis are large. The bands are narrower for p close to one and zero. The variability for the PPV of differential time to positivity is greater than that of a semiquantitative catheter segment and leads to wider intervals for a broad range of p.

This meta-analysis example assumes that sensitivity and specificity are constant in different studies. That is, the accuracy remains constant across different populations with different disease prevalence. However, in practice such an assumption may not be correct. Prevalence can influence the analysis of accuracy at three levels. First, low prevalence of disease may lead to a small number of diseased subjects included in a study sample. Thus, the estimator $\hat{se}$ may have a low precision relative to an estimator obtained from a population with a high disease prevalence. Second, the disease prevalence may potentially affect the number of correctly identified diseased subjects. One may assume that the occurrence of the disease is linked to the event that some underlying continuous variable crosses a threshold and that such a latent variable may also determine the binary classification of the diagnostic test. Such dependence may be explained by the disease spectrum being linked to prevalence, for example higher percentages of severe cases in populations with higher prevalence. Third, human factors may play a role. If the prevalence is high, as in a clinical setting, the medical staff who is judging an image may be more aware that the patient may be diseased than in a screening setting with few diseased, but thousands of disease-free individuals. This may result in a positive correlation between prevalence and sensitivity and a negative correlation between prevalence and specificity. More discussions can be found in Li and Fine (2011).

We next consider a specific clinical application of PPV and NPV. It is well understood that sensitivity and specificity may not reflect the clinical utility of a diagnostic test, which depends implicitly on the risk level of the population in which the test is to be used. In practice, the primary interest for an individual patient is the conditional probability of disease given the positive (negative) test. The PPV and NPV are therefore of high clinical importance (Kuk et al., 2010; Li et al., 2012). We consider the following clinical trial example.

Example (STAR*D Trial). For the STAR*D data, we now focus on a research question different from those studied in Chapter 2. Some patients with depression will respond to a particular medication, while others will require a different medication to achieve a response. Unfortunately, there are no clinically useful pretreatment assessments that can reliably recommend one antidepressant medication over another for a particular patient. Treatment efficiency will be increased if we can predict early on whether a medication is likely to be ineffective for an individual patient. Such predictions will enable patients to bypass ineffective medications and proceed to those more likely to be effective, thereby reducing the time and cost required to achieve treatment response and decreasing patient exposure to unnecessary medication.

At Level 1 of the STAR*D trial, investigators attempted to use the degree of depressive symptom improvement during the first few weeks of treatment to

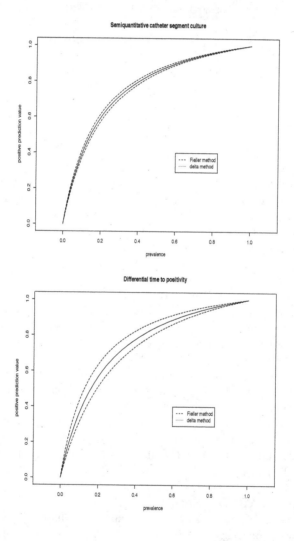

FIGURE 5.1: Positive predictive values of two diagnostic tests for IVDR BSI, with 95% confidence bands.

predict later outcomes (e.g., response after 6 weeks of treatment). Specifically, Kuk et al. (2010) considered using the baseline features and symptom change at two weeks to predict which patients will or will not respond to treatment at Level 1. Specifically, they developed a recursive subsetting algorithm which directly controls the NPV and ensures satisfactory prediction accuracy. The algorithm essentially searches among all possible categories of patients classified by week two symptom changes and baseline variables such as age, sex,

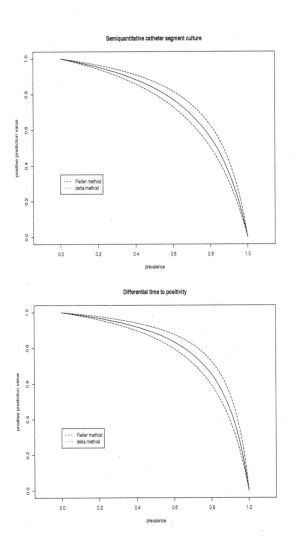

FIGURE 5.2: Negative predictive values of two diagnostic tests for IVDR BSI, with 95% confidence bands.

chronicity, and anxiety and selects those categories which have the largest number of subjects while still maintaining a satisfactory NPV level.

Define W_2 to be the proportion of reduction in the QIDS score from baseline to week two using the formula $W_2 = (S_2 - S_0)/S_0$, with S_0 representing the participant's baseline QIDS score and S_2 representing the participant's QIDS score at week two. Before the classification starts, a desired level of NPV for nonresponse at week six is set at 0.8.

Starting with the group least likely to respond (i.e., category 1): if the proportion of participants in this category (and, at the recursive stage, all other categories and subcategories previously labeled as nonresponding) who did not respond to treatment was at least 0.8, Kuk et al. (2010) predicted that everyone in this category would be nonresponders and moved to the next category. If the proportion who did not respond was <0.8, they partitioned the category into two subcategories according to the values of one of the baseline variables: gender, anxious status (present/absent), chronicity (present/absent), general medical condition (GMC, present/absent), baseline QIDS score, and age. All continuous variables were dichotomized at proper cut-offs. The variable used for partitioning and its cut-off were chosen to maximize the size of the subcategory (i.e., predict nonresponse for as many participants as possible based on the requisite NPV 0.8). The procedure stopped when a further partitioning to grow the set of participants with at least 80% actual nonrespondents was impossible.

Of the 2280 participants, 1058 (46%) responded by week six. The recursive subsetting algorithm eventually identified the following categories of participants as nonresponders: (1) those who have symptom worsening at week two (i.e., $W_2 < 0$); (2) those with a minimal improvement in the QIDS score after two weeks ($0 \leq W_2 < 1/12$) and also with a chronic condition; (3) those with a minimal improvement in QIDS score after week two ($0 \leq W_2 < 1/12$), and who are nonchronic, and who are > 50 years old; and (4) males who have a modest symptom improvement ($1/12 \leq W_2 < 1/6$). At an NPV level 80%, they successfully identified 505 nonresponders of which 404 were true nonresponders. The classification results are depicted with a tree graph in Figure 5.3.

5.1.2 Continuous Diagnostic Test

Many diagnostic tests give nonbinary results and require practitioners such as raters to choose a cut-off. For example, in the analysis results for STAR*D (Figure 5.3), the variables Age and W_2 are continuous measures. Furthermore, the binary variable Chronicity is also defined by thresholding the disease duration which is continuous. To study the accuracy of continuous tests, we usually employ a different set of measures.

5.1.2.1 ROC Curve and AUC

In general, we may consider the test Y to be distributed on a continuous scale with a support $\mathcal{Y} \subset \mathbb{R}$. For discrete but nonbinary tests, they can be similarly dealt with, and no significant methodological difference arises. Assume that a high value of Y leads to a positive diagnosis. For real applications where a low value Y defines a positive diagnosis, we can easily take a transformation $\tilde{Y} = -Y$ and use $\tilde{Y}$ in the calculation in the same way as follows.

The true positivity fraction (TPF) and false positive fraction (FPF) for the marker Y at a cut-off y can be defined as

$$\text{TPF}(y) = P(Y \geq y | S = 1),$$
$$\text{FPF}(y) = P(Y \geq y | S = 0).$$

The event $Y \geq y$ indicates that a positive diagnosis has been made, and the subject is announced to be diseased, while $Y < y$ indicates a negative diagnosis. Researchers also call TPF as sensitivity and $1-$ FPF as specificity, though we keep in mind that these terms now depend on the particular cut-off value y. At different y values, sensitivity and specificity of the test Y may be different.

The curve of $\text{TPF}(y)$ versus $\text{FPF}(y)$ across all y values is called the receiver operating characteristic (ROC) curve. Figure 5.4 provides an example where

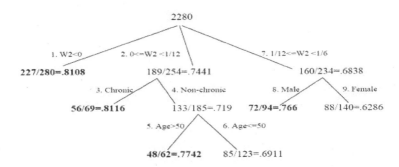

FIGURE 5.3: Early prediction of nonresponders with NPV=80% for STAR*D data (total sample size = 2280). For each node, the numerator is the number of true nonresponders, and the denominator is the number of predicted nonresponders. Nodes are numbered to reflect the recursive feature. W_2 is the week two symptom change. The categories with boldface numbers were chosen.

the ROC curve is an ascending curve from $(0,0)$ to $(1,1)$ in the unit square of the two-dimensional plane. It provides a visual inspection of the trade-off between sensitivity and specificity as we change the threshold of the test.

Mathematically, the ROC curve may be written as the composition of $TPF(y)$ and the inverse function of $FPF(y)$:

$$ROC(p) = TPF\{[FPF]^{-1}(p)\}, \qquad (5.20)$$

for $p \in (0,1)$, where

$$[FPF]^{-1}(p) = \inf\{y : FPF(y) \le p\}.$$

This formulation is especially useful in practice for selecting a proper threshold value y such that the resulting TRF and FRF are controlled at desirable levels.

An overall summary measure is the area under the ROC curve (AUC)

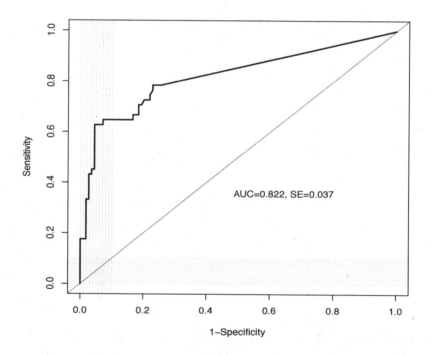

FIGURE 5.4: ROC curve for the cervical cancer example. The diagonal line corresponds to a random guess. AUC is the area under the ROC curve, and SE is the standard error for the AUC estimate.

which may be defined as

$$AUC = \int_0^1 ROC(p)\,dp = \int_0^1 \text{TPF}\{[\text{FPF}]^{-1}(p)\}\,dp$$
$$= \int_y \text{se}(y)\,d\{1 - \text{sp}(y)\}. \tag{5.21}$$

Denote the distributions of Y to be F_1 and F_0, corresponding to $S = 1$ and $S = 0$, respectively. The TPF and FPF at a threshold are simply the survival functions of the two distributions and equal to $1 - F_1$ and $1 - F_0$, respectively. The ROC curve may then be written as a function of these distributions

$$ROC(\text{p}) = 1 - F_1(F_0^{-1}(1 - \text{p})), \tag{5.22}$$

and the AUC may be written as

$$AUC = \int_0^1 1 - F_1(F_0^{-1}(1 - \text{p}))\,d\text{p}. \tag{5.23}$$

Another way of defining and interpreting the AUC is via probability, that is,

$$AUC = P(Y_1 > Y_2 | S_1 = 1, S_2 = 0), \tag{5.24}$$

where Y_i and S_i are the test value and disease status for subject i, respectively. This can be shown if we observe that (5.23) may be written as

$$\int_y 1 - F_1(y)\,d\{1 - F_0(y)\} = \int_y P(Y_1 > y | S_1 = 1) f_{Y_2}(y | S_2 = 0)\,dy$$
$$= \int_y P(Y_1 > y, Y_2 = y | S_1 = 1, S_2 = 0)\,dy.$$

This heuristic interpretation indicates that the AUC is equal to the probability that a randomly chosen diseased subject will have a marker value greater than that of a healthy subject. In other words, AUC is the chance that the two subjects from different categories are correctly ordered. The definition (5.24) turns out to be very handy when we make statistical inference for AUC.

Parametric Inference. Parametric methods are easy to implement and may lend support to simple interpretations. A common approach is to consider a bi-normal model. We assume that $F_i = N(\mu_i, \sigma_i)$ for $i = 1, 0$. The diagnostic accuracy measures introduced earlier can be described by the normal distribution functions. Specifically, the TPF and FPF are

$$\text{TPF}(y) = P(Y \geq y | S = 1) = 1 - \Phi\left(\frac{y - \mu_1}{\sigma_1}\right) \tag{5.25}$$

$$\text{FPF}(y) = P(Y \geq y | S = 0) = 1 - \Phi\left(\frac{y - \mu_0}{\sigma_0}\right), \tag{5.26}$$

where Φ is the standard normal distribution function. The ROC function based on the two normal distributions is

$$ROC(p) = \Phi(a + b\Phi^{-1}(p)), \qquad a = \frac{\mu_1 - \mu_0}{\sigma_1}, b = \frac{\sigma_0}{\sigma_1}. \qquad (5.27)$$

The AUC also has a simple form given by

$$AUC = \Phi\left(\frac{a}{\sqrt{1 + b^2}}\right). \qquad (5.28)$$

Under the above formulation, the estimation of various diagnostic accuracy measures boils down to the estimation of parameters involved in the normal distribution functions. In practice, sample means and sample variances for the diseased and healthy samples can be used to estimate the corresponding mean and variance parameters.

Now let us consider a more general parametric inference procedure. Assume that the distributions of Y given $S = 1$ or $S = 0$ both come from a location-scale family $\{Q_{(\mu,\sigma)} : \mu \in \mathbb{R}, \sigma > 0\}$, with different location parameter μ and scale parameter σ. The distribution function of a member in this family satisfies

$$Q_{(\mu,\sigma)}(z) = Q\left(\frac{z - \mu}{\sigma}\right),$$

for a known distribution function $Q(\cdot)$. The family of normal distributions is clearly an example of location-scale family. Other families include Gumbel distribution, logistic distribution, and Cauchy distribution. Sometimes these heavy-tailed distributions may be more appropriate to model the distribution of diagnostic tests than the thin-tailed normal distribution.

Assume that $Y|S = 1 \sim Q_{(\mu_1,\sigma_1)}$ and $T|S = 0 \sim Q_{(\mu_0,\sigma_0)}$. Under the assumed distributions, the parametric ROC curve can be represented as

$$ROC(p) = Q(a + bQ^{-1}(p)), \qquad (5.29)$$

where

$$b = \frac{\sigma_0}{\sigma_1}, \qquad a = \frac{\mu_1 - \mu_0}{\sigma_1}. \qquad (5.30)$$

Under the assumed parametric model, the AUC can be shown to be

$$AUC = E\left\{Q\left(\frac{Z - a}{b}\right)\right\}, \qquad (5.31)$$

where the expectation is taken with respect to Z which follows the standard Q distribution.

A Monte Carlo approach can be implemented to evaluate (5.31) numerically: simulate a large number of z_m ($m = 1, \cdots, M$) from Q and compute $\sum_{m=1}^{M} Q\left(\frac{z_m - a}{b}\right)/M$. When M is large enough, this finite summation gives a close approximation of (5.31).

Formula (5.31) may be applied with general Q for parameterized ROC curve. Under some specifications, this general formula may become familiar quantities that have been widely used in practice. For example, when Q is a symmetric distribution such as the standard normal distribution or the standard logistic distribution, expression (5.31) reduces to

$$E\left[Q\left(\frac{Z-a}{b}\right)\right] = Q\left(\frac{a}{\sqrt{1+b^2}}\right),\tag{5.32}$$

after simple algebra. We notice that (5.28) is a special case of (5.31).

The proofs of (5.29) and (5.31) are given below:

Proof 1 (Proof of 5.29) *For any threshold y,*

$$\begin{aligned}
\text{TPF}(y) &= P(Y > y|S=1) = 1 - Q\left(\frac{y-\mu_1}{\sigma_1}\right) \\
\text{FPF}(y) &= P(Y > y|S=0) = 1 - Q\left(\frac{y-\mu_0}{\sigma_0}\right).
\end{aligned}$$

For an FPF, we notice that $y = \mu_0 + \sigma_0 Q^{-1}(1 - \text{FPF})$ is the corresponding threshold for the test positive criterion. Hence

$$\begin{aligned}
ROC(p) = \text{TPF}(y) &= 1 - Q\left(\frac{y-\mu_1}{\sigma_1}\right) \\
&= 1 - Q\left(\frac{\mu_0 + \sigma_0 Q^{-1}(1-p) - \mu_1}{\sigma_1}\right) \\
&= 1 - Q(-a - bQ^{-1}(p)) \\
&= Q(a + bQ^{-1}(p)).
\end{aligned}$$

∎

Proof 2 (Proof of 5.31) *Assume that the density function $q(z)$ of $Q(z)$ exists. Then*

$$\begin{aligned}
AUC &= \int_0^1 \{1 - Q(a + bQ^{-1}(p))\}\,d\mathrm{p} \\
&= \int_0^1 \int_{a+bQ^{-1}(p)}^\infty q(z)\,dz\,d\mathrm{p} \\
&= \int_{-\infty}^\infty \int_0^{Q\left(\frac{z-a}{b}\right)} q(z)\,d\mathrm{p}\,dz \\
&= \int_{-\infty}^\infty Q\left(\frac{z-a}{b}\right) q(z)\,dz \\
&= E\left\{Q\left(\frac{Z-a}{b}\right)\right\}.
\end{aligned}$$

∎

If consistent parameter estimates $\hat{a}$ and $\hat{b}$ are available, we can estimate AUC consistently by

$$\widehat{AUC} = E\{Q(\frac{Z-\hat{a}}{\hat{b}})\}. \tag{5.33}$$

For many parametric models, maximum likelihood estimation (MLE) methods (Zhou et al., 2002) are widely used to obtain the estimates. The consistency of MLE thus lends support to the consistency of the estimated AUC. For normal distribution, familiar moment estimates such as sample mean and sample variance can be directly applied.

Furthermore, suppose that $(\hat{a}, \hat{b})$ are asymptotically normal, as is true with MLEs under fairly mild regularity conditions. This leads to the asymptotic normal distribution of $\widehat{AUC}$ with mean equal to the true AUC and variance equal to

$$\int_0^1 \int_0^1 \left[\text{var}(\hat{a}) + Q^{-1}(\text{p}_1)Q^{-1}(\text{p}_2)\text{var}(\hat{b}) + \{Q^{-1}(\text{p}_1) + Q^{-1}(\text{p}_2)\}\text{cov}(\hat{a}, \hat{b}) \right]$$
$$\times q(a + bQ^{-1}(\text{p}_1))q(a + bQ^{-1}(\text{p}_2)) \, d\text{p}_1 \, d\text{p}_2, \tag{5.34}$$

under the assumption that the density $q(z)$ of $Q(z)$ exists. The derivation of the above variance formula is via a simple delta method argument. To implement this formula in an MLE procedure, the asymptotic variances $\text{var}(\hat{a})$, $\text{var}(\hat{b})$ and covariance $\text{cov}(\hat{a}, \hat{b})$ can be estimated consistently from the observed inverse information matrix (Hessian matrix of the log-likelihood function).

Nonparametric Inference. In modern statistics, it is more robust to consider distribution-free approaches. Based on a sample $\{(Y_i, S_i) : i = 1, \cdots, n\}$, the distribution functions F_1 and F_0 are estimated empirically as

$$\hat{F}_1(y) = \sum_{i=1}^n I\{Y_i \leq y, S_i = 1\} / \sum_{i=1}^n I\{S_i = 1\}, \tag{5.35}$$

$$\hat{F}_0(y) = \sum_{i=1}^n I\{Y_i \leq y, S_i = 0\} / \sum_{i=1}^n I\{S_i = 0\}. \tag{5.36}$$

Using the above notations, the estimated ROC curve is given by

$$\widehat{ROC}(\text{p}) = 1 - \hat{F}_1(\hat{F}_0^{-1}(1 - \text{p})). \tag{5.37}$$

The so-constructed empirical ROC curve (e.g., Figure 5.4) is an increasing step function since empirical functions are always discrete step functions. Some authors have attempted to smooth this empirical curve by adopting a kernel method (Zou et al., 2011).

The AUC is estimated by

$$\widehat{AUC} = \int_0^1 1 - \hat{F}_1(\hat{F}_0^{-1}(1 - \text{p})) \, d\text{p}. \tag{5.38}$$

In practice, the integration can be exactly computed by summing up the areas of the trapezoids formed under the empirical ROC curve. The estimator may be slightly biased downward because the ROC curve is usually concave. Another alternative of evaluating the integral is via numerical integration methods.

We now consider the statistical properties of this estimator when sample sizes n_0 and n_1 are large. It has been shown in Hsieh and Turnbull (1996) that

$$\sup_{0<\mathbf{p}<1} |\widehat{ROC}(\mathbf{p}) - ROC(\mathbf{p})| \to 0 \qquad \text{almost surely} \qquad (5.39)$$

under the following technical conditions:

(a) $F_1(y)$ and $F_0(y)$ have continuous derivatives.

(b) The derivative of $ROC(\mathbf{p})$, taken with respect to $\mathbf{p}$, is bounded on any subinterval (a, b) of $(0, 1)$.

(c) $n_1/n_0 \to \lambda > 0$ as $n_0 \to \infty$ where $n_k = \sum_{i=1}^{n} I\{S_i = k\}$ (k=0, 1).

Under the same set of conditions, the stochastic process $\sqrt{n_1}\{\widehat{ROC}(\mathbf{p}) - ROC(\mathbf{p})\}$, indexed by p, converges weakly to a Brownian bridge process. A simple consequence of the weak convergence is the asymptotic multivariate normality for any finite dimensional distributions of this process. We may use such a result to build asymptotic confidence intervals for ROC curves at any $p \in (0, 1)$.

It is also straightforward to show that the AUC estimator (5.38) is strongly consistent to the true AUC if we notice that

$$\left| \int_0^1 \widehat{ROC}(\mathbf{p})d\mathbf{p} - \int_0^1 ROC(\mathbf{p})d\mathbf{p} \right| \leq \int_0^1 |\widehat{ROC}(\mathbf{p}) - ROC(\mathbf{p})|d\mathbf{p}, \quad (5.40)$$

and apply the dominated convergence theorem (Durrett, 2005) to the right-hand side.

Furthermore, following the results in Hsieh and Turnbull (1996), we have

$$\sqrt{n_1}(\widehat{AUC} - AUC) \to_d N(0, v^2), \qquad (5.41)$$

where

$$\begin{aligned}
v^2 &= \lambda\left[\int_0^1 \{F(\mathbf{p})\}^2 dROC(\mathbf{p}) - \left\{ \int_0^1 ROC(\mathbf{p})d\mathbf{p} \right\}^2 \right] + \\
&\quad \left[2\int_0^1 \int_{\mathbf{p}_2}^1 \mathbf{p}_2 dROC(\mathbf{p}_1)dROC(\mathbf{p}_2) - \left\{ \int_0^1 \mathbf{p}\,dROC(\mathbf{p}) \right\}^2 \right].
\end{aligned}$$

$$(5.42)$$

Another nonparametric AUC estimator can be obtained from noticing the

probabilistic interpretation (5.24). We may consider the following estimator
which is related to the Mann-Whiteney-Wilcoxon test statistic,

$$\widehat{AUC} = \frac{\sum_{i=1}^{n}\sum_{j=1}^{n} I\{Y_i < Y_j, S_i = 0, S_j = 1\}}{n_0 n_1}. \tag{5.43}$$

One apparent advantage of this estimator is that we no longer need to deal
with integrations. This is a two-sample U-statistic, and so its asymptotic distribution can follow from the famous Hoeffding's U-statistic theories (Hoeffding,
1948). The variance of this estimator is

$$v^2 = \frac{AUC(1 - AUC) + (n_0 - 1)(J_0 - AUC^2) + (n_1 - 1)(J_1 - AUC^2)}{n_0 n_1} \tag{5.44}$$

with $J_0 = P(Y_1 < Y_2, Y_3 < Y_2 | S_1 = 0, S_2 = 1, S_3 = 0)$ and $J_1 = P(Y_1 < Y_2, Y1 < Y_3 | S_1 = 0, S_2 = 1, S_3 = 1)$. The terms J_0 and J_1 can be estimated by
their empirical counterparts (another two U-statistics), and we can substitute
them into (5.44) along with the estimate of AUC to obtain the estimated
variance.

Example (Cervical Cancer). Diagnosis of cervical neoplasia hinges
upon microscopic inspection of cervical samples. Epigenetic modifications, for
example, DNA methylation, have been observed in the early stages of neoplas-
tic change, preceding gene mutations. The numeric value of DNA methylation
may be used as a test for cervical cancer status. In Lim et al. (2010), a total
of 165 samples were selected for analysis. Cytological examination determined
that 53 were normal, 49 were low-grade squamous intraepithelial lesion (LSIL),
and 63 were high-grade squamous intraepithelial lesion (HSIL). Based on a literature search and a trial run (Methylight) of shortlisted candidates, CCNA1,
PAX1, HS3ST2, DAPK1, and TFPI2 were chosen for analysis. The methylation measures of the five genes were able to segregate HSIL patients from
LSIL and normal patients, with AUC values ranging from 0.719 to 0.822 and
HS3ST2 registering the greatest AUC value. The ROC curve for HS3ST2 was
displayed in Figure 5.4, produced in R using the library Epi.

We note that the ROC curve and AUC may play important roles in many
other fields including financial forecasting and regression diagnostics. In SPSS,
ROC graphs are routinely generated together with AUC values for all fitted
logistic regression models as model evaluation tools.

5.1.2.2 Partial and Weighted AUC

AUC has been popular for summarizing the test accuracy. However, AUC
may be too *global* a summary measure, and sometimes may not reflect the
region of the ROC curve which is of the greatest interest. In some cases, it
may be of interest to assess tests across different ranges, corresponding to
different hypothetical operating scenarios for the test. For a particular clinical
application, a decision threshold may be chosen so that the diagnostic test

will have either satisfactory specificity (e.g., > 90%) or sensitivity. Under these circumstances, the average accuracy of the test over the specified range of specificities is more meaningful than the area under the entire ROC curve. For such settings, partial area under the ROC curve (PAUC) has been studied as a more appropriate summary measure of the diagnostic test's accuracy. A related measure is the PAUC index (Zhou et al., 2002; Dodd and Pepe, 2003), which is defined as the PAUC divided by the length of the specified interval of specificity. Normalized in this way, the PAUC index has a probabilistic interpretation similar to AUC.

Both AUC and PAUC (or more appropriately, the PAUC index) can be regarded as an integration of the ROC curve with uniform weights over $(0, 1)$ and its subinterval. We may consider flexible weight functions for AUC. Using notations defined earlier, we consider the weighted area under the ROC curve (WAUC)

$$
\begin{aligned}
WAUC &= \int_0^1 ROC(\mathbf{sp}) \cdot w(\mathbf{sp}) \, d\mathbf{sp}, \\
&= \int_0^1 \{1 - F_1(F_0^{-1}(\mathbf{sp}))\} \cdot w(\mathbf{sp}) \, d\mathbf{sp},
\end{aligned}
$$

where w is a density function on $[0, 1]$ such that $\int_0^1 w(x) \, dx = 1$. This measure can be reduced to some familiar measures for special choices of $w(\mathbf{sp})$. When $w(\mathbf{sp}) = 1$ for $\mathbf{sp} \in [0, 1]$, WAUC is simply the ordinary AUC. When $w(\mathbf{sp}) = \frac{1}{b-a} I\{\mathbf{sp} \in [a, b]\}$, WAUC is the partial AUC index restricted in $[a, b]$.

The specification of such general weight functions can be helpful if one conceptualizes the decision threshold in a particular application as being random across patients, with the weight in the WAUC corresponding to the probability distribution of the threshold. We note that different weights correspond to different beliefs regarding the relative importance of different portions of the ROC curve.

In general, WAUC may be interpreted as the weighted average of sensitivity with weights emphasizing specificity of interest. When w is a probability density function, the WAUC corresponds to the expected sensitivity under the assumed specificity distribution, with a WAUC value >0.5 indicating that in the long run the test classifies diseased patients correctly more often than not under f. However, one should realize that the minimum acceptable WAUC is generally different from .5 if the weight function is not symmetrical around .5. A useless diagnostic test whose distributions for diseased and nondiseased population are the same has $\mathbf{se} = 1 - \mathbf{sp}$ and the WAUC value

$$
W_0 = 1 - \int_0^1 \mathbf{sp} \cdot w(\mathbf{sp}) d\mathbf{sp}.
$$

Such a null value is $1 - (a + b)/2$ for f being uniform on $[a, b]$. Standard AUC has $W_0 = 0.5$ while PAUC index on the interval $[a, 1]$ has $W_0 = (1 - a)/2$. A test can only be considered useful if its WAUC exceeds W_0.

The WAUC offers a sensible measure of diagnostic performance when a clinically relevant range of specificity has been specified. The relative importance of the specificity values is characterized by the weight function in the computation of WAUC. Instead of assigning noninformative equal weights to all possible values in $(0, 1)$, a general weight function can be any probability density or weight function defined within this range. Depending on the focus of study, we can choose right-skewed, symmetric, or left-skewed distributions.

The class of beta density functions (Gupta and Nadarajah, 2004) provides abundant choices for the weight function. The beta(α, β) density is given by

$$f(x) = \frac{\Gamma(\alpha + \beta)}{\Gamma(\alpha)\Gamma(\beta)} x^{\alpha-1}(1 - x)^{\beta-1},$$

for $\alpha > 0, \beta > 0$ and $0 \leq x \leq 1$, where the boundary values at $x = 0$ or $x = 1$ are defined by continuity (as limits). The relative importance can be adjusted by modifying the parameter values (α, β) in the density function.

When assessing his or her knowledge about the specificity weight (or equivalently, the distribution of the random threshold in the population), one naturally tends to anchor on the uniform distribution and begin to establish departures from uniformity. We notice that the uniform distribution on $(0, 1)$ is a special case of the beta density when $\alpha = \beta = 1$. We therefore propose the following weight elicitation procedure, following a similar idea in Chaloner and Duncan (1983).

Step 1 Choose the distribution mode which is the most likely value of specificity in the weighting scheme.

Step 2 Choose the distribution variance which describes the relative likelihood of mode to adjacent values. When only vague information is available for the spread pattern, one chooses the value close to that of the uniform distribution.

Step 3 Choose a sensible beta density that best satisfies the two conditions.

The mode elicitation method in Step 1 has an intuitive appeal to data analysts and excludes certain beta densities with no modes for $\mathbf{sp} \in (0, 1)$. The first two steps effectively mimic human cognitive processing and create a distribution that can reflect one's view of the relative importance of specificity values. We note that the mode and variance of the beta density are given by $(\alpha - 1)/(\alpha + \beta - 2)$ and $\alpha\beta/[(\alpha + \beta + 1)(\alpha + \beta)^2]$, respectively. Given desired values of the mode and variance at Steps 1 and 2, we may solve a system of two equations to find the two parameters α and β exactly. Finally for the sake of interpretability we usually have to adjust the solved values to sensible parameters to strike a balance between mathematical precision and practical utility.

We prefer the mode of the weight to be near large values of the specificity, say 85% or 90%. This suggests that α has to be much greater than β, and both parameters must be above one. As for the variance, we imagine that the overall distribution displays a comparable variation relative to the noninformative uniform distribution which is the standard weight for the ordinary AUC. This requirement implies that α and β may not be too extreme. We note that if these two parameters are too large, the variance will be very small and the distribution will be indistinguishable from a degenerate point mass at the mode. This kind of weight simply gives the sensitivity of the test at a particular specificity and may be too restrictive for a meaningful comparison of the test accuracy. To qualify the two demands, we suggest $Beta(8,2)$ be used in real data analysis.

We remark that using integer parameters in a beta distribution is not a necessary requirement in the final step. Sometimes the exact solutions obtained from solving the equations set at the first two steps may involve complex numbers. Therefore, the adjustment at the final step ensures that the chosen parameters are those best satisfying the conditions specified in the first two steps. Using whole numbers is just a relatively simple approach to do the adjustment. It is always acceptable to consider noninteger parameters.

In addition to the beta weight function, one may also consider a class of conceptually simple weight functions by keeping a uniform distribution in the interval [0.9,1] and decreasing weights gradually below 0.9. Herein we consider a linearly decreasing weight between 0.9 and 0.5. The resulting distribution after proper rescaling is a trapezoid density function. The advantage of this weight function is that it does not downweight the area of extremely large specificity. We compare this weight function in the real examples and find similar performance to the beta weight function.

We realize that in practice assessors may choose different weight distributions and obtain different WAUC values for a test. This may be due to varying degree of belief among assessors. The conclusion from each weighted ROC analysis thus needs to be construed with respect to the weight function considered by the assessor. A relatively objective approach is to obtain independent estimates from different assessors and combine their results with a method called recoiling quantitative opinions proposed by Berger (1985, pp. 272–286). The aggregated WAUC value is then representative of how people in the general population perceive the relative importance of the specificity values. Such aggregated analyses may be carried out by noting that the WAUC is a linear functional of the ROC curve and the implied weight function for the aggregated WAUC is thus just a weighted average of the weight functions for the different assessors.

The idea of weighting in the AUC has precedent in Wieand et al. (1989) to streamline the theoretical properties of nonparametric estimation of AUC and PAUC. However, not much practical attention has been placed on the development of appropriate statistical procedures about general weighted AUC measures. Bandos et al. (2005) used a specific weighting system when estimating AUC to incorporate the clinical importance (utility) when evaluating the accuracy of a diagnostic test. Their method is a special case of the WAUC with weight function depending on the observed test results. Li and Fine (2010) provided a general framework on how to conduct a diagnostic accuracy study with WAUC and supply necessary asymptotic results as the basis for statistical inference. Specifically, parametric and nonparametric inferences for WAUC can be carried out by plugging in the AUC estimate and following similar procedures as introduced in the preceding section.

5.1.2.3   ROC Surface

Traditional ROC analysis relies on the existence of a binary gold standard for the true disease status. While there are some medical tests where binary gold standards are appropriate, there are many medical diagnostic situations in which the true disease status has more than two dimensions. For such a diagnostic situation, one naive solution is to dichotomize the gold standard so that the methods introduced in preceding sections for binary classification can be applied. However, Obuchowski (2005) noted that the unnatural dichotomization of nonbinary gold standards can induce a rather substantial bias in the estimation of the test diagnostic accuracy. Dichotomizing nonbinary gold standards will bias the estimated AUC values toward 1.0, meaning that the diagnostic accuracy appears to be greater than it actually is.

There are numerous examples of medical diagnostic tests that do not easily fit into the binary disease classification structure, and many medical diagnoses appear to have a natural ordinal gold standard of disease status. These diagnoses have an ordered gradation of illness from disease-absent to seriously ill. For example, cognitive function declines from normal function, to mild impairment, to severe impairment and/or dementia. Another example is the stage of cancer progression at the time of detection, from localized cancer through distant metastases already present. We consider statistical methods for the assessment of diagnostic accuracy when the true disease status is multicategory.

Consider diagnostic tests which distinguish subjects each from one of three or more classes. The accuracy of such diagnostic tests can be statistically evaluated by constructing three-dimensional ROC surfaces or multidimensional ROC manifolds. Scurfield (1996) brought out the mathematical definition of ROC measures for more than two categories. However, rigorous statistical inferences arrived later, motivated by sophisticated multicategory biostatistical research problems. Mossman (1999) introduced the concept of three-way ROC analysis into medical decision making. Nakas and Yiannoutsos (2004) considered the estimation of volume under the ROC surface (VUS) for the ordered three-class problem. Li and Fine (2008) further proposed the estimation of VUS for unordered classification by following the probabilistic interpretation and applied VUS as a model selection criterion in microarray studies. Li and Zhou (2009) discussed the estimation of three-dimensional ROC surfaces. Zhang and Li (2011) considered combining multiple markers to improve diagnostic accuracy for three-way ROC analysis.

Let us first introduce some key concepts in multidimensional ROC analysis. Scurfield (1996) proposed the three-dimensional ROC surface which extends the two-dimensional ROC curve. Suppose that each subject in the sample is from one of three different categories, and the sample sizes for the three categories are $n_1, n_2,$ and n_3, respectively. The total sample size is thus $n = n_1 + n_2 + n_3$. We employ a continuous diagnostic test to differentiate the three classes. Let $Y_1, Y_2,$ and Y_3 denote the test result for a subject randomly selected from Classes I, II, and III, respectively. All individuals have undergone the examination, and their test values have been recorded. The test results $Y_{1,i}$ ($i = 1, \cdots, n_1$) are i.i.d. with a distribution G_1; the test results $Y_{2,j}$ ($j = 1, \cdots, n_2$) are i.i.d. with a distribution G_2; and the test results $Y_{3,k}$ ($k = 1, \cdots, n_3$) are i.i.d. with a distribution G_3. $G_1, G_2,$ and G_3 are all continuous probability distributions on $\mathbb{R}$. The three kinds of test results $Y_1, Y_2,$ and Y_3 are independent to each other as they are obtained from different subjects.

Suppose that *we know $Y_1, Y_2,$ and Y_3 tend to take high, intermediate, and low values*, respectively. A diagnostic decision for each subject is based on the following rule (R_I) by specifying two-ordered decision thresholds $c_1 < c_2$:

1. IF $Y \leq c_1$ THEN decision is "Class III."

2. ELSE IF $c_1 < Y \leq c_2$ THEN decision is "Class II."

3. ELSE decision is "Class I."

The correct classification probability for the three classes are

$$d_1 = P(Y_1 > c_2) = 1 - G_1(c_2),$$

$$d_2 = P(c_1 < Y_2 \leq c_2) = G_2(c_2) - G_2(c_1)$$

and

$$d_3 = P(Y_3 < c_1) = G_3(c_1),$$

respectively. We plot (d_1, d_2, d_3) for all possible $(c_1, c_2) \in \mathbb{R}^2$ and obtain an ROC surface (e.g., Figure 5.8) in the three-dimensional space. We note that when the disease status is binary, we only need one threshold and evaluate a pair of correct classification probabilities which are equivalent to sensitivity and specificity, respectively.

By writing the correct classification probability for the intermediate class d_2 as a function of d_1 and d_3, the equation for defining an ROC surface for the test may be given as

$$ROC(u,v) = \begin{cases} G_2(G_1^{-1}(1-u)) - G_2(G_3^{-1}(v)) & \text{if } G_3^{-1}(v) \leq G_1^{-1}(1-u), \\ 0 & \text{otherwise.} \end{cases}$$

$$(5.45)$$

Thus defined, the map $(u,v) \mapsto ROC(u,v)$ is monotone nonincreasing for both arguments. Just as the ROC curve for binary diagnostics represents the trade-off between sensitivity and specificity which are the correct classification probability for the two classes (diseased and healthy), the ROC surface represents the three-way trade-off among the correct classification probabilities for the three classes.

The volume under the ROC surface (VUS) is usually used to characterize the overall accuracy of the test (Mossman, 1999). It is defined as the following double integral

$$\text{VUS} = \int_0^1 \int_0^1 ROC(u, v) \, d\, u\, d\, v. \tag{5.46}$$

VUS is mathematically equivalent to the probability $P(Y_1 > Y_2 > Y_3)$. This immediately suggests one empirical method to estimate VUS nonparametrically with a three-sample U-statistic:

$$\widehat{VUS} = n_1^{-1} n_2^{-1} n_3^{-1} \sum_{i_1=1}^{n_1} \sum_{i_2=1}^{n_2} \sum_{i_3=1}^{n_3} I(Y_{1i_1} > Y_{2i_2} > Y_{3i_3}). \tag{5.47}$$

Statistical properties of this estimator have been discussed in Nakas and Yiannoutsos (2004) and Li and Zhou (2009). One can construct confidence intervals for VUS based on the asymptotic approximation methods described therein.

The italic sentence in the third paragraph above may be questionable for a practical data set. At the beginning of a study it is necessary to justify this statement by finding the correct order. In fact, Scurfield (1996) introduced six VUS measures which correspond to six different triple-comparison probabilities $P(Y_{p1} > Y_{p2} > Y_{p3})$ where (p_1, p_2, p_3) is a permutation of $(1, 2, 3)$. Among the six VUSs, only the largest one is a reasonable measure of the accuracy of the test. The estimator (5.47) is only sensible if $P(Y_1 > Y_2 > Y_3)$ is the largest among the six versions. Otherwise, the estimation method must be adjusted to preserve the correct order specified in the definition of the largest VUS.

If we intend to find the order of three classes and obtain the correct VUS, we need to compute $3! = 6$ VUSs for comparison. In M-category problem $(M > 3)$, the concept of ROC surface extends to ROC manifold in a high dimensional space (Li and Fine, 2008). The hypervolume under the ROC manifold (HUM) becomes a meaningful measure similar to AUC and VUS. The estimator of HUM can be computed as an M sample U-statistics, similar to (5.47), after the order of the M classes are determined. For a general M-category problem, we need to evaluate $M!$ such HUM measures to identify the largest HUM. From our experiences, the number of categories in a biomedical study rarely exceeds six, and the extensive search for a correct HUM definition is applicable in practice.

When the order of the categories is known, Nakas and Yiannoutsos (2004) and Li and Zhou (2009) derived the asymptotic distribution of the estimator (5.47) and provided the basis for statistical inference. The former reference used U-statistics theory, while the latter used the Brownian bridge technique. We note that even if we do not know the order of the categories in advance and only sort the categories after an exhaustive computation, we may still carry out inferences based on the same asymptotic results. In fact, denote

$$\hat{h}_\tau = (n_1 n_2 \cdots n_M)^{-1} \sum_{i_1, i_2, \cdots, i_M} I\{Y_{t_1, i_1} < Y_{t_2, i_2} < \cdots < Y_{t_M, i_M}\} \quad (5.48)$$

where $\tau = (t_1, \cdots, t_M)$ represents a permutation of $(1, 2, \cdots, M)$. Let $\mathcal{T}$ be the collection of all possible τ. We note that the index set $\mathcal{T}$ is finite with a total of $M!$ elements. Then, the sensible HUM estimator we report in practice is $\hat{H} = \max_{\tau \in \mathcal{T}} \hat{h}_\tau$. For each $\hat{h}_\tau$, previous authors have established (Nakas and Yiannoutsos, 2004; Li and Fine, 2008) asymptotically $\hat{h}_\tau \sim N(h_\tau, \sigma_\tau)$. Notice that $\max_{\tau \in \mathcal{T}} \hat{h}_\tau = \lim_{\lambda \to \infty} (\sum_{\tau \in \mathcal{T}} \hat{h}_\tau^\lambda)^{1/\lambda}$. We then define $a(\hat{h}_\tau, \lambda) = (\sum_{\tau \in \mathcal{T}} \hat{h}_\tau^\lambda)^{1/\lambda}$. For any fixed λ, we can show that $a(\hat{h}_\tau, \lambda)$ is asymptotically normal with the delta method since this is a smoothing mapping of $\hat{h}_\tau$. Noting that for any positive λ the function $a(\hat{h}_\tau, \lambda)$ is bounded between $(0, 1)$, we then let $\lambda \to \infty$ and obtain the asymptotic distribution of $\hat{H}$. The distribution of $\hat{H}$ derived in this way turns out to be identical to $\hat{h}_{\tau_0}$ where τ_0 corresponds to the particular known order that gives the maximum of $\hat{h}_\tau$. Subsequently statistical inference follows.

Finally, we present a more general framework of ROC analysis for unordered multicategory classification. This requires us to consider classification rules different from R_I given above.

Consider M subjects, each from one of the M classes, with probability ratings $p^{(1)}, p^{(2)}, \cdots, p^{(M)}$, respectively. Each probability rating $p^{(m)}$ is a vector $(E_{m1}, E_{m2}, \cdots, E_{mM})$, where $E_{m1}, E_{m2}, \cdots, E_{mM} > 0$ and $E_{m1} + E_{m2} + \cdots + E_{mM} = 1$. Each component E_{mk} in the vector indicates the likelihood that the mth subject is from the kth category. In practice, the probability assessment vectors can be indirectly derived from the continuous diagnostic test. The simplest way to generate such vectors is to fit a multinomial logistic regression model by using the multiple category indicator variable as the response and using the diagnostic test as the predictor and evaluate the model-based prediction on the probability scale. In R, fitting multinomial logistic regression can be realized by adopting the library **nnet**. Alternatively, one can choose multicategory support vector machine techniques (Lee et al., 2004) or classification trees (Brieman et al., 1984), among many other classifiers documented in the literature (Dudoit et al., 2002).

Let v_m ($m = 1, 2, \cdots, M$) be an M-dimensional vector whose elements are all 0 except that the mth element equals 1. Now we consider the following classification rule (R_{II}) based on the probability assessment vectors of M subjects: assign subjects to class $k_1, k_2, \cdots, k_M$ such that

$$\|p^{(1)} - v_{k_1}\|^2 + \|p^{(2)} - v_{k_2}\|^2 + \cdots + \|p^{(M)} - v_{k_M}\|^2$$

is minimized among all possible assignments $k_1 \neq \ldots \neq k_M$, where $\|\cdot\|$ is the Euclidean distance.

Let $CR(p^{(1)}, p^{(2)}, \cdots, p^{(M)})$ be 1 if all M subjects are classified correctly, and 0 otherwise. The probability $\Pr\{CR(p^{(1)}, p^{(2)}, \cdots, p^{(M)}) = 1\}$ is the HUM according to its probabilistic interpretation. Let p_{mj}, $j = 1, \ldots, n_i$, be the probability assessment vectors for the n_i subjects from class $m = 1, \ldots, M$. The estimator of HUM is given by

$$\frac{1}{\prod_{m=1}^{M} n_m} \sum_{k_1=1}^{n_1} \sum_{k_2=1}^{n_2} \cdots \sum_{k_M=1}^{n_M} CR(p_{1k_1}, p_{2k_2}, \cdots, p_{Mk_M}).$$

Following U-statistic theories, we can show that this quantity is asymptotically normal and its variance can be estimated consistently. For $M > 3$, the variance estimator is complicated, involving summations of order $(\prod n_i)^2$. We recommend using the bootstrap for inference.

The calculation of HUM for unordered multicategory outcomes with $M = 3$ and $M = 4$ has been implemented in an R program, and the code is freely downloadable from the following Web site:

$$http://www.stat.nus.edu.sg/\sim stalj$$

Users may prepare data in the right format and paste the code in R to obtain the HUM values.

Remarks. There are still many open research questions for HUM and the ROC surface. For example, it is widely accepted that the ROC curve should rise above the diagonal line of the unit square and thus produce an AUC greater than $1/2$. The lower bound for an M class HUM is actually $1/M!$, corresponding to a silly classifier sorting M categories by random guess. The interpretation of an HUM value is more difficult than AUC in applications. For two-class problems, an AUC value lower than 75% usually indicates that the test has very low discrimination ability. However, a meaningful benchmark threshold for HUM in multiclass problems has yet arrived in the literature.

5.1.2.4 Applications in Genetics

Since 1990s, AUC and other diagnostic accuracy measures have been effectively used in genetic studies. Investigators usually aim at finding genes with differential ability, that is, being able to discriminate diseased subjects from healthy subjects. All genes under examination may preserve some classification power, and their accuracy can be reflected by the corresponding AUC values. A natural strategy to identify the most differentially expressed genes is to rank all genes using their AUC values. In fact, using the two-sample Mann-Whitney-Wilcoxon rank sum statistic (which is equivalent to AUC) to compare the diagnostic performance of different genes has been widely accepted in clinical practice. Pepe et al. (2003) proposed to use PAUC to rank genes when the comparison of discrimination accuracy must be conditional on having acceptable specificity levels.

We now consider two examples from genetic studies. The first example illustrates weighted ROC analysis while the second example illustrates mutli-category ROC analysis.

Example (Lymphoma). We conduct an ROC analysis using a well-known publicly available lymphoma dataset (Shipp et al., 2002). The data set contains 77 samples, 58 of which came from diffuse large B-cell lymphoma (DLBCL) patients and 19 follicular lymphoma from a related germinal center B-cell lymphoma. The gene expression data have been obtained using Affymetrix human 6800 oligonucleotide arrays.

The full data set has been downloaded from the Broad Institute Web site:

$$http://www.genome.wi.mit.edu/MPR/lymphoma$$

We compute the WAUC values under three weights for all 7129 genes and rank them. The three weights are U(0,1), U(0.9,1), and Beta(8,2). U(0,1) gives the ordinary AUC. U(0.9,1) leads to a PAUC index over a range of high specificity values. Beta(8,2) gives a negatively skewed weight for WAUC with an emphasis on high specificity values. The results for the top 10 genes under different weights are summarized in Table 5.2. The pairwise correlation coefficients between the estimated WAUCs are 0.044 (U(0,1) and U(0.9,1)), 0.217 (U(0,1) and Beta(8,2)), and 0.974 (U(0.9,1) and Beta(8,2)), respectively. WAUCs under the U(0.9,1) and Beta(8,2) weights are thus closely correlated and yield similar rankings of genes for their ability to classify DLBCL while retaining relatively high specificity for follicular lymphoma.

The results for WAUC under the U(0.9,1) weight (or PAUC index) indicate that a total of 409 genes have perfect WAUC values. The second highest WAUC value is .992 which is achieved by the next 280 genes. In this case, using such a WAUC produces an extremely large number of ties when ranking genes.

TABLE 5.2: Gene Number and Estimated Weighted Area under the ROC Curve under the Three Different Weights for the Top Genes from the Lymphoma Data Set (The standard errors for WAUC are given in parentheses.)

	U(0,1)			U(.9,1)			Beta(8,2)	
Rank	Gene	WAUC (SE)	Rank	Gene	WAUC (SE)	Rank	Gene	WAUC (SE)
1	506	.964 (.017)	1	23	1 (0)	1	2064	.988 (.008)
2	972	.942 (.031)	...	...	...	2	87	.987 (.017)
3	2137	.936 (.035)	1	7113	1 (0)	3	1984	.986 (.011)
4	4028	.927 (.033)	410	80	.992 (.019)	4	4194	.985 (.017)
5	605	.926 (.042)	...	...	...	5	6325	.984 (.018)
6	2789	.921 (.038)	410	7125	.992 (.019)	6	5124	.983 (.013)
7	6179	.920 (.041)	690	122	.984 (0.021)	7	3409	.982 (.023)
8	4292	.919 (.041)	...	...	...	8	2006	.982 (.013)
9	6815	.918 (.039)	...	...	...	9	5998	.981 (.013)
10	4372	.915 (.038)	...	...	...	10	5867	.981 (.027)

It is difficult to judge the relative accuracy for genes with the same WAUC values.

Eyeballing the empirical distribution (Figure 5.5) of gene 23 and 53 (obtained by smoothing the histogram), both of which have perfect WAUC under the U(0.9,1) weight, we notice that there is no DLBCL subject whose values

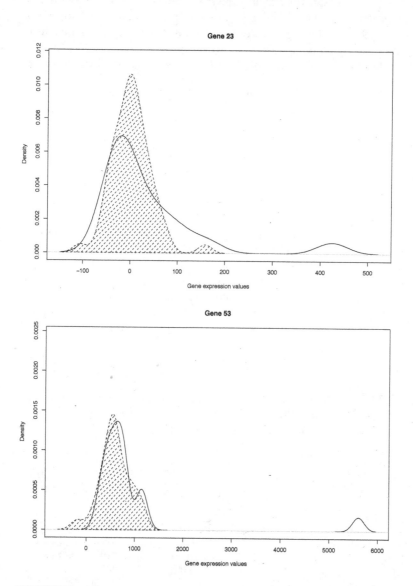

FIGURE 5.5: Distribution of gene expressions for the diseased and normal populations from the DLBCL data set. The area under the density for the DLBCL patients is shaded.

fall into the upper 10th percentile of the follicular lymphoma group. Therefore, none from the DLBCL group is mislabeled under such a truncated region. However, the overall classification performance of these genes is not great. In most of the domain, the two distributions overlap, indicating very weak differential power for the gene. The overall AUCs (or WAUC under the uniform weight on $(0,1)$) for these two genes are 0.548 and 0.590, respectively, only slightly better than $W_0 = 0.5$.

It may be better to look at the WAUC under the beta weight. For example, if we care about the performance of genes under the high specificity requirement, we may use Beta(8,2). Using such a criterion, we may have a more reasonable comparison among all the genes. We notice that WAUC under this weight ranks genes 23 and 53 at 520 and 1485 with corresponding values 0.910 and 0.865, respectively.

The distribution of the top two genes under weight Beta(8,2) are shown in Figure 5.6. The overall AUC for these two genes are 0.889 and 0.758, respectively. These two also have WAUC to be 1 under $U(0.9,1)$. Clearly, not all genes with high ranks under WAUC with the $U(0.9,1)$ weight deserve the same amount of attention. WAUC with $U(0.9,1)$ may mislead us to believe too many genes to have the same accuracy while in fact they do not.

If our purpose is to select genes according to their overall performance on the general population, AUC may be good to provide rankings. However, in a genetic screening study, having people without disease undergo unnecessary work-up procedures may produce inconvenience and stress since the procedure may be costly and invasive. Because of such a concern, it is better to focus the ROC curve over the high specificity range. In fact, genes with the top three AUC values (No. 506, 972, and 2137) only have WAUC 0.781, 0.786, and 0.789 under the $U(.9,1)$ weight. These genes may not be the best candidates to screen out diseased subjects when high specificity is desired.

The sorted WAUC values for the 7129 genes under the three different weights are shown in Figure 5.7. While the WAUC values under $U(0,1)$ and Beta(8,2) are decreasing strictly, those under $U(0.9,1)$ decrease as a step function. Thus, while the WAUC under $U(0.9,1)$ may be successful in some cases to rank genes for the aforementioned goal, it can potentially create a large number of ties due to the truncated nature of the estimator. It is more flexible to consider weights like Beta(8,2) when we want to focus on high specificity regions of the ROC curve and still give strictly ordered rankings for all genes.

The choice of Beta$(8,2)$ entails that the mean of the specificity distribution is around $8/(8+2) = 0.8$. If an investigator has an even stronger opinion on putting a high emphasis on large specificity values, she or he may consider a larger α parameter with β fixed at 2. For example, one may consider Beta$(18,2)$ which places the mean of the specificity distribution to be $18/(18+2) = 0.9$. This will result in slightly different rankings from what we report for Beta$(8,2)$. Six among the original top ten genes, namely 2064, 1984, 6325, 5124, 2006, and 5867, are still among the top ten genes under Beta$(18,2)$. Gene 87 is ranked at 22, but has a very high WAUC value (0.997).

Furthermore, the correlation of rankings from these two beta distributions is 0.986, indicating a close agreement on rankings when we shift the mean of beta distribution from 0.8 to 0.9. As we advised earlier, it is important to interpret the results based on what weight function is chosen. When different

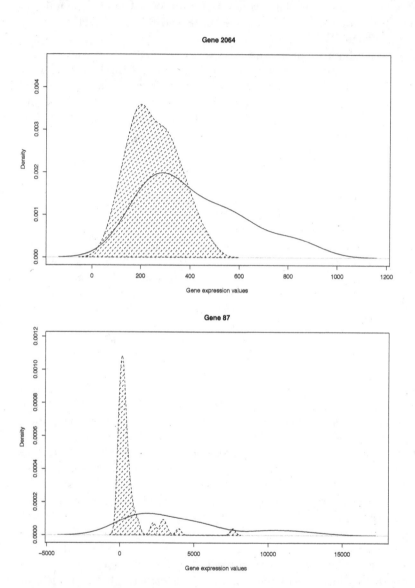

FIGURE 5.6: Distribution of gene expressions for the diseased and normal populations from the DLBCL data set. The area under the density for the DLBCL patients is shaded.

expert opinions have to be considered jointly, a fair conclusion may be drawn from taking an average of different votes.

In addition to the AUC measures reported in Table 5.2, we may rank genes by using p-values from paired two-sample t-tests or Wilcoxon rank sum tests. Essentially a significant p-value from a test suggests a location (mean) difference between the diseased and healthy groups. A T-test heavily depends on the data normality assumption, and thus the nonparametric Wilcoxon rank sum test is usually preferred. As noted earlier, the test statistic for the Wilcoxon test is mathematically equivalent to the unweighted AUC.

Example (Proteomic Study). We next look at a mass spectrometry data set for the detection of Glycan biomarkers for liver cancer (Ressom et al., 2008). The investigators included 203 participants from Cairo, Egypt, among whom there were 73 hepatocellular carcinoma cases (denoted by HC), 52 patients with chronic liver disease (denoted by QC), and 78 healthy individuals (denoted by NC). In the following, we denote the three categories as: 1 for HC, 2 for NC, and 3 for QC. The spectra were generated by matrix-assisted laser desorption/ionization time-of-flight (MALDI-TOF) mass analyzer (Applied Biosystems Inc., Frammingham, Massachusetts). Each spectrum consisted of approximately 121,000 m/z values with the corresponding intensities in the mass range of 1,500-5,500 Da. The data set analyzed here has been downloaded from:

http : //microarray.georgetown.edu/ressomlab/index_downloads.html

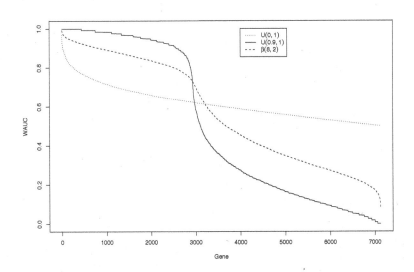

FIGURE 5.7: Sorted WAUC values for the 7129 genes in the lymphoma data.

and contains a total of 484 peaks after extensive preprocessing of the raw data (Ressom et al., 2007). Here, the term "peaks" refers to intensity measurements of protein segments or peptides. In the literature, researchers have used the terms "peaks," "biomarkers," "tests," and "intensities" interchangeably. Note that "peaks" and "intensities" may have broader biological meanings. In this book and many statistical research papers, researchers have specifically focused on their numerical attributes and used the magnitude of peaks or intensities to predict disease status. They thus serve the same function as a traditional diagnostic test.

In this study, the disease status is a categorial variable taking three distinct values. The most appropriate accuracy summary measures should be three-way ROC surface and VUS. By using the methods described in the preceding section, we have computed the VUS exhaustively for six versions of probability definitions and identified the largest value to be the correct VUS for each marker. The VUSs of 112 peaks are greater than 0.4, indicating that they are able to correctly sort three subjects, each from one of the three categories in four out of ten runs.

TABLE 5.3: **Top 20 Peaks Ranked by the VUS Values for the Liver Cancer Data** (μ_i and e_i are the means and relative effects for the ith class ($i = 1, 2, 3$). Classes 1, 2, and 3 are HC, NC, and QT, respectively. (SE) is the standard error of the estimated VUS.)

Rank	VUS (SE)	Definition	$\hat{\mu}_1$	$\hat{\mu}_2$	$\hat{\mu}_3$
1	0.647(0.0406)	$P(X_1 < X_2 < X_3)$	896.611	1326.071	2732.444
2	0.632(0.0423)	$P(X_1 < X_2 < X_3)$	651.121	985.067	1372.388
3	0.623(0.0430)	$P(X_1 < X_2 < X_3)$	1452.321	2010.886	4829.766
4	0.584(0.0427)	$P(X_1 < X_2 < X_3)$	124.784	286.132	412.497
5	0.563(0.0435)	$P(X_1 < X_2 < X_3)$	481.267	697.342	988.530
6	0.558(0.0413)	$P(X_1 < X_2 < X_3)$	544.353	748.769	1122.159
7	0.533(0.0417)	$P(X_1 < X_2 < X_3)$	314.048	401.839	607.952
8	0.529(0.0409)	$P(X_1 < X_2 < X_3)$	150.320	366.533	553.001
9	0.524(0.0428)	$P(X_1 < X_2 < X_3)$	10552.81	21490.65	26878.118
10	0.513(0.0409)	$P(X_1 < X_2 < X_3)$	413.769	526.830	772.249
11	0.509(0.0417)	$P(X_1 < X_2 < X_3)$	1014.861	1245.426	1928.783
12	0.504(0.0424)	$P(X_1 < X_2 < X_3)$	785.593	854.669	1577.023
13	0.503(0.0415)	$P(X_1 < X_2 < X_3)$	229.566	285.739	428.804
14	0.502(0.0421)	$P(X_1 < X_2 < X_3)$	171.408	226.698	322.948
15	0.501(0.0425)	$P(X_1 < X_2 < X_3)$	170.734	281.202	364.334
16	0.499(0.0415)	$P(X_1 < X_2 < X_3)$	85.506	126.392	189.421
17	0.498(0.0408)	$P(X_2 < X_3 < X_1)$	567.211	198.593	227.376
18	0.496(0.0418)	$P(X_1 < X_2 < X_3)$	114.326	170.651	363.178
19	0.495(0.0431)	$P(X_1 < X_2 < X_3)$	660.858	949.397	1331.265
20	0.491(0.0428)	$P(X_1 < X_2 < X_3)$	333.676	425.174	741.058

We report the results for 20 peaks with the highest VUS values in Table 5.3 along with standard errors computed using a nonparametric bootstrap method. Different peaks seem to maintain the same ordering definition except for the 17th peak. For most peaks, healthy subjects (NC) tend to have an intermediate value. A large value tends to lead to chronic liver disease (QT), while a low value leads to hepatocellular carcinoma (HC). The 17th peak behaves differently from the other peaks where HC patients tend to have greater peak values than the other two groups. Identification of such order information may bring more insights to mass spectrometry studies. In all these 20 cases, it is interesting to note that the orders of three categories for VUS definitions are correctly detected from the orders of sample means of the three categories. In fact, it can be shown that when the distribution is normal, the order of the means of the three categories can be correctly prescribed by the order of the three classes in the definition of VUS (Zhang, 2010b). Alternatively, the unordered ROC analysis by using the method in Li and Fine (2008) yields the same VUS values for the peaks and ranking results as in Table 5.3.

The ROC surface for the peak with the largest VUS is plotted in Figure 5.8. One can choose the appropriate cut-off values c_1 and c_2 for a particular decision to satisfy the required correct classification probabilities (d_1, d_2, d_3) by locating the corresponding values on this operating surface. The distributions of this peak for the three classes are shown in Figure 5.9. The overall shapes of the three empirical density curves are quite close to normal distributions.

When the number of categories goes beyond three, it may be hard to visually display the ROC manifold. Nevertheless the calculation of HUM may still be carried out in the same manner as in the above example. We may also note that the VUS values for individual markers tend to be small in many applications. In the above protemoic study, the maximum VUS is only roughly 65%. It is often advised in similar bioinformatics studies to combine more than one marker to form a stronger "composite marker" with more satisfactory accuracy (McIntosh and Pepe, 2002; Li and Fine, 2008; Zhang and Li, 2011).

When there are thousands of markers as in typical genetic studies, it is often observed that the AUC value for an individual marker is quite low. Even if we choose the marker with the highest AUC, its differential ability can still be unsatisfactorily low (e.g., $< 80\%$). Since we have abundant information from enormous markers, we may consider combining multiple markers to form a more accurate composite marker.

Specifically, denote Y_k to be the kth marker, $k = 1, 2, \cdots, K$, where K is the total number of markers. Write $\mathbf{Y} = (Y_1, \cdots, Y_K)^T$. We aim at finding a K-vector $\boldsymbol{\beta} = (\beta_1, \cdots, \beta_K)^T$ so that a linear combination $Y = \sum_{k=1}^{K} \beta_k Y_k = \mathbf{Y}^T \boldsymbol{\beta}$ becomes a marker with an improved accuracy.

There are many methods to estimate the coefficient vector $\boldsymbol{\beta}$. The first method is based on fitting a logistic regression model (or another parametric generalized linear model)

$$\log \frac{P(S=1)}{P(S=0)} = \beta_0 + \mathbf{Y}^T \boldsymbol{\beta} \tag{5.49}$$

by using the observed disease status data S and then outputting the estimated regression coefficient $\hat{\boldsymbol{\beta}}$, resulting from maximum likelihood estimation. McIntosh and Pepe (2002) demonstrate certain optimality of this approach. Since logistic regression analysis is available in various software, many authors have adopted this simple method in applications.

The second method is to directly solve the following maximization problem:

$$\operatorname{argmax}_{\boldsymbol{\beta}} \sum_{i=1}^{n_0} \sum_{j=1}^{n_1} \mathsf{J}(\mathbf{Y}_j^T \boldsymbol{\beta} - \mathbf{Y}_i^T \boldsymbol{\beta}), \tag{5.50}$$

where the subscripts i and j index disease-absent and disease-present membership, respectively, and $\mathsf{J}(x) = 0$ when $x \leq 0$, and $\mathsf{J}(x) = 1$ when $x > 0$. The resulting composite marker $\mathbf{Y}$ can ensure a maximum empirical AUC value among all possible linear combinations. The estimated coefficients have been referred to as the maximum rank correlation estimator (Han, 1987; Sherman, 1993; Cavanagh and Sherman, 1998; Abrevaya, 1999; Pepe and Thompson, 2000). An iterative marginal algorithm has been proposed by Wang (2006) to

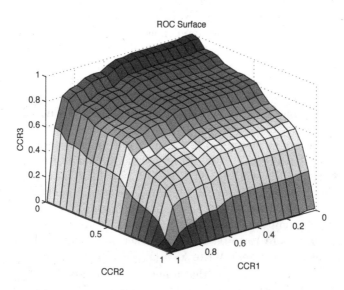

FIGURE 5.8: ROC surface for the peak with the largest VUS. The three coordinates are the correct classification probabilities for the three classes (d_1, d_2, d_3) defined in Section 5.1.2.3.

improve the computation speed. The computation for this approach is indeed nontrivial for some data sets since the objective function has a nondifferentiable mass.

The third method is to approximate the nondifferentiable function $J(x)$ with a differentiable one $S(x)$. We can then consider solving the following maximization problem

$$\text{argmax}_{\boldsymbol{\beta}} \sum_{i=1}^{n_0} \sum_{j=1}^{n_1} S(\mathbf{Y}_j^T \boldsymbol{\beta} - \mathbf{Y}_i^T \boldsymbol{\beta}). \tag{5.51}$$

In Ma and Huang (2005), it is proposed that $S(x) = 1/(1 + \exp(-x))$. This smooth function approximates $J(x)$ closely, and the resulting coefficient estimator $\hat{\boldsymbol{\beta}}$ can thus achieve similar optimality as those obtained in the second method. This method is appealing for practitioners since derivative-based optimization algorithms can be implemented. Theoretical results are obtained in Ma and Huang (2007).

The first two methods can be easily extended for multicategory classifi-

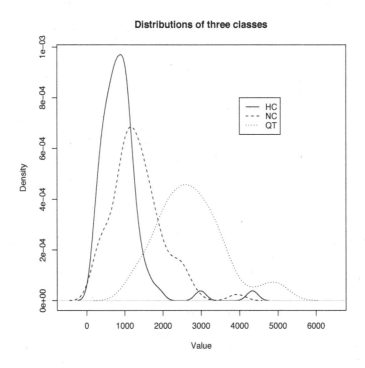

FIGURE 5.9: The distributions of the peak with the largest VUS for the three groups.

cation problems as well. For example, Li and Fine (2008) consider the first method, while Zhang and Li (2011) consider the second method.

We note that when the multivariate normality assumption can be imposed on $\mathbf{Y}$, an optimal linear combination can be easily constructed (Su and Liu, 1993). Specifically, if $\mathbf{Y}_i$ i.i.d. $\sim N(\boldsymbol{\mu}_0, \Sigma)$ $(i = 1, \cdots, n_0)$ and $\mathbf{Y}_j$ i.i.d. $\sim N(\boldsymbol{\mu}_1, \Sigma)$ $(j = 1, \cdots, n_1)$, the best linear combination for maximizing the binormal AUC is $\boldsymbol{\beta} = \Sigma^{-1}(\boldsymbol{\mu}_1 - \boldsymbol{\mu}_0)$. Sample means and covariance matrices can be plugged in for estimation. In practice, such a rigid distribution assumption is hardly valid for genetic data, and we seldom take this parametric approach.

Other than combining multiple markers, the aforementioned methods may also be used to adjust for confounding variables such as gender and age in epidemiology studies (Enrique et al., 2004; Yu et al., 2012). The adjustment method is to regard the confounders as biomarkers as well and use the above procedures.

Finally, we want to stress that gene selection is not a task entirely equivalent to model selection in machine learning. In addition to numeric consideration from its expression value, other biological and medical evidences are needed to support the claimed disease differential function for a particular gene. It is also necessary to conduct meta-analysis to reduce sampling variability and firmly establish a reproducible genetic effect.

More discussion on ROC analysis and its applications in genetics will be covered in the next chapter.

5.1.3 Accuracy Improvement Analysis

In this section we consider the problem of assessing the improvement of diagnostic accuracy with multicategory classification. It is quite common that we use multiple biomarkers to predict the outcome. These biomarkers are usually placed in a well-structured model, such as the logistic regression model, and from the model we may obtain the prediction based on an evaluated probability vector for each category. The overall model accuracy is pertinent to how often the probability assessment leads to the correct membership for the sample and can be accounted by simple correct proportions or ROC-based measures. Sometimes when adding a new biomarker into the model, investigators may question how much accuracy gain it can bring. When the assessed gain is sufficient, the new marker may be warranted to enter the model. Otherwise, it is unnecessary to enroll this marker into the existing model.

5.1.3.1 Accuracy Improvement Parameters

Improvement of accuracy can be described as a difference (or occasionally ratio) of accuracy measures between two models. All the accuracy measures introduced in the previous sections can be considered. The most relevant measure may be AUC (or its multidimensional counterpart HUM) since the model-based predictions are continuous quantities. However, it is noticed by many epidemiologists that the increase of AUC values is not sensitive enough to detect a meaningful accuracy increment, especially when the two AUCs are both very large. Two metrics have been introduced to complement AUC (Pencina et al., 2008, 2011). To make our scope more general, we present these two metrics for multidimensional classification. The results reduce to the common two-category problem easily.

Consider a set of predictors $\Omega = \{Y_1, \cdots, Y_K\}$. Suppose that we have a sample of subjects with measurements of all Y_k ($k = 1, \cdots, K$). The goal is to use them to construct a meaningful statistical model for predicting the multicategory outcome S which takes discrete values from $\mathcal{S} = \{1, 2, \cdots, M\}$. We define the binary random variable $S_m = I(S = m)$ and let the prevalence for the mth category be $\rho_m = E(S_m) = P(S = m)$.

Suppose that a model $\mathcal{M}_1$ is constructed based on a set of predictors $\Omega_1 \subset \Omega$. The model $\mathcal{M}_1$ can generate a probability vector $\mathbf{p}(\mathcal{M}_1) = (p_1(\mathcal{M}_1), \cdots, p_M(\mathcal{M}_1))$ for each subject, where $\sum_{m=1}^{M} p_m(\mathcal{M}_1) = 1$. Decision makers may assign a subject to one of the M categories according to the greatest component in the probability vector. One may quantify the accuracy of $\mathcal{M}_1$ based on Ω_1 by the following multicategory correct classification probability (CCP):

$$CCP = \sum_{m=1}^{M} \rho_m CCP_m, \tag{5.52}$$

where each CCP_m is the correct classification probability for the mth-category. For the model $\mathcal{M}_1$, we write

$$CCP_m(\mathcal{M}_1) = P\{p_m(\mathcal{M}_1) = \max \mathbf{p}(\mathcal{M}_1) | S = m\}. \tag{5.53}$$

Now suppose that more variable(s) are included and we construct a model $\mathcal{M}_2$ which is based on a set of predictors $\Omega_2 \supset \Omega_1$. The newly constructed model $\mathcal{M}_2$ generates another probability vector $\mathbf{p}(\mathcal{M}_2) = (p_1(\mathcal{M}_2), \cdots, p_M(\mathcal{M}_2))$ for each subject, where $\sum_{m=1}^{M} p_m(\mathcal{M}_2) = 1$. Again, decision makers may assign the subject according to the greatest value of this probability vector, and the mth-category accuracy of $\mathcal{M}_2$ based on Ω_2 can be quantified by

$$CCP_m(\mathcal{M}_2) = P\{p_m(\mathcal{M}_2) = \max \mathbf{p}(\mathcal{M}_2) | S = m\}. \tag{5.54}$$

The overall accuracy improvement from $\mathcal{M}_1$ to $\mathcal{M}_2$ may be summarized as

$$\mathsf{T} = \sum_{m=1}^{M} w_m \{CCP_m(\mathcal{M}_2) - CCP_m(\mathcal{M}_1)\}, \tag{5.55}$$

where w_m are positive weights for the mth category. When $M = 2$, the two

CCPs are the sensitivity and specificity of the model-based test. The T measure thus quantifies the overall increase of the weighted sum of sensitivity and specificity. When equal weights are used for the two categories, T is simply the difference of Youden's index between the two models. We call (5.55) the *net reclassification improvement* (NRI). When $M = 2$ and $w_m = 1/2$, NRI is equivalent to the original NRI proposed in Pencina et al. (2008). NRI indicates the probability that added markers in $\mathcal{M}_2$ lead to correct classification of subjects who are incorrectly classified using the smaller model $\mathcal{M}_1$.

We now introduce another accuracy improvement measure based on R^2 for categorical response. The interpretation and computation of R^2, also called a coefficient of determination, has been discussed for binary logistic regression models. Simply speaking, the value of R^2 is the fraction of the total variation explained by the model. For linear regression models, R^2 is closely related to the correlation coefficient and the ANOVA F-test, while for binary regression, it is closely connected to the probabilities of correct classification. Let $\mathbf{R}^2(\mathcal{M}_j) = (R_1^2(\mathcal{M}_j), \cdots, R_M^2(\mathcal{M}_j))$ be an M-dimensional vector with

$$R_m^2(\mathcal{M}_j) \;=\; \frac{\mathrm{var}(S_m) - E\{\mathrm{var}(S_m|\mathcal{M}_j)\}}{\mathrm{var}(S_m)} = \frac{\mathrm{var}\{p_m(\mathcal{M}_j)\}}{\rho_m(1-\rho_m)}. \quad (5.56)$$

The second equality follows because $E(Y_m|\mathcal{M}_j) = p_m(\mathcal{M}_j)$. It has been shown in Pepe et al. (2008) that the increase in R^2 for binary classification ($M = 2$) from model $\mathcal{M}_1$ to model $\mathcal{M}_2$ is equivalent to the *integrated discrimination improvement* (IDI) proposed in Pencina et al. (2008). A natural adaptation of the R^2 definition of IDI to the multicategory set-up is

$$\mathsf{R} = \sum_{m=1}^{M} w_m \{R_m^2(\mathcal{M}_2) - R_m^2(\mathcal{M}_1)\}. \quad (5.57)$$

We refer to (5.57) as IDI in this book, similarly to the binary case. The multicategory IDI (5.57) reduces to that in Pencina et al. (2008) when $M = 2$ and equal weights $w_1 = w_2 = 1/2$ are used.

The IDI for two-category classification given in Pencina et al. (2008) is in the following integration form:

$$IDI = \int \{\mathrm{TPF}_{\mathcal{M}_2}(y) - \mathrm{FPF}_{\mathcal{M}_2}(y)\}\, dy - \int \{\mathrm{TPF}_{\mathcal{M}_1}(y) - \mathrm{FPF}_{\mathcal{M}_1}(y)\}\, dy \quad (5.58)$$

where $\mathrm{TPF}_{\mathcal{M}_j}(y)$ and $\mathrm{FPF}_{\mathcal{M}_j}(y)$ ($j = 1, 2$) are the sensitivity and 1 minus specificity defined in (5.25) and (5.26) for a marker based on model $\mathcal{M}_j$ at a threshold y. For example, suppose our modeling device is the widely used logistic regression, and $\mathcal{M}_1$ and $\mathcal{M}_2$ involve $\{Y_1, Y_2\}$ and $\{Y_1, Y_2, Y_3\}$, respectively. The propensity scores from the models can be extracted as

$$p(\mathcal{M}_1) \;=\; 1/(1 + \exp\{-(\beta_{11}Y_1 + \beta_{12}Y_2)\}) \quad (5.59)$$
$$p(\mathcal{M}_2) \;=\; 1/(1 + \exp\{-(\beta_{21}Y_1 + \beta_{22}Y_2 + \beta_{23}Y_3)\}), \quad (5.60)$$

and the IDI can then be expressed as

$$IDI = \int_0^1 \{P(p(\mathcal{M}_2) > y|S = 1) - P(p(\mathcal{M}_2) > y|S = 0)\}\, dy$$

$$- \int_0^1 \{P(p(\mathcal{M}_1) > y|S = 1) - P(p(\mathcal{M}_1) > y|S = 0)\}\, dy. \quad (5.61)$$

Interesting, the measure can also be regarded as the improvement of the integrated Youdon's index.

The choice of weights in the definitions of NRI and IDI may depend on the goal and design of the study. When aiming for the overall test accuracy to differentiate multiple classes, it is natural to weigh all categories equally; on the other hand, as pointed out in Pencina et al. (2011), sometimes it is useful to reward some categories with higher weights when savings associated with correct classification of such categories outweigh other categories. When cost-efficiency information is available, we can incorporate them easily in the inference for weighted NRI and IDI. There are also other practical considerations that invoke unequal weights, and one can run a Bayesian prior elicitation procedure to construct reasonable weights (Li and Fine, 2010).

Remark. The traditional accuracy improvement value based on the difference between AUCs is still widely used. However, simulation studies in Pencina et al. (2012) show that when the baseline old model has an AUC value equal to $0.50, 0.60, 0.70, 0.80$, and 0.90, the corresponding AUC increase for adding an uncorrelated useful marker is $0.214, 0.132, 0.080, 0.045$, and 0.019, respectively. The larger the baseline AUC, the smaller the AUC increment for adding this same marker. On the other hand, IDI remains rather stable at $0.064, 0.066, 0.071, 0.071$, and 0.058, correspondingly, when the prevalence is about 10%. The NRI is even more stable and equals a constant 0.622 for all cases. Such an experiment clearly illustrates the advantage of using NRI and IDI to select additional useful biomarkers regardless of the original accuracy level.

5.1.3.2   Estimation Theory

Suppose that we obtain a sample $\{Y_{i1}, \cdots, Y_{ip}, S_i : i = 1, \cdots, n\}$. We denote the class sample size as $n_m = \sum_{i=1}^n I(S_i = m)$ for the mth category. We assume that $n \to \infty$ and $n_m/n \to \rho_m > 0$ for all m. One may fit the candidate models $\mathcal{M}_1$ and $\mathcal{M}_2$ from the preceding section using one's method of choice. The main requirement is that the method provides estimated probability assessment vectors for each model. Using the fitted models, one may then estimate the NRI by

$$\hat{\mathsf{T}} = \sum_{m=1}^M \frac{w_m}{n_m} \sum_{i=1}^n \{I(\hat{p}_{mi}(\mathcal{M}_2) = \max \hat{\mathbf{p}}_i(\mathcal{M}_2), Y_i = m)$$

$$- I(\hat{p}_{mi}(\mathcal{M}_1) = \max \hat{\mathbf{p}}_i(\mathcal{M}_1), Y_i = m)\}, \quad (5.62)$$

where $\hat{\mathbf{p}}_i(\mathcal{M}_j) = (\hat{p}_{1i}(\mathcal{M}_j), \hat{p}_{2i}(\mathcal{M}_j), \cdots, \hat{p}_{Mi}(\mathcal{M}_j))$ is the estimated membership probability for the ith subject based on the jth model. In practice, they can be obtained from fitting a multinomial logistic regression model to the data and then outputting the predicted probabilities from the fitted model. We note that when the models are consistently estimated, $\hat{\mathbf{p}}_i(\mathcal{M}_j)$ is consistent to $\mathbf{p}_i(\mathcal{M}_j)$ and therefore by the law of large numbers $\hat{\mathsf{T}}$ is consistent to T. Furthermore, by using the central limit theorem, we can show that $\sqrt{n}(\hat{\mathsf{T}} - \mathsf{T}) \to_d N(0, \sigma_T^2)$, where

$$
\sigma_T^2 = \sum_{m=1}^{M} \frac{w_m^2}{\rho_m} (a_m + b_m - 2c_m)
$$

$$
- \sum_{i=1}^{M} \sum_{j=1}^{M} w_i w_j (a_i - b_i)(a_j - b_j), \tag{5.63}
$$

with $a_m = CCP_m(\mathcal{M}_1)$, $b_m = CCP_m(\mathcal{M}_2)$, $c_m = P(p_m(\mathcal{M}_1) = \max \mathbf{p}(\mathcal{M}_1), p_m(\mathcal{M}_2) = \max \mathbf{p}(\mathcal{M}_2)|Y = m)$.

The multicategory IDI can be estimated by using the following formula

$$
\hat{\mathsf{R}} = \sum_{m=1}^{M} \frac{w_m}{n_m(1 - n_m/n)} \sum_{i=1}^{n} \{[\hat{p}_{mi}(\mathcal{M}_2) - \overline{\hat{p}_m(\mathcal{M}_2)}]^2 - [\hat{p}_{mi}(\mathcal{M}_1)
$$

$$
- \overline{\hat{p}_m(\mathcal{M}_1)}]^2\}, \tag{5.64}
$$

where $\overline{\hat{p}_m(\mathcal{M}_j)} = n^{-1} \sum_{i=1}^{n} \hat{p}_{mi}(\mathcal{M}_j)$.

We can also show that (5.64) is consistent to R for a large n by noting that for a large sample, $\hat{p}_{mi}(\mathcal{M}_j)$ is consistent to $p_{mi}(\mathcal{M}_j)$ and the average squared distance to the mean $n^{-1} \sum_{i=1}^{n} \{\hat{p}_{mi}(\mathcal{M}_2) - \overline{\hat{p}_m(\mathcal{M}_2)}\}^2$ is consistent to $\mathrm{var}\{p_m(\mathcal{M}_j)\}$. The consistency then follows from the law of large numbers. As with NRI, one may further show that $\sqrt{n}(\hat{\mathsf{R}} - \mathsf{R}) \to_d N(0, \sigma_R^2)$, where

$$
\sigma_R^2 = \sum_{j=1}^{M} \sum_{k=1}^{M} \frac{w_j w_k}{\rho_j \rho_k (1 - \rho_j)(1 - \rho_k)} \Big\{ E\big[((p_j(\mathcal{M}_1) - \mu_j(\mathcal{M}_1))^2
$$

$$
- (p_j(\mathcal{M}_2) - \mu_j(\mathcal{M}_2))^2)((p_k(\mathcal{M}_1) - \mu_k(\mathcal{M}_1))^2 - (p_k(\mathcal{M}_2) - \mu_k(\mathcal{M}_2))^2)\big]
$$

$$
- E\big[(p_j(\mathcal{M}_1) - \mu_j(\mathcal{M}_1))^2 - (p_j(\mathcal{M}_2) - \mu_j(\mathcal{M}_2))^2\big]
$$

$$
\times E\big[(p_k(\mathcal{M}_1) - \mu_k(\mathcal{M}_1))^2 - (p_k(\mathcal{M}_2) - \mu_k(\mathcal{M}_2))^2\big] \Big\}, \tag{5.65}
$$

and $\mu_m(\mathcal{M}_j) = E(p_m(\mathcal{M}_j))$. All the moments involved in the variance expression can be readily estimated by using empirical moment estimators. The variance can then be estimated by the plug-in method.

The above parameter estimation and variance estimation formula are implemented in R by Binyan Jiang, and the code is downloadable at: $http://www.stat.nus.edu.sg/ \sim stalj$. Though the variance formula (5.63) and (5.65) look complicated in the above presentation, our experiences with simulation and real data analysis suggest that they can be evaluated instantly following the point estimation by using our code. These formulas allow inferences to be carried out much faster than a resampling-based approach which is also asymptotically valid. An advantage of the resampling method is that the sampling variability in estimation of the probability vector may be formally accounted for in inference.

5.2 Diagnostics for Survival Outcome under Diverse Censoring Patterns

We may extend the diagnostic accuracy measures for dichotomous and polychotomous outcomes to continuous outcomes, in particular survival time outcomes. At each time point, survival data is equivalent to binary data with status being alive or dead. Thus, the ROC approaches can be applied at each time point. The accuracy measures for binary classification can be applied, and an overall accuracy measurement can be obtained by integrating the AUC over time. Of note, the time-dependent ROC analysis is much more complicated than an ordinary ROC analysis because of censoring and the time-dependent nature of the study cohort. Censoring causes the status of censored subjects not well defined at certain time points. Thus, at a fixed time point, not all observations are equally "informative." In addition, unlike with cross-sectional studies and categorical outcomes, the study cohort varies as time passes. Thus, the reliability of AUC may change over time. Instead of a simple integration, a weighted integration with time- and data-dependent weights may be needed.

Let $T \in \mathcal{T}$ be the time to a definitive event where $\mathcal{T} \subset \mathbb{R}^+$. The event can be the onset of a disease or the occurrence of death, and T is referred to as the survival time or failure time. Denote the continuous marker to be $Y \in \mathcal{Y} \subset \mathbb{R}$. Assume that T and Y are random variables defined on proper measurable spaces. Their distributions can be continuous, discrete, or a mixture of continuous and discrete components.

We now introduce time-dependent accuracy measures with similar definitions as those given in the preceding section. Out of respect to the convention in diagnostic medicine, assume that a high value of Y leads to a positive diagnosis. For a cut-off y, the true positivity fraction (TPF) and false positive fraction (FPF) are both functions of time and defined as

$$\text{TPF}_t(y) = P\{Y > y | T \leq t\}, \quad \text{FPF}_t(y) = P\{Y > y | T > t\}.$$

The event $T \leq t$ indicates that the failure outcome happens before t and corresponds to a disease-present status while $T < t$ corresponds to a disease-free status. $Y \geq y$ indicates that a positive diagnosis has been made, and the subject is declared to be diseased, while $Y < y$ indicates a negative diagnosis. Hereafter, we use the subscript t to emphasize the dependence on time. In the literature, TPF and $1-$FPF have also been referred to as sensitivity and specificity, respectively.

At a given time point t, the corresponding ROC curve is the two-dimensional plot of $\text{TPF}_t(y)$ versus $\text{FPF}_t(y)$ across all possible values of y. Since both functions are time-dependent, the resulting ROC curve is also time-dependent. The ROC curve may also be written as the composition of $\text{TPF}_t(y)$ and the inverse function of $\text{FPF}_t(y)$:

$$ROC_t(p) = \text{TPF}_t\{[\text{FPF}_t]^{-1}(p)\} \tag{5.66}$$

for $p \in (0,1)$, where

$$[\text{FPF}_t]^{-1}(p) = \inf\{y : \text{FPF}_t(y) \le p\}.$$

At time t, the diagnostic performance of marker Y can be summarized with the time-dependent AUC, which is defined as

$$AUC(t) = \int_0^1 ROC_t(p)\,dp = \int_0^1 \text{TPF}_t\{[\text{FPF}_t]^{-1}(p)\}\,dp. \tag{5.67}$$

Another way of defining and interpreting the AUC is via the following probability statement,

$$AUC(t) = P\{Y_1 > Y_2 | T_1 \le t, T_2 > t\}, \tag{5.68}$$

where Y_1 and Y_2 are the marker values of two randomly chosen subjects, and T_1 and T_2 are their corresponding failure times. In another word, the time-dependent AUC is equal to the probability that a marker value of a subject with failure time no greater than t exceeds that of a subject with failure time longer than t.

The $AUC(t)$ defined above can quantify the prognostic performance at a fixed time point. Its integration over time can measure the overall prognostic performance. Consider the follow-up time period $[\tau_1, \tau_2]$. The integrated summary measure is defined as

$$C = \int_{\tau_1}^{\tau_2} AUC(t) \times w(t)\,dt. \tag{5.69}$$

Here, $w(t)$ is a weight function. Different weights can be used to emphasize and reflect the varying reliability within different time intervals. Two weight functions are of special interest. When $w(t) = (\tau_2 - \tau_1)^{-1}$, C is simply the algebraic mean $AUC(t)$ value over the time period $[\tau_1, \tau_2]$. Heagerty and Zheng (2005) propose another weight $w(t) = 2f(t)S(t)$, where f and S are the density and survival functions of the failure time distribution, respectively. Under this weight function, C is closely related to the concordance measure (see Section 4.2.1) and can be interpreted as

$$P(Y_1 > Y_2 | T_1 < T_2), \tag{5.70}$$

where Y_i and T_i are the test value and failure time for subject i from the population. This measure is especially useful for studying the interrelationship between Y and T. More specifically, if $C > 1/2$, a subject who develops the disease earlier is more likely to have a larger marker value than a subject who develops the disease later. Unlike $AUC(t)$, the integrated measure C is time-independent and more appropriate for comparing the overall prognostic performance of different markers. There have been a series of discussions and applications of C in survival analysis. See for example Harrell et al. (1982, 1984, 1996), Pencina and D'Agostino (2004), and Gonen and Heller (2005).

Another set of diagnostic accuracy measures are the predictive values which may be more clinically relevant. They are defined similarly as their time-independent versions

$$\text{PPV}_t(y) = P(T \leq t | Y \geq y),$$
$$\text{NPV}_t(y) = P(T > t | Y < y).$$

These two prospective accuracy measures may be of more interest to the end users of the test, since they provide a quantification of the subject's risk of an outcome by time t, given a positive or negative test result. More discussions can be found in Heagerty and Zheng (2005) and Zheng et al. (2008).

5.2.1 Accuracy Measures under Diverse Censoring Patterns

Consider censored survival data, where the event times of some or even all subjects are not accurately observable. As can be seen in Heagerty et al. (2000), Heagerty and Zheng (2005), the time-dependent ROC curve is constructed from known binary classification of disease status over time. Recall that in the preceding section the time-dependent TPF and FPF are defined as

$$\text{TPF}_t(y) = P\{Y > y | t \geq T\}, \quad \text{FPF}_t(y) = P\{Y > y | t < T\},$$

where the disease status is a conditional event whether the survival time passes t. Because of censoring, the disease status of some subjects may not be well defined at certain time points. In the following we introduce some relatively convenient notations to facilitate intuitive estimation.

Consider the scenario where the event time is only known to lie in the interval $[L, R]$ where $0 \leq L \leq R \leq \infty$. It includes the following special cases: (a) when $L = R < \infty$, the event time is accurately observed and uncensored; (b) when $L = 0$ and $R < \infty$, the observation is left censored, and the true event time is only known to be earlier than R; (c) when $L > 0$ and $R = \infty$, the observation is right censored, and the true event time is only known to be later than L; (d) when $0 < L < R < \infty$, the observation is interval censored. In summary, the disease status is known to be 0 before L, 1 after R, but undetermined between L and R.

Consider TPF_t and FPF_t, which are defined on the marker values of diseased and healthy subjects. Under censoring when the event time is only known to lie in $[L, R]$, a subject is known *for sure* to be diseased if and only if $t \geq R$ and to be healthy if and only if $t \leq L$. Accordingly, we may write

$$\text{TPF}_t(y) = P\{Y > y | t \geq R\}, \quad \text{FPF}_t(y) = P\{Y > y | t < L\}.$$

That is, an uncensored observation contributes to the computation of TPF and FPF at any time t. A right censored observation only contributes to the computation of FPF. A left censored observation only contributes to the computation of TPF. And an interval censored observation contributes to the computation of FPF only in the time interval of $[0, L)$ and TPF only in the time interval of (R, ∞).

To quantify the loss of information caused by censoring, define the undetermined fraction (UF) as

$$\text{UF}_t = P\{L \leq t \leq R\}.$$

Such a proportion does not provide any useful information on the classification power of marker Y.

When UF_t is nonzero, the estimation of TPF_t, FPF_t, and other diagnostic accuracy measures (to be given in the next section) is affected by two aspects. First, less data for diseased or healthy subjects are available. This leads to an inflation of the variances for the estimates. Second, the undetermined subjects may be distributed unevenly between the diseased and healthy categories. Unable to reflect the exact trade-off between TPF_t and FPF_t in the presence of such missing information, the estimation of the ROC curve thus may potentially be biased. One needs to be cautious at stating the conclusion of a study when UF_t is too large.

5.2.2 Estimation and Inference

Consider data which may be composed of subjects under various censoring schemes. As described in the preceding section, all observations can be written in an "interval-censored" format. Assume a sample of n i.i.d. observations $\{(Y_i, L_i, R_i) : i = 1, \cdots, n\}$. In the following, we will investigate estimation and inference for the quantities defined for censored survival time outcomes.

5.2.2.1 Estimation of TPF and FPF

Denote $N_L(t) = \sum_{i=1}^{n} I\{L_i > t\}$ and $N_R(t) = \sum_{i=1}^{n} I\{R_i \leq t\}$, where I is the indicator function. These are the actual sample sizes for healthy and diseased subjects at time t, resembling n_0 and n_1 in Section 5.1. TPF and

FPF can be estimated by their empirical correspondences

$$\widehat{\text{TPF}}_t(c) = \frac{\sum_{i=1}^n I\{Y_i > c, R_i \le t\}}{N_R(t)}, \tag{5.71}$$

$$\widehat{\text{FPF}}_t(c) = \frac{\sum_{i=1}^n I\{Y_i > c, L_i > t\}}{N_L(t)}. \tag{5.72}$$

Note that the estimates for TPF_t and FPF_t given here appear different from the estimates given in the next section under right censoring. In general numerical results of the above estimators are asymptotically equivalent to those in the next section when data is only right censored and the familiar Kaplan-Meier (KM) curve can be constructed. However, in a finite sample, KM estimates may lead to slightly biased estimates. In comparison, our estimates given here are conditionally unbiased in any finite sample. These estimates have variances

$$\text{var}\{\widehat{\text{TPF}}_t(c)\} = \frac{\text{TPF}_t(c)(1 - \text{TPF}_t(c))}{N_R(t)},$$

$$\text{var}\{\widehat{\text{FPF}}_t(c)\} = \frac{\text{FPF}_t(c)(1 - \text{FPF}_t(c))}{N_L(t)},$$

which can be estimated by

$$\widehat{\text{var}}\{\widehat{\text{TPF}}_t(c)\} = \frac{\widehat{\text{TPF}}_t(c)(1 - \widehat{\text{TPF}}_t(c))}{N_R(t)},$$

$$\widehat{\text{var}}\{\widehat{\text{FPF}}_t(c)\} = \frac{\widehat{\text{FPF}}_t(c)(1 - \widehat{\text{FPF}}_t(c))}{N_L(t)}.$$

As $n \to \infty$, for a time point t such that $P(N_R(t) \ge 1), P(N_L(t) \ge 1) \to 1$, consistency and asymptotic normality of the above estimators can be established using the law of large number and central limit theorem. For finite sample data, the above estimates are reliable unless the censoring rate is extremely high. Asymptotic confidence intervals can be constructed following the normality result.

5.2.2.2 Estimation of Time-Specific AUC

Consider $AUC(t)$ at a fixed time point t, where $\min_i R_i < t < \max_i L_i$. At this time point, the ROC curve can be estimated as

$$\widehat{ROC}_t(p) = \widehat{\text{TPF}}_t\{[\widehat{\text{FPF}}_t]^{-1}(p)\} \qquad p \in (0, 1). \tag{5.73}$$

Accordingly, $AUC(t)$ can be estimated with

$$\widehat{AUC}(t) = \int_0^1 \widehat{ROC}_t(p)\, dp. \tag{5.74}$$

The above estimator is conceptually straightforward. However, the ROC estimator involves the functional estimation of TPF and FPF and inversion of FPF, and the AUC estimator further involves an integration. A computationally more feasible estimator can be based on the probabilistic interpretation of the AUC in (5.68). Specifically, consider the following rank sum type estimator

$$\widehat{AUC}(t) = \frac{\sum_i \sum_j I\{Y_i > Y_j, t \geq R_i, t < L_j\}}{\sum_i \sum_j I\{t \geq R_i, t < L_j\}}. \qquad (5.75)$$

As $n \to \infty$, when the denominator is bounded away from zero, consistency of the above estimate can be established using the law of large number.

For inference, one possibility is to use the U-statistic theories (Hoeffding, 1948) to establish the asymptotic normality of (5.75). Investigating such an approach closely, we may find that the asymptotic variance of this U-statistic may involve complicated functionals that are not easy to estimate accurately in practice. As an alternative solution, we suggest the following m-out-of-n bootstrap approach:

(a) Randomly sample m subjects *without replacement*.

(b) Compute $\widehat{AUC}_t$ (5.75) using those m subjects.

(c) Repeat Steps (a)–(b) B (e.g., 500) times.

(d) Compute the sample variance, rescale, and obtain an estimate of $var(\widehat{AUC})$.

This version of nonparametric bootstrap approach has been proposed earlier to address inference with censored observations (Bickel and Freedman, 1981; Bickel et al., 1987). Following Theorem 5.3 of Gine (1997), it can be proved that this bootstrap procedure is valid when $m/n \to 0$ as $n \to \infty$ and $m \to \infty$.

For right censored survival data, many authors have contributed different approaches to directly estimate $AUC(t)$ without the preliminary estimation of TPF, FPF, or ROC (Antolini et al., 1982; Heagerty and Zheng, 2005; Chambless and Diao, 2006; Chiang and Hung, 2010a,b; Hung and Chiang, 2010). The construction in their works is to derive $AUC(t)$ for each event time t as an analytic function of survival and conditional survival functions. The estimation of $AUC(t)$ is then translated into the estimation of quantities pertinent to the survival experiences. Nonetheless, the introduction of additional functionals does not reduce the overall computational cost, and the precision of the final estimate of $AUC(t)$ depends on the performance of such terms, usually acquired nonparametrically and hence of low efficiency themselves. It is therefore appealing to consider implementing the simple estimator (5.75) for right censored data.

5.2.2.3 Estimation of Integrated AUC

As with the time-specific AUC, there are also two ways to compute the integrated AUC. The first approach is based on the definition in (5.69), where we can estimate C with

$$\hat{C} = \int_0^\tau \widehat{AUC}(t) \times \hat{w}(t)\, dt.$$

Here, $\hat{w}(t)$ is the estimate of the weight function $w(t)$.

Consider, for example, the concordance measure in (5.70). As pointed out by some authors (Heagerty and Zheng, 2005), among the many possible integrated AUC measurements, the concordance is of special interest. We have its estimate

$$\hat{C} = \int_0^\tau \widehat{AUC}(t) \cdot 2\hat{f}(t)\hat{S}(t)\, dt. \tag{5.76}$$

In (5.76), $\widehat{AUC}(t)$ has been estimated by using either (5.74) or (5.75). For unknown functions $\hat{f}$ and $\hat{S}$, we may consider the following estimation approach. First, rewrite the density as

$$f = \exp(h).$$

Assume that (a) h belongs to the Sobolev space indexed by the order of derivative s_0; and (b) $|h| \leq M < \infty$. It is easy to see that

$$S = \int f = \int \exp(h).$$

We may estimate the function h as

$$\hat{h} = \text{argmax}_h \left(\sum_{i=1}^n \log(S(R_i) - S(L_i)) \right) - \lambda_n \int (h^{s_0})^2 dt, \tag{5.77}$$

and $\hat{f} = \exp(\hat{h})$ and $\hat{S} = \int \hat{f}$. In (5.77), λ_n is the data-dependent tuning parameter and can be chosen using cross-validation. We use the penalty on smoothness, which has been commonly adopted in spline studies, to control the estimate of h. For practical data, the computation of $\hat{h}$ can be realized using a "basis expansion + Newton maximization" approach. Following Ma and Kosorok (2006a), we can prove that (a) $\hat{f}$ and $\hat{S}$ are $n^{s_0/(2s_0+1)}$ consistent; and (b) $\int \left(\hat{f}^{(s_0)} \right)^2 = O_p(1)$ and $\int \left(\hat{S}^{(s_0)} \right)^2 = O_p(1)$. These results, together with the $\sqrt{n}$ consistency of $\widehat{AUC}_t$, can lead to the consistency of $\hat{C}$. For inference, we propose using the bootstrap approach (see for example, Liu et al., 2005), which supports an intuitive interpretation and has satisfactory empirical performance.

This approach can generate reasonably satisfactory numerical results, when there is a moderate to large sample size so that f and S can be estimated well. However, the computational cost is rather substantial. As an alternative, consider the following computationally more affordable approach.

The second approach for estimating the time-integrated AUC has been motivated by the probabilistic interpretation in (5.70). Specifically, note that the concordance measure can be interpreted as the following probability

$$P(Y_1 > Y_2 | T_1 < T_2) = 2P(Y_1 > Y_2, T_1 < T_2).$$

See the Appendix of Heagerty and Zheng (2005) for a proof.

Define $\mathcal{I} = \{(i, j) : \text{without any ambiguity, } T_i < T_j \text{ or } T_i > T_j \}$. Consider two event time intervals $[L_1, R_1]$, and $[L_2, R_2]$. This pair of subjects may belong to $\mathcal{I}$ when, for example, $R_1 < L_2$ and it is for sure that $T_1 < T_2$. We estimate (5.70) with the following rank-based estimator

$$
\begin{aligned}
\hat{C} &= \frac{\sum_{(i,j) \in \mathcal{I}} I\{Y_i > Y_j, T_i < T_j\}}{\sum_{(i,j) \in \mathcal{I}} I\{T_i < T_j\}} \\
&= \frac{\sum_{i=1}^{n} \sum_{j=1}^{n} I\{Y_i > Y_j, R_i < L_j\}}{\sum_{i=1}^{n} \sum_{j=1}^{n} I\{R_i < L_j\}}.
\end{aligned}
\tag{5.78}
$$

Consistency of this estimate can be established using large sample theories, when the denominator is bounded away from zero as $n \to \infty$.

For statistical inference, the first possible option is based on the U-statistic theories. Direct application of Hoeffding's theory can lead to the asymptotic normality of (5.78). However, the variance may involve complicated unknown functionals. Therefore, we suggest adopting the bootstrap approaches again. The first kind of bootstrap is still the m-out-of-n nonparametric bootstrap described above. This approach requires that $m/n \to 0$ as $n \to \infty$. When n is not large and m is even smaller, the size of the index set $\mathcal{I}$ can be rather small and hence fail to support the validity of the bootstrap. To provide a better remedy, researchers have proposed the semiparametric bootstrap approach that consists of the following steps.

(a) Construct $\hat{S}_X(t)$, an estimate of the conditional survival function (given X).

(b) For $i = 1 \ldots n$, simulate survival time T_i^* from the conditional survival function $\hat{S}_{X_i}(t)$ and regard T_i^* as the true survival time.

(c) Compute $\hat{C}^* = \frac{\sum_i \sum_j I\{Y_i > Y_j, T_i^* < T_j^*\}}{\sum_i \sum_j I\{T_i^* < T_j^*\}}$.

(d) Repeat Steps (b)–(c) B (for example, > 500) times.

(e) Compute the sample variance and obtain an estimate of $var(\hat{C})$.

In Step (a), the conditional distribution $\hat{S}_X(t)$ is attainable via fitting an accelerated failure time (AFT) model with a specified parametric distribution.

Compared to the nonparametric bootstrap, the semiparametric bootstrap can provide a variance estimate with less variability, especially when the sample size is not large. However, there is a risk of model misspecification and consequently the variance estimate may be biased. A careful model justification must be accompanied with this approach. In terms of computational cost, the semiparametric bootstrap involves the computation of $\hat{S}_X$ and is hence more expensive.

Remark. Using the U-statistic approach to estimate the integrated AUC is computationally easier. However, it is limited to certain weight functions where C has simple probabilistic interpretations. We note that since the concordance measure is perhaps the only integrated AUC extensively used, this limitation is not of serious concern. In contrast, the integral estimator (5.76) is more generically applicable for any weight choices, at the price of a higher computational burden.

For right censored survival data, there exists abundant literature on the estimation of concordance measures (Harrell et al., 1982, 1984, 1996; Pencina and D'Agostino, 2004; Gonen and Heller, 2005). Usually, the PH or AFT models are used to characterize the dependence of T on Y, and the estimation of concordance can be based on the fitted regression models. Some of those approaches may be viewed as a special case of the formulation given above, which can accommodate right censored data as well as interval censored data.

We now look at a real example that illustrates part of the methods introduced in this section.

Example (Calcification Study). The calcification study investigated calcification after implantation of hydrogel intraocular lenses, which is an infrequently reported complication of cataract treatment (Yu et al., 2001). In this study, patients were examined by an ophthalmologist to determine the status of calcification at a random time ranging from 0 to 36 months after implantation of the intraocular lenses. Thus, all observations were "case I" interval censored. At the examination, the severity of calcification was graded on a discrete scale ranging from 0 to 4, with severity ≤ 1 classified as "not calcified." The data set contains 379 records, among which one has missing measurements. Among the 378 subjects with complete measurements, 48 experienced calcification during follow-up. The markers of interest include age, incision width, and incision length.

For the three markers, we compute their time-integrated AUCs (along with 95% bootstrap confidence intervals) as Age: 0.503 [0.494, 0.511]; incision width: 0.658 [0.647, 0.667]; incision length: 0.805 [0.795, 0.814].

Among the three markers, age has almost no prognostic power and may be considered as independent to failure time. Incision length has the largest AUC and can make around 80% correct discrimination between the calcified and normal subjects. The accuracy of incision width is intermediate. We plot the ROC curves as a function of time in Figure 5.10.

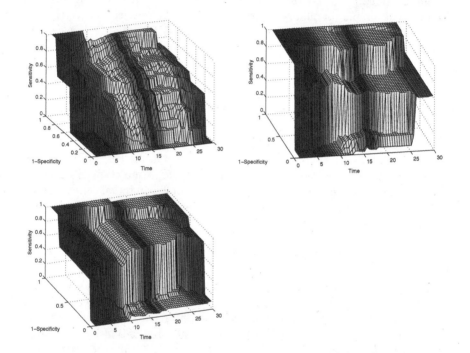

FIGURE 5.10: Analysis of calcification study: Estimated time-dependent ROC curves for three continuous markers: age (upper-left panel), incision width (upper-right panel), and incision length (lower panel).

5.3 Diagnostics for Right Censored Data

5.3.1 Estimation of Accuracy Measures

Since the early 2000s, diagnostic accuracy studies for survival time outcomes have been initiated and drawn extensive attention from biostatisticians. The pioneer works were mostly confined to right censored survival data and borrowed established results from survival analysis as we introduce in the earlier chapters of this book. We summarize some of those results in this section as right censoring is the most popular type of censoring. In what follows, we use the same notations as in Section 5.2 and adopt the same definitions for various time-dependent diagnostic accuracy measures as introduced earlier. We focus more on TPF, FPF, PPV, NPV, and ROC curves in this section. The study of AUC and concordance measure for right censored data is not different from what have been presented in the preceding section.

Suppose that we have a cohort of n individuals followed prospectively for a clinical event of interest. The censoring time is denoted as C. Due to right censoring, we observe a bivariate vector (X, δ), where $X = \min(T, C)$ and $\delta = I(T \leq C)$. Let $D = \{(X_i, \delta_i), i = 1, \cdots, n\}$ be the full data. Assume that T and C are independent conditional on Y, the diagnostic marker of interest.

We use the following notations for the joint and marginal survival distributions:

$$S_{Y,T}(y,t) = P(Y > y, T > t), \qquad (5.79)$$

$$S_Y(y) = P(Y > y), \qquad (5.80)$$

$$S_T(t) = P(T > t). \qquad (5.81)$$

Also denote $F_Y(y) = 1 - S_Y(y)$ and $F_T(t) = 1 - S_T(t)$. In order to estimate the accuracy parameters defined in Section 5.2, we need to consistently estimate the three survival distribution functions (5.79), (5.80), and (5.81). In fact, time-dependent TPF, FPF, PPV, and NPV can all be expressed as functions of the three functions by writing out the conditional probabilities and using Bayes formula. Specifically, we have

$$\text{TPF}_t(y) = \frac{S_Y(y) - S_{Y,T}(y,t)}{1 - S_T(t)},$$

$$\text{FPF}_t(y) = \frac{S_{Y,T}(y,t)}{S_T(t)},$$

$$\text{PPV}_t(y) = \frac{S_Y(y) - S_{Y,T}(y,t)}{S_Y(y)},$$

$$\text{NPV}_t(y) = \frac{S_T(t) - S_{Y,T}(y,t)}{1 - S_Y(y)}.$$

We thus seek estimating survival functions first. Once their estimates are available, the accuracy parameters can be estimated by substitution. The plug-in estimators are given by

$$\hat{\text{TPF}}_t(y) = \frac{\hat{S}_Y(y) - \hat{S}_{Y,T}(y,t)}{1 - \hat{S}_T(t)},$$

$$\hat{\text{FPF}}_t(y) = \frac{\hat{S}_{Y,T}(y,t)}{\hat{S}_T(t)},$$

$$\hat{\text{PPV}}_t(y) = \frac{\hat{S}_Y(y) - \hat{S}_{Y,T}(y,t)}{\hat{S}_Y(y)},$$

$$\hat{\text{NPV}}_t(y) = \frac{\hat{S}_T(t) - \hat{S}_{Y,T}(y,t)}{1 - \hat{S}_Y(y)}.$$

Other time-dependent diagnostic accuracy measures such as the ROC curve, AUC, and integrated AUC can then be similarly estimated based on these estimates.

The two marginal distributions (5.80) and (5.81) can be estimated consistently using the empirical distribution function and Kaplan-Meier (KM) estimator, respectively, and we denote them by $\hat{S}_Y(y)$ and $\hat{S}_T(t)$. The nontrivial issue is the estimation of the joint distribution function (5.79). To this end, we consider the following three statistical approaches.

Approach 1 (Conditional KM Estimate). By using the Bayesian theorem, we can rewrite TPF and FPF as

$$\begin{aligned}
\text{TPF}_t(y) &= P(Y > y | T \leq t) \\
&= \frac{P(Y > y)P(T \leq t | Y > y)}{P(T \leq t)} \\
&= \frac{S_Y(y)\{1 - S_T(t | Y > y)\}}{1 - S_T(t)}, \\
\text{FPF}_t(y) &= P(Y > y | t < T) \\
&= \frac{P(Y > y)P(T > t | Y > y)}{P(T > t)} \\
&= \frac{S_Y(y)S_T(t | Y > y)}{S_T(t)}.
\end{aligned}$$

Noticing the structure above, we can estimate all the survival functions regarding T using the KM estimation method. Specifically, we estimate $S_T(t)$ using the KM method for the whole sample and estimate $S_T(t | Y > y)$ using the KM method for the subsample with $Y > y$. The distribution $S_Y(y)$ can be estimated using the empirical distribution function.

This naive approach is very easy to implement but has received criticisms. First, the sample size for the subsample with $Y > y$ may be rather low and lead to a very inefficient estimate for the conditional survival function. Second, the conditional KM estimate may sometimes yield nonmonotone TPF or FPF sequences over y and render a nonmonotone ROC curve. This can be easily fixed by taking the following simple transformation

$$\hat{\text{TPF}}_t^*(y) = \min_{y'>y} \hat{\text{TPF}}_t(y'), \qquad \hat{\text{FPF}}_t^*(y) = \min_{y'>y} \hat{\text{FPF}}_t(y')$$

for all the numerical implementation. Such a redistribution method has been widely practiced in monotone functional estimation (see for example, Zhang et al., 2008).

Approach 2 (Conditional Smoothing Estimate). For the second approach, we rewrite TPF and FPF as

$$
\begin{aligned}
\text{TPF}_t(y) &= P(Y > y | T \leq t) \\
&= \frac{S_Y(y) - \int_y^\infty P(T > t | Y = u)\, d\, F_Y(u)}{1 - S_T(t)}, \\
\text{FPF}_t(y) &= P(Y > y | t < T) \\
&= \frac{\int_y^\infty P(T > t | Y = u)\, d\, F_Y(u)}{S_T(t)}.
\end{aligned}
$$

We may then direct attention to estimating the conditional survival function $S_T(t|y) = P(T > t | Y = y)$.

To estimate the function $S_T(t|y)$ without imposing any parametric assumption, we consider the nonparametric smoothing approach. Many smoothing methods have been developed in the literature of nonparametric density estimation. The most popular choices include the nearest neighbor method, Nadaraya-Watson kernel method, and local polynomial method. In what follows, we present a construction based on the kernel smoothing method.

Consider the conditional Nelson-Aalen estimation method. Specifically, we estimate the conditional cumulative hazard function $H_T(t|y) = -\log S_T(t|y)$ by

$$\hat{H}_T(t|y) = \int_0^t \frac{d\hat{N}_T(u|y)}{\hat{\pi}_Y(u|y)},$$

where

$$\hat{N}_Y(t|y) = n^{-1} \sum_{i=1}^n K_\varsigma(Y_i - y) I(X_i \leq t)\delta_i,$$

$$\hat{\pi}_T(t|y) = n^{-1} \sum_{i=1}^n K_\varsigma(Y_i - y) I(X_i \geq t),$$

$K_\varsigma(\cdot) = K(\cdot/\varsigma)/\varsigma$, ς is the bandwidth, and $K(\cdot)$ is a user-selected kernel function. As a result, $S_T(t|y)$ can be estimated as

$$\hat{S}_T(t|y) = \exp(-\hat{H}_Y(t|y)).$$

The joint survival function $S(y,t)$ may then be estimated by

$$\hat{S}(y,t) = \int_y^\infty \hat{S}_T(t|u)\, d\hat{F}_Y(u),$$

where $\hat{F}_Y$ is the empirical distribution function of Y.

There are many possible choices for the kernel function $K(\cdot)$, including the Gaussian kernel, Epanechnikov kernel, and nearest neighbor kernel. It is acknowledged in the literature that the Epanechnikov kernel has the best performance among all possible kernel functions. The bandwidth ς is a user-controlled design parameter. Many rules have been developed to select an optimal bandwidth for estimating the functions. One commonly adopted method is cross-validation.

This smoothing approach is now popularly employed in nested case-control (NCC) studies (Cai and Zheng, 2011) for diagnostic accuracy evaluation. An NCC study involves a cost-efficient sampling scheme to collect exposure data in observational studies where large cohorts are needed to observe enough events of interest, making it impractical to collect all exposure data for the complete cohort. In an NCC study, measurements of some or all risk factors of interest are obtained only for a subset of the original study cohort. Typically, the risk factor information is collected on all cases and on a sample of the study cohort that acts as a control group. Operationally speaking, the risk set $R(t_i)$ in the partial likelihood (2.30) is modified to be $\tilde{R}(t_i)$, a subset of $R(t_i)$ consisting of one case and m controls. In practice, little added estimation efficiency is realized with m greater than six (Goldstein and Langholz, 1992).

For an NCC study, the nonparametric estimators of survival functions given above must be modified by incorporating the sampling weights. Cai and Zheng (2011) proposed nonparametric inverse probability weighted estimators for accuracy parameters. Such a weighting method is common for NCC studies (Saarela et al., 2008; Salim et al., 2009). Specifically, the joint survival function can be estimated by

$$
\begin{aligned}
\hat{S}(y,t) &= \int_y^\infty \hat{S}_T(t|u)\, d\hat{F}_Y(u), \\
&= \frac{\sum_{i=1}^n \hat{S}_T(t|Y_i)\hat{w}_i I(Y_i > y)}{\sum_{i=1}^n \hat{w}_i},
\end{aligned}
$$

where $\hat{w}_i$ is the inverse probability weight estimated for the ith subject. Similarly the marginal survival function of Y needs to be modified as

$$\hat{S}_Y(y) = \frac{\sum_{i=1}^n \hat{w}_i I(Y_i > y)}{\sum_{i=1}^n \hat{w}_i}.$$

The two aforementioned conditional approaches for accuracy estimation are based on the conditional survival function estimation, with either a conditional product limit construction (Campbell, 1981) or a nonparametric smoothing approach (Akritas, 1994). Such conditional approaches suffer a major criticism that they do not return to the empirical distribution function in the absence of censoring. Furthermore, these approaches also have a slow convergence rate and may not be practical for data with moderate sample sizes. The third approach is to directly estimate the bivariate joint survival function.

Approach 3 (Dabrowska Estimate). The approach presented herein can even incorporate the situation where Y is also censored. Sometimes the marker Y itself may be subject to the same censoring C and we observe (Z, ξ), where $Z = \min(Y, C)$, $\xi = I(Y \leq C)$. Let $\mathcal{D} = \{(X_i, \delta_i, Z_i, \xi_i), i = 1, \cdots, n\}$ be the full data. Assume that T and C are independent, and that Y and C are independent.

We can rewrite TPF and FPF as

$$
\begin{aligned}
\text{TPF}_t(y) &= P(Y > y | T \leq t) \\
&= \frac{S_Y(y) - S_{Y,T}(y, t)}{1 - S_T(t)}, & (5.82) \\
\text{FPF}_t(y) &= P(Y > y | t < T) \\
&= \frac{S_{Y,T}(y, t)}{S_T(t)}. & (5.83)
\end{aligned}
$$

For some applications where a low marker value Y defines a positive diagnosis, we can modify the above definitions to be

$$\text{TPF}_t^*(y) = P(Y \leq y | T \leq t) \tag{5.84}$$

$$\text{FPF}_t^*(y) = P(Y \leq y | T > t). \tag{5.85}$$

One may argue that a transformation of $\tilde{Y} = -Y$ can lead to equivalent definitions as (5.82) and (5.83). When Y is right censored, such a transformation leads to a left censored $\tilde{Y}$. The statistical estimation methods for left censored data are less convenient, and hence we choose to not flip the sign of Y but use alternative definitions (5.84) and (5.85). We call (5.82) and (5.83) negative-direction definitions, while (5.84) and (5.85) are referred to as positive-direction definitions. To present the estimation methods in this chapter, we focus on the negative-direction definitions (5.82) and (5.83), as they are more commonly encountered for uncensored diagnostic tests.

The two marginal survival functions can be estimated by the familiar product limit estimates. To facilitate notations for Approach 3, we write

$$\mathcal{H}(z,x) = n^{-1}\sum_{i=1}^{n} I(Z_i > z, X_i > x),$$

$$K_1(z,x) = n^{-1}\sum_{i=1}^{n} I(Z_i > z, X_i > x, \xi_i = 1),$$

$$K_2(z,x) = n^{-1}\sum_{i=1}^{n} I(Z_i > z, X_i > x, \delta_i = 1),$$

and denote

$$\Lambda_{10}(y,t) = \int_0^y \frac{K_1(du,t)}{\mathcal{H}(u-,t)}, \qquad \Lambda_{01}(y,t) = \int_0^t \frac{K_2(y,dv)}{\mathcal{H}(y,v-)}.$$

The familiar KM estimates for the marginal survival functions may be written as

$$\hat{S}_Y(y) = \prod_{u \le y} \{1 - \Lambda_{10}(\Delta u, 0)\}$$

$$\hat{S}_T(t) = \prod_{v \le t} \{1 - \Lambda_{01}(0, \Delta v)\},$$

where $\Lambda(\Delta u) = \Lambda(u) - \Lambda(u-)$ and $u-$ is the time prior to u.

For the bivariate joint survival function, consider the estimator proposed by Dabrowska (Dabrowska, 1988, 1989). Further denote

$$K_3(z,x) = n^{-1}\sum_{i=1}^{n} I(Z_i > z, X_i > x, \xi_i = 1, \delta_i = 1),$$

$$\Lambda_{11}(y,t) = \int_0^y \int_0^t \frac{K_3(du,dv)}{\mathcal{H}(u-,v-)}.$$

The Dabrowska estimator for $S_{Y,T}(y,t)$ can be written as

$$\hat{S}_{Y,T}(y,t) = \hat{S}_Y(y)\hat{S}_T(t) \prod_{u \le y, v \le t} \{1 - L(\Delta u, \Delta v)\},$$

where

$$L(\Delta u, \Delta v) = \frac{\Lambda_{10}(\Delta u, v)\Lambda_{01}(u, \Delta v) - \Lambda_{11}(\Delta u, \Delta v)}{\{1 - \Lambda_{10}(\Delta u, v)\}\{1 - \Lambda_{01}(u, \Delta v)\}}.$$

Similar to the Kaplan-Meier estimator for one-dimensional survival function, Dabrowska estimator for a two-dimensional survival function can automatically incorporate tied observations and thus is suitable for continuous and/or discrete distributions. The estimator (5.86) reduces to a bivariate empirical survival function in the absence of censoring.

Subsequently the time-specific ROC curve can be estimated by

$$\hat{ROC}_t(p) = \hat{\text{TPF}}_t\{\hat{\text{FPF}}_t^{-1}(p)\}.$$

However, the estimated ROC curve is only well defined for $p \leq 1 - \hat{\text{FPF}}_t(z^U)$, where $z^U = \max\{Z_i : \xi_i = 1\}$ is the largest uncensored marker value. Consequently, when there are censored marker values greater than z^U, the ROC curve may not be closed.

Similarly, for positive-direction definitions, the estimated ROC curve is

$$\hat{ROC}_t(p) = \hat{\text{TPF}}_t^*\{\hat{\text{FPF}}_t^{*-1}(p)\},$$

where

$$\hat{\text{TPF}}_t^*(y) = \frac{1 - \hat{S}_T(t) - \hat{S}_Y(y) + \hat{S}_{Y,T}(y,t)}{1 - \hat{S}_T(t)},$$

$$\hat{\text{FPF}}_t^*(y) = \frac{\hat{S}_T(t) - \hat{S}_{Y,T}(y,t)}{\hat{S}_T(t)}.$$

Such an estimated ROC curve is only defined for $p \geq 1 - \hat{\text{FPF}}_t^*(z^U)$.

Therefore, the time-specific AUC may not be evaluatable because of the incomplete ROC graph and we may only report a partial AUC in practice.

Remark (Other Bivariate Estimators). In addition to the Dabrowska estimator for the joint survival function, we may also choose Prentice and Cai's estimator (1992) which is based on the Peano series. It is noticed from many numeric studies that these two approaches agree very well in many settings, and we will not elaborator more on the second method in this book. Another theoretically sound estimator is the nonparametric maximum likelihood estimation (NPMLE) for the bivariate survival function proposed by van der Laan (1996a). It is the most efficient method for the bivariate estimation problem. However, the semiparametric efficient score function in van der Laan (1996a) is usually difficult to implement and thus prohibits us from conducting further computationally intensive inferences such as bootstrap.

Cheng and Li (2012) derived the asymptotic distribution results for Approach 3 using empirical process theories. Statistical inference based on normal approximation can follow from such theoretical results. However, the asymptotic distributions involve some unknown functions in the covariance structures which are difficult to estimate accurately in practice. Therefore, we consider the bootstrap inference, whose validity is justified in Cheng and Li (2012).

Specifically, we may repeatedly take random samples from the original sample $\mathcal{D} = \{(X_i, \delta_i, Z_i, \xi_i), i = 1, \cdots, n\}$ with replacement to obtain B bootstrap samples $\mathcal{D}^b$ ($b = 1, \cdots, B$). For each bootstrap sample $\mathcal{D}^b$, we may evaluate the bootstrap estimates $\hat{\text{se}}_t^b(y), \hat{\text{sp}}_t^b(y), \hat{\text{PPV}}_t^b(y), \hat{\text{NPV}}_t^b(y), \hat{R}_t^b(u), \hat{A}^b(t)$, and

$\hat{S}^b$. The standard error for each accuracy estimate is obtained from the sample standard deviation of the bootstrap estimates. The $(1 - \alpha)100\%$ asymptotic confidence intervals for all the accuracy measures can then be constructed from the normal approximation method. For example, a 95% asymptotic confidence intervals for S may be attained as $\hat{S} \pm z_{.025} SD(\hat{S}^b)$, where $z_{.025}$ is the upper .025 quantile of the standard normal distribution and $SD(\hat{S}^b)$ is the standard deviation of the bootstrap estimates $\{\hat{S}^b, b = 1, \cdots, B\}$.

To improve empirical coverage rates, we may construct confidence intervals based on the logarithm transformation, for example, $\exp\{\log \hat{S} \pm z_{.025} SD(\log \hat{S}^b)\}$.

Example (Transplant). We apply the accuracy measures to a bone marrow transplant study containing 137 patients (Copelan et al., 1991) using Approach 3. This data set has been an instructive example in survival analysis to illustrate the concept of time-dependent covariates (Klein and Moeschberger, 2003, p. 297). We refer to Section 2.3.1.4 of this book. Specifically, the outcome of interest is leukemia relapse-free survival. Predictors of interest include time to the first platelet recovery, time to acute graft-versus-host (GVH) disease, time to chronic GVH disease, and baseline characteristics such as patient age. In a typical Cox model, the effects of the three intermediate event times have been studied by forming binary time-varying covariates which jump from zero to one at the intermediate event times. Their relationship with the survival outcome can be examined using the hazard ratio. In our analysis, we use the information from observed event times directly as opposed to the derived binary indicators. We also provide alternative interpretations on how accurately the intermediate event times can reflect the disease-free survival outcome. In this example, age is uncensored, while the times to the three intermediate events are all right censored.

In Figure 5.11, we present four ROC curves evaluated at $t = 6$ months (roughly the median of the observed event times) for the three intermediate event times and patient age. Analysis results from Cox regression suggest that times to platelet recovery and chronic GVH disease as well as patient age are all negatively correlated with the outcome, while the development of acute GVH is positively associated with the outcome. Hence we adopt the usual negative-direction definitions of sensitivity and specificity for the platelet recovery, chronic GVH and age, and use the positive-direction definitions for the acute GVH time in our calculation.

Upon inspection, we note that the ROC curve for time to platelet recovery always lies on top of that for patient age. Their corresponding AUCs at this particular time are 0.71 and 0.56, respectively, and the 2-marker bootstrap test yields a p-value of 0.051, which is marginally significant at level 0.05. We also calculate the integrated AUCs between 107 days (.2th quantile of the observed event or censoring time) and 1491 days (.8th quantile). The integrated AUC is 0.68 for time to platelet recovery and 0.55 for age, and the p-value for testing their difference is 0.053. Both the AUCs at six months

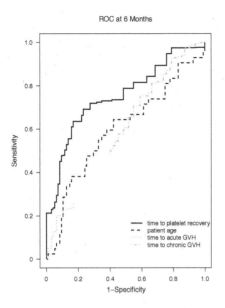

FIGURE 5.11: ROC curves for the three intermediate event times and patient age for the transplant example.

and the integrated AUCs suggest that time to platelet recovery may be more accurate than patient age in predicting relapse-free survival. In Klein and Moeschberger (2003), time to platelet recovery is a highly significant marker in the proportional hazards model. Our analysis further confirms its importance. The ROC curves across time for this marker are displayed in Figure 5.12. AUCs at earlier time points seem to be slightly smaller than those at later time points.

It is interesting to note that the ROCs for times to acute and chronic GVH diseases are incomplete in Figure 5.11 due to heavy censoring of these two time measures. There are 101 subjects that were censored after the largest observed acute GVH event time (88 days), and 33 subjects that were censored after the largest observed chronic GVH event time (487 days). These open ROC graphs present a quite unique phenomenon associated with censored predictors. We can thus only discuss their accuracy within a limited range and evaluate the partial areas from their curves. Based on the Cox regression model, times to acute and chronic GVH diseases are not significant (Klein and Moeschberger, 2003). It is not clear whether this lack of significance is due to insufficient observations or an intrinsic weak relationship between a GVH disease and relapse-free survival. In contrast, ROC analysis for both censored predictor and event time allows us to evaluate the prediction accuracy of these two intermediate events within observable regions. Both ROC curves are inferior

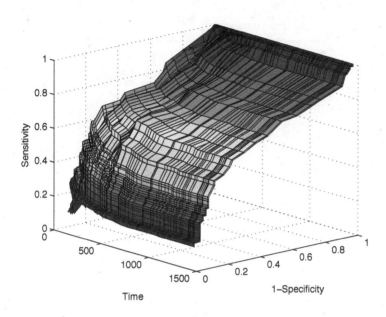

FIGURE 5.12: Time dependent ROC curves for time to platelet recovery in the transplant example.

in the observed range, comparing to time to platelet recovery for predicting six months relapse-free survival. Similar trends are observed for three months and one year relapse-free survival (figures not presented). Therefore, we may conclude that these are weak markers with inadequate accuracy and will not recommend their use in practice.

5.3.2 Additional Applications

Applications of time-dependent ROC analysis in genetics can be conducted similarly, as we illustrate in Section 5.1.2.4 for time-independent ROC analysis. In the literature, Ma and Song (2011) considered ranking genes for predicting failure time outcomes with applications to a breast cancer data set; Yu et al. (2012) considered adjusting for confounder effects when ranking genes using AUC by adopting a model-based approach. The next chapter illustrates and exemplifies with greater details.

For the accuracy improvement measures discussed in Section 5.1.3.1 for categorical outcomes, we may similarly define their counterparts for failure time outcomes. More results are provided in Pencina et al. (2011) and Shi et al. (2012a), where the first reference provides a general conceptual basis for an accuracy incremental index, and the second reference supplies estimation techniques for data with independent censoring and competing-risk censoring. The R code to compute NRI and IDI can be downloaded from: $http://www.stat.pitt.edu/yucheng/software:html$, where the authors also provide examples to demonstrate the usage. This area is currently expanding rather rapidly, and what has been covered here may need to be updated and expanded soon.

Remarks. Diagnostic tests and biomarkers are usually expected to predict the survival outcomes and can be easily formulated as predictor variables in regression models. The analysis focused on in this chapter is significantly different from regression. The strength of association between a marker and a survival outcome is characterized by the actual prediction accuracy at a probability scale, as opposed to by the relative effect on instantaneous failure rate. The interpretation of analysis results is thus also different from the usual hazards regression that only stresses the significance of regression coefficients. Some researchers have attempted to interweave the two types of approaches by proposing covariate-adjusted ROC methods or other regression-type analysis for accuracy measures (Pepe, 1998, 2003). When evaluating conditional accuracy measures specific to certain covariate values (such as age or gender group), the available information from the sample becomes limited, and the resulting estimates are not as efficient as those based on the full sample. We realize that the regression procedure is essentially equivalent (though not literally carried out in this way) to divide the whole sample into small strata according to covariate values and evaluate diagnostic accuracy within each stratum. The problem is even more severe for continuous covariates since they can only attain zero mass. Another potential weakness for covariate adjustment is inference: when too many accuracy measures are computed with the same data, the problem of multiple comparisons may arise, and any significant finding has to be reconsidered after more sophisticated calculation.

5.4 Theoretic Notes

Consistency of the Empirical ROC Curve. We now show that the estimator (5.37) is uniformly consistent to the true ROC curve. First, observe that

$$\sup_p |\widehat{ROC}(p) - ROC(p)|$$

$$\leq \sup_p |\hat{F}_1(\hat{F}_0^{-1}(1-p)) - F_1(\hat{F}_0^{-1}(1-p))|$$

$$+ \sup_p |F_1(\hat{F}_0^{-1}(1-p)) - F_1(F_0^{-1}(1-p))|.$$

The first term converges to zero almost surely by applying the Glivenko-

Cantelli theorem. The second term converges to zero almost surely by applying Bahadur's representation of the sample quantiles (Bahadur, 1966).

Weak Convergence of the Empirical ROC Curve. The Brownian bridge process $B(p)$ is a special Gaussian process with mean zero and covariance $\operatorname{cov}(B(p_1), B(p_2)) = p_1(1 - p_2)$ for $0 \leq p_1 \leq p_2 \leq 1$.

To show that $\sqrt{n_1}(\widehat{ROC}(p) - ROC(p))$ converges weakly to a Brownian bridge process, it suffices to show the following results.

(a) The finite-dimensional distribution

$$\sqrt{n_1}(\widehat{ROC}(p_1) - ROC(p_1), \cdots, \widehat{ROC}(p_k) - ROC(p_k))$$

converges in distribution to the multivariate normal distribution.

(b) The process $\sqrt{n_1}(\widehat{ROC}(p) - ROC(p))$ is tight.

(a) can be proved by a straightforward calculation. (b) can be proved using the techniques in Billingsley (1999) for the proof of Donsker's theorem. The detailed proof is omitted. Alternatively, one can apply results from the empirical process theory.

5.5   Exercises

1. TRUE or FALSE?

    (a) An ROC function is nondecreasing and bounded between $[0, 1]$.

    (b) The estimated sensitivity and specificity are negatively correlated.

    (c) PPV and NPV remain the same across different populations with different disease prevalence.

2. If $\mathbf{Y}_i$ i.i.d. $\sim N(\boldsymbol{\mu}_0, \Sigma)$ ($i = 1, \cdots, n_0$) and $\mathbf{Y}_j$ i.i.d. $\sim N(\boldsymbol{\mu}_1, \Sigma)$ ($j = 1, \cdots, n_1$), show that the best linear combination for maximizing the bi-normal AUC is $\boldsymbol{\beta} = \Sigma^{-1}(\boldsymbol{\mu}_1 - \boldsymbol{\mu}_0)$. Please also compute the optimal AUC value.

3. Show that (5.69) is equivalent to (5.70).

4. What is the advantage of an NCC study for survival analysis?

Chapter 6

Survival Analysis with High-Dimensional Covariates

All statistical techniques introduced in the previous chapters are limited to data with a relatively small number of covariates. When the sample size n goes to infinity while keeping the number of covariates p fixed, standard likelihood- and estimating equation-based methods can be used straightforwardly for estimation and inference. Since the late 1990s, advancements in biomedical technologies have generated a large number of "large p, small n" data sets, where the number of covariates is comparable to or even much larger than the sample size. For example, in a typical microarray gene expression study, the number of subjects n is usually no greater than 1000, while the number of genes profiled can be more than several thousands, and all their expression values are recorded to be the p covariates. With such data, standard survival analysis techniques are no longer directly applicable. Suppose that we attempt to fit a Cox regression model with the number of covariates larger than the sample size. Mathematically it can be shown that multiple or even infinite many maximizers of the partial likelihood function exist, with most or all of them being unreasonable. Some existing software packages would fit a model using only the first few covariates, while setting the estimated regression coefficients to zero for the rest. Such results are unreasonable in that they depend on the order of covariates set in a computing code.

Consider a study with a well-defined censored survival outcome. In addition to the outcome observation, a large number of covariates are measured for each subject. With such data, researchers may be interested in conducting the following two types of analyses.

- The first type of analysis is to identify covariates that are associated with survival in a univariate sense. With each individual covariate, quantification of the strength of association can be easily carried out using standard procedures introduced in Chapter 2. However, when inference on a large number of covariates is carried out for the same data (i.e., fitting a large number of models for the same group of subjects), simply using, for example, 0.05 as the p-value cut-off for significance may lead to a substantial amount of false discoveries. We will introduce statistical techniques that can properly adjust for multiple comparisons and so control the false discovery rate.

- For the second type of analysis, our goal is to develop an adequate multivariate regression model that can describe the relationship between survival and covariates and be used to predict survival for a new subject. Such analysis would demand the incorporation of multiple covariates in a single statistical model. Properly reducing the dimensionality of covariates is needed along with estimation. In this chapter, we will introduce several widely used dimension reduction and variable selection techniques that can properly accommodate high-dimensional covariates in survival model fitting.

Materials introduced in this chapter may not be limited to censored survival time data. The screening, dimension reduction, and variable selection methods are equally applicable to Gaussian response variables, such as financial transaction data, weather time series data, or binary outcomes, such as disease status data (disease-present or -absent, see Section 5.1). There are limited operational obstacles for practitioners to borrow techniques covered in this chapter to analyze other types of data.

This chapter is organized as follows. In Section 6.1, we describe several applications that may generate survival data with high-dimensional covariates and motivate methodological development. In Section 6.2, we describe the procedure to identify important covariates associated with survival using univariate analysis methods. In Section 6.3, we describe dimension reduction and variable selection techniques that can be used to construct multivariate regression models. In Section 6.4, we describe univariate and multivariate methods that can properly incorporate the hierarchical structure of covariates. Such hierarchical structure may also exist with low-dimensional covariates. However, it is more critical with high-dimensional covariates. In Section 6.5, we consider the problem of analyzing multiple heterogeneous high-dimensional survival data. Such analysis can effectively increase sample size and so performance of multivariate analysis methods. Theoretical notes and some further developments are described in the Appendix of this chapter.

6.1 Applications

6.1.1 Gene Expression Study on Mantle Cell Lymphoma

Genes are segments of an individual's DNA sequence that encode proteins which carry out all basic functions of living cells. In different disease states, the same gene may have different levels of expressions, that is, a different amount of mRNA (which is an intermediary along the way to protein production) is present. Gene expression studies have been extensively conducted using microarray techniques, identifying previously unknown disease subtypes, classifying new patients into groups with different prognostic patterns, and selecting optimal treatment regimens. There has been a large amount of literature on microarray and gene expression studies, see for example Knudsen (2006), Lesk (2002), and Wong (2004).

A lymphoma gene expression study was reported in Rosenwald et al. (2003), which used microarray gene expression analysis in mantle cell lymphoma (MCL). Among 101 untreated patients with no history of previous lymphoma, 92 were classified as having MCL based on established morphologic and immunophenotypic criteria. Survival times of 64 patients were available, and 28 patients were censored. The median survival time was 2.8 years (range: 0.02–14.05 years). Lymphochip DNA microarrays were used to quantify mRNA expressions in the lymphoma samples from the 92 patients. Data that contains the expression values of 8810 cDNA elements is available at: $http://llmpp.nih.gov/MCL$.

In this example, the sample size $n = 92$ is much smaller than the number of covariates $p = 8810$. It is infeasible to apply traditional methods described in Chapter 2 to study how the covariates affect the survival outcome. This

MCL data will be visited repeatedly in this chapter to illustrate the statistical methods.

6.1.2   Genetic Association Study on Non-Hodgkin Lymphoma

In genetic association studies, researchers search for single DNA mutations (single nucleotide polymorphisms, or SNPs) that may be associated with the development and progression of diseases. It is now possible to assay hundreds of thousands of SNPs for an individual at a given time. For example, the Affymetrix Genome-Wide Human SNP array 6.0 features 1.8 million genetic markers, among which are more than 906,600 SNPs. By 2009, about 600 human genetic association studies have been conducted, examining 150 diseases and traits, and finding 800 SNP associations (Johnson, 2009). Almost all common SNPs have only two alleles so that one records a variable z_{ij} on individual i taking values in $\{0, 1, 2\}$ depending upon how many copies of, say, the rare allele one individual has at location j. The problem is then to decide whether or not a quantitative trait, such as survival time, has a genetic background. In order to scan the entire genome for signal, we need to screen about 1 million SNPs. If the trait has a genetic background, it will be typically regulated by only a very small number of genes (SNPs).

Between 1996 and 2000, a population-based NHL (non-Hodgkin lymphoma) study was conducted in Connecticut women including 601 histologically confirmed incident NHL cases (Han et al., 2010). Subjects were interviewed, and information on anthropometrics, demographics, family history of cancer, smoking and alcohol consumption, occupational exposure, medical conditions and medication use, and diet were collected. Of the 601 cases, 13 could not be identified in the CTR (Connecticut Tumor Registry) system, and 13 were found to have a history of cancer prior to the diagnosis of NHL, leading to a prognostic cohort of 575 NHL patients. Among the 575 patients, 496 donated either blood or buccal cell samples for genotyping. DNA extraction and genotyping was performed at the Core Genotyping Facility of National Cancer Institute. A total of 1462 tag SNPs from 210 candidate genes related to immune response were genotyped using a custom-designed GoldenGate assay. In addition, 302 SNPs in 143 candidate genes previously genotyped by Taqman assay were also genotyped. There were thus a total of 1764 SNPs measured. Vital status was abstracted from the CTR in 2008. Other follow-up information was also abstracted, including date of death, date of most recent follow-up, date of treatments, treatment regimens, date of first remission, dates of relapse, date of secondary cancer, B-symptoms, serum LDH levels, and tumor stage.

This data serves another example of $n = 496$ being much smaller than $p = 1764$. We mostly exemplify our methods with gene expression data such as MCL in this book and place less weight on applications of SNPs. Nonetheless, the methods used for gene expressions can be similarly implemented for SNPs with minor modifications.

6.1.3 Epigenetic Study on Adult Acute Myeloid Leukemia

In biological systems, methylation is catalyzed by enzymes. It can be involved in the modification of heavy metals, regulation of gene expression, regulation of protein function, and RNA metabolism. DNA methylation in vertebrates typically occurs at CpG sites. It results in the conversion of cytosine to 5-methylcytosine.

A DNA methylation study was conducted investigating survival of AML (adult acute myeloid leukemia) patients (Bullinger et al., 2010). A total of 182 DNA samples derived from 98 peripheral blood and 84 bone marrow specimens from adult AML patients were provided by the German-Austrian AML Study Group (AMLSG). Patients were enrolled between February 1998 and November 2001. The median follow-up time was 534 days overall (1939 days for survivors). The 74 validation set samples were enrolled under the same protocol and between February 1998 and May 2004. MALDI-TOF-MS-based DNA methylation analysis was performed. Genes subject to methylation analysis in this study have been selected based on previous gene expression profiling studies in AML. Researchers selected the top 48 genes associated with outcome represented by 50 genomic regions. They also included an additional 32 genes known to be aberrantly methylated in cancer or to be associated with leukemogenesis represented by 42 genomic regions. The promoter regions of these 80 genes (represented by 92 genomic regions; i.e., amplicons) included more than 2170 CpG sites for each sample. The CpG sites were analyzed in 1320 informational units that contained either individual CpG sites or short stretches of subsequent CpG sites (CpG units).

Covariates in this data are $p = 1320$ informational units, and the goal is to examine how they can be used to predict the survival of AML patients. The training sample size is only 182 (with another 74 validation samples), much less than p.

6.1.4 Remarks

Besides cancer, high-throughput profiling studies have also been extensively conducted on cardiovascular diseases, diabetes, mental disorders, and others. In addition to genomic and epigenetic studies, genetic and proteomic studies also generate data with the "large p, small n" characteristic. The profiling techniques used to generate such data can be widely different, leading to covariates with significantly different characteristics. For example, a typical microarray profiling study generates $\sim$40,000 continuous measurements, whereas a typical genetic association study may generate $\sim$1,000,000 categorical measurements with at most three levels.

Despite the remarkable differences in technologies and covariate characteristics, the survival analysis problems involved and hence analytic techniques needed are similar. Analysis methods introduced in this chapter are "insensitive" in that they do not heavily depend on the specific type of covariates. Let T denote the survival time of interest and $\mathbf{X} = (X_1, \ldots, X_p)^T$ denote the p-dimensional covariates. The following two types of problems will be studied in this chapter.

- The first type of analysis is to identify which X_js ($j = 1, \ldots, p$) are significantly associated with T in a univariate sense. For example, in the MCL study, it is of interest to identify which of the $p = 8810$ measured genes are significantly associated with disease-specific survival. This type of problem demands analyzing each X_i separately and then ranking across all genes. Statistically speaking, this is a "hypothesis testing + multiple comparisons" problem.

- The second type of analysis is to construct a multivariate regression function $f(X_1, \ldots, X_p)$, which can be used to predict the survival time or failure risk of a new subject. With p fixed and n sufficiently large, this problem has been addressed in Chapter 2. With $p >> n$, this becomes a dimension reduction or variable selection problem. In this chapter, we will focus on linear predictive models where $f(X_1, \ldots, X_p) = f(\mathbf{X}^T \boldsymbol{\beta})$ with $\boldsymbol{\beta} = (\beta_1, \ldots, \beta_p)^T$ being the regression coefficient. Nonlinear predictive models are more flexible than linear models and have been investigated in several recent publications. However, such models may incur prohibitively high computational cost and lack lucid interpretations, and will not be considered in this chapter.

Sometimes the two procedures are conducted in a sequential manner where investigators use the first type of analysis to conduct a rough screening and remove "noisy" covariates and then build a statistical model with the remaining covariates by conducting the second type of analysis. However, many or most studies may only carry out one of the two analyses. The same data set may be analyzed independently by using one of the two approaches in different studies. Published studies have shown that these two types of analyses can lead to quite different results for the same data set. In fact, they describe different aspects of covariates and thus complement each other: the first type of analysis studies the *marginal effect* of each X_j on the survival outcome, while the second type of analysis examines the *conditional effect* of each covariate given other covariates presented in the model.

In the remainder of this chapter, we will use the cancer microarray gene expression study as an example because of its practical importance and popularity. The statistical techniques introduced are generically applicable for studies with similar characteristics.

6.2   Identification of Marginal Association

Consider survival time T, which can be progression-free, overall, or other types of survival. Denote C to be the censoring time. Although in Chapters 1–4 we have introduced various censoring patterns, we will focus on right censored data in this chapter. It is expected that other censoring patterns (which are significantly less frequently encountered) can be studied in a similar manner. Under right censoring, one observation consists of $(Y = min(T, C), \Delta = I(T \leq C), \mathbf{X})$. Identification of covariates marginally associated with survival demands first ranking covariates based on their strength of associations, which proceeds as follows. For $j = 1, \ldots, p$,

1. Describe the relationship between covariate X_j and survival time T using the model $T \sim f(X_j \beta_j)$, where β_j is the unknown regression coefficient and f is a known link function. Here f is a parametric or semiparametric survival model as described in Chapter 2, particularly including for example the Cox model, additive risk model, and AFT model.

2. A statistic measuring the strength of association for X_j is computed. Examples of the statistic include the magnitude of the estimate of β_j, significance level (p-value) of the estimate, likelihood of the model, and others. Of note, as the statistical models have only a single covariate, standard estimation and inference approaches described in Chapter 2 are directly applicable and can be realized easily using many existing software packages.

The p covariates are then ranked based on the statistics obtained above.

This procedure only provides a relative ranking of covariates. It is useful when, for example, researchers are interested in investigating a fixed number of top-ranked covariates. However, this procedure itself does not contain rigorous inference, and it cannot rule out the case where even covariates with the highest ranks are not significantly associated with survival.

For rigorous statistical inference, take the p-value as the ranking statistic. For p markers, the jth p-value p_j is obtained for testing the null hypothesis: $H_0 : \beta_j = 0$, where β_j is the regression coefficient for the jth marker in a, say Cox PH or AFT, model $(j = 1, \cdots, p)$. In Chapter 2, covariates with p-values less than $\alpha = 0.05$ are identified as having significant associations with survival. With a large number of covariates, for example, $p \sim 40{,}000$ covariates as in a typical microarray study, using 0.05 as the cut-off for significance may lead to a large number of incorrect significant results. Consider the scenario where there is no association between T and X, and X_i and X_j are independent for $i \neq j$. Then it can be proved that the 40,000 p-values obtained using the procedure described above follow a $Uniform[0, 1]$ distribution. Thus, using 0.05 as the cut-off will lead to approximately 2000 $(= 40{,}000 \times 0.05)$ "discoveries" on average whose significance can be completely due to chance.

TABLE 6.1:   Possible Outcomes from p Hypothesis Tests

	# declared nonsignificant	# declared significant	Total
# true null	U	V	m_0
# true alternative	T	S	$p - m_0$
Total	$p - R$	R	p

Note: The FDR is defined as $E\left(\frac{V}{R}I(R > 0)\right)$.

Classic multiple testing procedures suggest the consideration of a *family-wise type I error rate* (FWER) defined as

$$P(S_1 \cup S_2 \cup \cdots \cup S_p)$$

where $S_j = \{p_j < \alpha | \beta_j = 0\}$ is the event that the jth null hypothesis is incorrectly rejected ($j = 1, \cdots, p$). Using the Bonferroni inequality, we can show that

$$P(S_1 \cup S_2 \cup \cdots \cup S_p) \leq \sum_{j=1}^{p} P(S_j).$$

Bounding individual type I error rates $P(S_j)$s does not imply the same bound for the familywise type I error rate. In fact, when $\alpha = 0.05$ is used across all tests, the upper bound for the FWER is $0.05p$ for p hypotheses. When $p > 20$, this is equivalent to assert that the type I error is controlled to be at most 1, a correct but useless statement.

A computationally simple remedy is to use $\alpha = 0.05/p$ (as supposed to 0.05) as the preplanned cut-off for significance and hence retain the overall type I error rate under 0.05. The Bonferroni adjustment approach has been extensively used in multivariate data analysis (Johnson and Wichern, 2001) and experimental design (Wu and Hamada, 2000). There are other inequality-based adjustment methods introduced in the history for a similar goal, including the Dunnett method, Scheffe method, Tukey method, and Sidak method, among others. These methods all enjoy simple mathematical forms and can be implemented easily. However, these crude correction methods are not favored by analysts of ultrahigh-dimensional data, as they often lead to overly conservative results. That is, too few true positives can be discovered. Bonferroni correction is the most conservative among these methods while others are to some degree less conservative but may demand additional data assumptions.

An alternative approach, which is less conservative than the Bonferroni correction, is to control the false discovery rate (FDR). Consider simultaneous inference with a set of p hypotheses (covariates). The possible outcomes are displayed in Table 6.1. Traditional approaches, including the Bonferroni correction, provide control of familywise type I error rate $P(V \geq 1) = 1 - P(V = 0)$. Controlling this error rate essentially requires the elimination of false positives with a very high probability. This is practically hard to achieve for genetic studies and leads some authors to consider the less stringent error rate $P(V \geq k)$ for a $k > 0$ (Dudoit et al., 2004). As an alternative, the FDR approach targets at controlling

$$E\left(\frac{V}{R}I(R > 0)\right),$$

which is the *proportion* of significant discoveries that are incorrect. Allowing V to be nonzero and controlling the size of FDR is reasonable and in fact necessary. In, for example, whole-genome cancer studies, the number of genes surveyed can be extremely large, and multiple genes are expected to be involved in the pathogenesis of cancer. At an exploration stage, researchers may be willing to accept a list of candidate genes that contains a relatively small number of (more than one) false positives. In general FDR $\leq$ FWER, and the equality holds only when $p = V$.

Denote $p_1, \ldots, p_p$ as the p p-values generated using the procedure described above. The FDR approach described in Benjamini and Hochberg (1995) and Benjamini and Yekutieli (2001) proceeds as follows:

1. Set the target FDR level to be q. In the literature, commonly adopted values include $0.05, 0.1$, and 0.2.

2. Order the p-values $p_{(1)} \leq \cdots \leq p_{(p)}$.

3. Let r be the largest i such that $p_{(i)} \leq \frac{i}{p} \times \frac{q}{c(p)}$. When it is reasonable to assume that different p-values are independent, $c(p) = 1$. When it is suspected that different p-values are correlated but the correlation structure is unknown, a conservative choice is $c(p) = \sum_{i=1}^{p} \frac{1}{i}$.

4. Covariates corresponding to $p_{(1)}, \ldots, p_{(r)}$ are concluded as significantly associated with survival.

At Step 3, the sum $\sum_{i=1}^{p} \frac{1}{i}$ may be approximated by $\log(p) + \gamma$ where $\gamma \approx 0.5772$ is the Euler-Mascheroni constant. The procedure given above guarantees that the FDR is no greater than q (Benjamini and Hochberg, 1995; Storey and Tibshirani, 2003). Practically we can claim that $(1-q)100\%$ of the rejected cases are true discoveries. Note that this kind of claim is not available for a significance test: setting $\alpha = 0.05$ for a hypothesis testing does not mean that a rejection of the null is correct with a 95% chance.

In R, the FDR approach can be implemented using packages `locfdr` and `LBE`. In addition, functions `Threshold.FDR` (library `AnalyzeFMRI`) and `compute.FDR` (library `brainwaver`) can also be used for such a purpose.

Example (Follicular Lymphoma). A study was conducted to determine whether the survival risks of patients with follicular lymphoma can be predicted by the gene-expression profiles of the tumors (Dave et al., 2004;

Yu et al., 2012). Fresh-frozen tumor-biopsy specimens from 191 untreated patients who had received a diagnosis of follicular lymphoma between 1974 and 2001 were obtained. The median age at diagnosis was 51 years (range: 23 to 81), and the median follow-up time was 6.6 years (range: less than 1.0 to 28.2). The median follow-up time among patients alive at last follow-up was 8.1 years. Records with missing measurements are excluded from the analysis. The analysis includes 156 subjects. Affymetrix U133A and U133B microarray genechips were used to measure gene expression levels. A log2 transformation was first applied to the Affymetrix measurements. As genes with higher variations are of more interest, we filter the 44,928 gene expression measurements with the following criteria: (1) the max expression value of each gene across 156 samples must be greater than the median max expressions; and (2) the max-min expressions should be greater than their median. Out of 44,928 genes, 6506 pass the above unsupervised screening.

We fit $p = 6506$ univariate Cox PH models on all the gene expressions and obtain p-values from Wald tests. We then take the FDR approach with $q = 0.20$ and obtain 10 significant genes (Figure 6.1). The solid line in the right panel is $0.2 \cdot j/p$ versus the order j. The largest j for which the p-value $p_{(j)}$ falls below the line occurs at $j = 10$, indicated by the vertical line. Using FWER control with a Bonferroni correction at significance level $\alpha = 0.05$, we cannot identify any significant gene.

Other versions of FDR have been proposed in the literature, including the conditional FDR (Tsai et al., 2003), positive FDR (Storey, 2002), local FDR

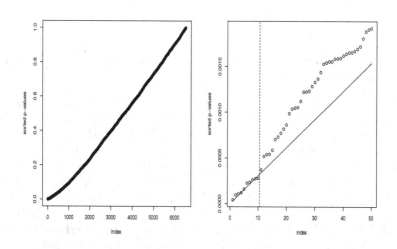

FIGURE 6.1: Plots of the ordered p-values in follicular lymphoma study. The left panel is the plot of $p = 6506$ ordered p-values. The right panel is the plot of the 50 smallest p-values $p_{(j)}$ and a solid line $0.2 \cdot j/p$.

(Efron et al., 2001), and two-dimensional local FDR (Ploner et al., 2006), among others. They may all be implemented using similar algorithms. These different methods usually lead to similar results in gene screening. Some studies have suggested controlling the FNR (false negative rate), which is defined as $E\left(\frac{T}{p-R}\right)$.

Using the terminology in Section 5.1, it is not hard to realize that the estimated FDR is equivalent to one minus the estimated PPV in the setting of a diagnostic test, and the estimated FNR is one minus the estimated NPV. However, there is a crucial difference between the construction entries of the 2×2 Tables 5.1 and 6.1. The $n_0 + n_1$ subjects under diagnostics are all independent, and therefore Table 5.1 easily invokes a binomial-based inference. On the other hand, the p p-values in this section are computed from fitting models on the same sample and cannot be reasonably assumed to be independent. Because of the intrinsic correlations of p-values, inference techniques for PPV and NPV introduced in Section 5.1.1 are not directly applicable for FDR and FNR. The dependence does not affect the point estimation of FDR and FNR but leads to inflated estimation variability.

For a more comprehensive review of the FDR methodology, we refer to Benjamini and Yekutieli (2001), Storey et al. (2007), Storey and Tibshirani (2003), and Efron (2010).

The methods introduced above are not limited to censored survival response. In fact, when facing enormous p-values produced from linear or logistic regression models, we can also go through the same screening procedures by controlling FWER or FDR.

6.2.1   Nonparametric Concordance Statistic

The procedure described in the previous section needs to assume a specific parametric or semiparametric model f. An advantage of this approach is that the ranking statistic usually has lucid interpretations and can be easily computed using existing software. On the other hand, a drawback is that the validity of ranking depends on the validity of model assumption. As described in previous chapters, determining the validity of model assumption is already difficult with low-dimensional covariates. With extremely high-dimensional covariates, although some *ad hoc* approaches have been implemented, to the best of our knowledge, there is no generically applicable approach for validating the model assumption. Here we describe a concordance measure, which is advocated by Ma and Song (2011) in the context of cancer prognosis studies with gene expression measurements. It relies on weaker assumptions and hence provides a more robust way of ranking covariates.

For covariate j $(= 1, \ldots, p)$, assume that $E(T|X_j) = \eta_j(X_j)$, where η_j is an *unknown* monotone function. Note that this model is generic and includes many existing parametric and semiparametric models as special cases. Without loss of generality, assume that η_j is an increasing function. A recoding

$X_j \to -X_j$ can be conducted if necessary. Assume that there are n i.i.d. observations. We use (Y_i, δ_i) to denote the observed time and event indicator of the ith subject, and $X_{i,j}$ to denote the jth component of X measured on the ith subject. Intuitively, if covariate j is associated with survival, then the order of $\{T_1, \ldots, T_n\}$ should be similar to that of $\{X_{1,j}, \ldots, X_{n,j}\}$. Thus, the strength of association between a covariate and survival can be evaluated using the *concordance* between the ranking of event times and the ranking of covariate values.

Particularly, for the jth covariate, the nonparametric concordance measure is defined as

$$\tau_j = P(X_{i,j} < X_{k,j} | T_i < T_k, i \neq k).$$

This measure has roots in the time-dependent ROC (see Section 5.1.2.1) techniques. As described in Chapter 5, the ROC approaches have been extensively used in evaluating the diagnostic and prognostic accuracy of markers. Using the ROC curve, we can summarize the diagnostic and prognostic power by the AUC (defined in Equation 5.23). A larger AUC value indicates higher prognostic accuracy of the marker. For covariate j at time t, the AUC for the incident ROC curve is

$$AUC_j(t) = P(X_{i,j} < X_{k,j} | T_i = t, T_k > t, i \neq k).$$

The concordance measure τ_j is related to AUC through the formula

$$\tau_j = \int_0^\infty w(t) \times AUC_j(t) dt,$$

where $w(t) = 2f(t)S(t)$, and $f(t)$ and $S(t)$ are the density and survival functions of the survival time T. Thus the concordance measure can be viewed as a weighted average of the AUC over time. A larger τ_j value indicates that the order of marker values is more likely to concord with the order of their failure times, thus implying more accurate prediction of the failure outcome. Unlike $AUC_j(t)$ which is a function of time, τ_j is time-independent, thus, it can summarize the association between a covariate and survival in an easier way and facilitate the comparison of multiple covariates.

We note that τ_j can be rewritten as

$$\tau_j = \frac{P(X_{i,j} < X_{k,j}, T_i < T_k)}{P(T_i < T_k)}, \quad i \neq k.$$

If $P(C \geq T) > 0$, then $E\left(\frac{\delta_i}{S_C^2(T_i)} I(Y_i < Y_k)\right) = P(T_i < T_k)$. Here, S_C is the survival function of the censoring time C. In addition, if C is independent of X, $E\left(\frac{\delta_i}{S_C^2(T_i)} I(X_{i,j} < X_{k,j}, Y_i < Y_k)\right) = P(X_{i,j} < X_{k,j}, T_i < T_k)$. Thus, τ_j can be estimated consistently by

$$\hat{\tau}_j = \frac{\sum_{i=1}^n \sum_{k=1}^n \frac{\delta_i}{\hat{S}_C^2(Y_i)} I(X_{i,j} < X_{k,j}, Y_i < Y_k)}{\sum_{i=1}^n \sum_{k=1}^n \frac{\delta_i}{\hat{S}_C^2(Y_i)} I(Y_i < Y_k)}.$$

TABLE 6.2: Analysis of MCL Data: Number of Genes Identified as Significantly Associated with Survival with FDR = 0.1 (The numbers in the diagonal are the number of genes identified by the methods while the numbers in the off-diagonal are the numbers of overlapping genes identified by two methods.)

	Nonparametric	Cox	AFT	Additive
Nonparametric	88	52	0	23
Cox		61	0	24
AFT			0	0
Additive				25

Here, $\hat{S}_C$ is the Kaplan-Meier estimate of S_C. It can be proved that

$$n^{1/2}\left(\hat{\tau}_j - \tau_j\right) = n^{-1/2} \sum_{i=1}^{n} \varphi_{i,j} + o_p(1).$$

For each j, $\varphi_{i,j}, i = 1 \ldots n$ are i.i.d. random variables. Thus, under mild regularity conditions, $\hat{\tau}_j$ is $\sqrt{n}$ consistent and asymptotically normal. Inference can be based on this asymptotic normality result. The definition of $\varphi_{i,j}$ and outline of the proof of asymptotic properties are provided in the Appendix of this chapter. Calculation of $\hat{\tau}_j$ and its significance test can be easily realized using existing software packages.

Example (MCL). We now analyze the MCL data described in Section 6.1. In real data analysis with a large number of covariates like the MCL data, we often conduct data preprocessing before implementing the statistical methods introduced in this chapter. We outline these procedures herein and recommend readers to check for relevant references for more details on these procedures. First, we need to fill in missing expression levels. Imputation is conducted in this case by using medians across samples. In the literature, a few model-based, computationally expensive imputation approaches have been proposed. We note that the impact of different imputation approaches has been less investigated. Our limited experience suggests that when the missingness is not severe, imputation may have a small impact. Second, since usually researchers are more interested in genes with a higher degree of variation of expressions, we conduct an unsupervised screening and select 2000 genes with the largest variances of expressions. Finally, we normalize the gene expressions to have marginal median zero and variance one.

In the following analysis, we set FDR = 0.1 and apply the approach described in preceding sections. In particular, we consider ranking using the nonparametric concordance measure as well as under three semiparametric regression models, namely the Cox, AFT, and additive risk models. Under the Cox model, the partial likelihood based approach is used for estimation and inference; under the additive risk model, we adopt the approach in Lin and

TABLE 6.3: Analysis of MCL Data: The Kendall Tau Rank Correlation of Rankings Using Different Approaches.

	Nonparametric	Cox	AFT	Additive
Nonparametric	1	0.604	0.546	0.603
Cox		1	0.687	0.902
AFT			1	0.576
Additive				1

Ying (1994a); under the AFT model, we assume unknown error distribution and use the weighted least squares approach in Stute (1993) for estimation and inference. The numbers of genes identified using different approaches and their overlaps are shown in Table 6.2. The nonparametric concordance measure identifies the most number of genes, while the AFT model identifies zero genes. The additive risk model highly agrees with the nonparametric method and Cox model as it identifies 25 genes of which 23 are identified by the nonparametric method and 24 are identified by the Cox model.

We also attain the complete ranking of all the 2000 genes and evaluate the (dis)similarity among the rankings using the Kendall's tau rank correlation coefficient. This measure is adopted as the rankings are discrete quantities. Results are shown in Table 6.3. The rankings from the Cox model and additive risk model are similar with a 90% rank correlation while the rankings between any other pairs of methods are only moderately correlated. This analysis suggests that the numbers of genes identified and rankings of genes highly depend on the ranking methods and survival models. We advise that practitioners consider multiple candidate approaches (particularly using parametric and semiparametric models) and conduct thorough examination and comparison. We have experimented with a few other data sets. In general, the discrepancy in ranking exists (between different methods). However, the degree of discrepancy depends on data.

Remark. Compared to (semi)parametric model-based approaches, the nonparametric concordance is less sensitive to model specification. However, like other nonparametric approaches, it may have lower efficiency. With a practical data set, we suggest conducting analysis under (semi)parametric models and also reporting the nonparametric concordance measure. If a specific (semi)parametric model generates results similar to that under the concordance measure, then this model, which is more efficient and interpretable, can be adopted in downstream analysis.

A taxonomy of dimension reduction and variable selection.

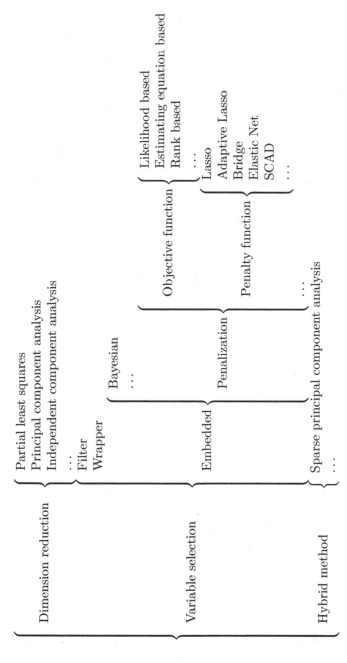

6.3   Multivariate Prediction Models

In this section, we investigate constructing multivariate predictive models $f(X_1,\ldots,X_p)$ which can describe the relationship between T and X and can be used to predict the survival risk of a new subject. With $p >> n$, standard survival analysis techniques are not directly applicable, as saturated models may have infinite many unreasonable estimates. The dimensionality of X needs to be reduced prior to or along with estimation. Statistical methodologies that can be used to reduce dimensionality can be roughly classified into three categories, namely dimension reduction methods, variable selection methods, and hybrid methods. We present a taxonomy of those methods below. Similar figures are also presented in Ma and Huang (2008) and Saeys et al. (2007).

All three classes of methods share the same spirit in that they all search for a new set of low-dimensional covariates to represent the effects of original ones. Dimension reduction methods construct new input covariates using linear combinations of all original covariates. Representative examples are partial least squares and principal component analysis. In contrast, variable selection methods search for a subset of the original covariates and use them as the new input covariates. Simple, well-known variable selection approaches include step-wise and best-subset approaches. More complicated and effective approaches include penalization and Bayesian approaches and others. Both dimension reduction and variable selection methods have a long history in survival analysis and other statistical research areas (Hastie et al., 2009). The third class of methods, the hybrid methods, are a "combination" of dimension reduction and variable selection methods. For example, the sparse principal component analysis constructs sparse loadings (that is, variable selection) in principal component analysis (which is a dimension reduction method). Such methods have been proposed in recent literature and have not been extensively used in practical survival analysis yet.

6.3.1   Dimension Reduction Approaches

Dimension reduction approaches construct a new set of covariates $\{X_j^* = b_{j1}X_1 + \ldots b_{jp}X_p, j = 1,\ldots,p^*\}$, where $p^* << p$. Usually, it is also taken that $p^* < n$. After transforming original high-dimensional covariates to low dimension, we can apply standard model-fitting approaches described in Chapter 2 to this new set of covariates. Thus, with dimension reduction approaches, the key step is the construction of $\{(b_{j1},\ldots,b_{jp}) :\ j = 1,\ldots,p^*\}$, which are referred to as "loadings" in the literature.

6.3.1.1 Partial Least Squares

Partial least squares (PLS) was first introduced in the field of chemometrics (Marten and Naes, 1989). With high-dimensional covariates, PLS was first successfully applied to data with categorical response variables (Nguyen and Rocke, 2002a). Nguyen and Rocke (2002b) later demonstrated that PLS also has satisfactory performance with right censored survival data under the Cox model.

The objective criterion for constructing loadings (and hence the new covariates) is to *sequentially* maximize the covariance between the response variable T and a linear combination of $\boldsymbol{X}$ (an $n \times p$ design matrix). Thus, we search for the loading vector $b_j = (b_{j1}, \ldots, b_{jp})^T$ satisfying the following objective criterion

$$b_j = \mathrm{argmax}_b Cov^2(\boldsymbol{X}b, T) = \mathrm{argmax}_b Corr^2(\boldsymbol{X}b, T)Var(\boldsymbol{X}b),$$

subject to $b^T b = 1$ and the orthogonality constraint

$$b_j^T Cov(\boldsymbol{X})b_k = 0, \quad \text{for all } j \neq k.$$

It is worth noting that the loadings are nonlinear functions of both T and $\boldsymbol{X}$, which is different from some other dimension reduction approaches, such as principal component analysis to be introduced in the next section. There are also some alternative ways of introducing this method.

The computation of PLS can be easily carried out for a Gaussian response T with complete observations. It is not entirely transparent how to handle censored nonnormal variables such as failure times. In the literature, three *ad hoc* methods are available:

1. The first approach is to consider censored failure times as missing observations and use an imputation algorithm to fill in unobserved failure times. The rest of the calculation of PLS is standard (Datta et al., 2007). One crucial assumption of this method is that the imputed quantity has approximately the same mean function as its unobserved counterpart. This method has only been used for AFT models.

2. Nguyen and Rocke (2002a) propose to consider (Y, δ), the observed survival time and survival indicator, as a bivariate response and subsequently conduct a bivariate PLS. The authors provided an SAS macro code which is available upon request. This method has been applied in a few genomic studies but comes with limited theoretical justification.

3. From an uncanny perspective, fitting the Cox model can be regarded as fitting a generalized linear model (GLM), specifically, a Poisson model (Whitehead, 1980). The estimation for GLM is via an iterative weighted least squares algorithm. Nygard et al. (2008) transplanted PLS to the Cox model by replacing the least squares step in the iterative calculation of GLM with a PLS. Their method has been implemented in a MATLAB code downloadable from: *http://www.med.uio.no/*

imb/english/research/networks/bmms/software/. It seems that performance of this method is the most stable from simulations and data analyses. However, there is still a lack of solid theoretical support in the literature.

The number of components, p^*, can be viewed as a tuning parameter. In general, a larger value of p^* may lead to better model fitting, however at the same time a higher risk of overfitting. In Nguyen and Rocke (2002b), two approaches have been proposed to choose p^*. The first is an *ad hoc* approach and specifies p^* as a small predefined number, for example, $p^* = 2$. The second approach, which may be more objective but computationally more expensive, uses cross-validation and searches for p^* that minimizes the predicted residual sum of squares. Nguyen and Rocke (2002b) analyzed a B-cell lymphoma study with gene expression measurements (Alizadeh et al., 2000) (sample size 40) and a breast carcinoma study reported in Sorlie et al. (2001) (sample size 49). It is shown that PLS with a very small value of p^* has satisfactory model fitting and prediction properties.

6.3.1.2   Principal Component Analysis

Principal component analysis (PCA) is one of the oldest dimension reduction approaches (Jolliffe, 1986). In PCA, the new input covariates $\{X_j^* : j = 1 \ldots p^*\}$ are referred to as "principal components" (PC). Unlike in PLS, in PCA, the loadings are constructed in an *unsupervised* manner. That is, the response variable T is not used in the construction of loadings.

To make covariates more comparable, we first conduct preprocessing so that the sample means of X_j ($j = 1, \ldots, p$) are all equal to zero and the sample variances are all equal to one. Denote $Cov_n(\boldsymbol{X})$ as the $p \times p$ sample variance-covariance matrix computed based on n i.i.d. observations. In PCA, eigenvalues and eigenvectors of $Cov_n(\boldsymbol{X})$ are computed. When $n > p$, this can be realized using the R function `princomp`. When $n < p$, PCA can be realized via singular value decomposition (R function `svd`). PCs are defined as the eigenvectors with nonzero eigenvalues and sorted by the magnitudes of corresponding eigenvalues, with the first PC having the largest eigenvalue.

There are two obvious reasons for promoting PCA in dimensional reduction analysis: (i) PCs sequentially capture the maximum variability among $\boldsymbol{X}$ and hence protect the procedure with the least loss of information (in terms of explaining variation); (ii) PCs are uncorrelated by construction, and we may refer to one PC without worrying about others. A major criticism, though, is that each PC is a linear function of all p variables X_j ($j = 1, \cdots, p$) with all loadings being nonzero. It is then difficult to give a sensible interpretation on the derived PCs. Two different sparse PCA approaches have been developed along with theoretical properties in the literature. The first approach (Johnstone and Lu, 2009) assumes a spiked population model where the first $M(< p)$ decreasingly ordered eigenvalues of $Cov_n(\boldsymbol{X})$ are greater than σ^2 and the remaining $p - M$ eigenvalues are all σ^2. The PCA is subsequently performed on a submatrix of $Cov_n(\boldsymbol{X})$ corresponding to selecting a subset of $\boldsymbol{X}$. The second approach (Jolliffe et al., 2003; Zou et al., 2006) employs a penalization idea to be discussed in Section 6.23. The upshot is that this approach gives zero loadings for the constructed PCs when sparsity is expected. An R package `elasticnet` can be adapted. We do not further elaborate on these extensions and focus on the ordinary PCA in the following presentation.

As PCA is performed on the matrices of correlation coefficients, data should satisfy certain assumptions. We refer to Chapter 6 of Hatcher and Stepanski (1994) for detailed discussions. Particularly for theoretical validity, it is assumed that data is normally distributed. This assumption is intuitively reasonable considering that when the mean is not of interest, the normal distribution is fully specified by the variance structure. In practice, for example, in gene expression analysis, data may not satisfy the normality assumption. In theory it is possible to transform covariates and achieve normality, although this is rarely done in practice because of high computational cost and a lack of appropriate methods for extremely high-dimensional data. Our literature review shows that PCA has been conducted with data obviously not having a normal distribution, and empirical studies demonstrate satisfactory performance, although there is a lack of theoretical justification for such results.

The PCs have the following main statistical properties: (a) $Cov(X_i^*, X_j^*) = 0$ if $i \neq j$. That is, different PCs are orthogonal to each other. In survival analysis with high-dimensional covariates, it is common that some covariates are highly correlated. PCA can effectively solve the collinearity problem. (b) The number of PCs is less than or equal to $min(n, p)$. In studies like the MCL, the dimensionality of PCs can be much lower than that of covariates. Thus, the PCs may not have the high-dimensionality problem encountered by gene expression and other high-dimensional data and have much lower computational cost. (c) Variation explained by PCs decreases, with the first PC explaining the most variation. Often the first few, say, three to five, PCs can explain the majority of variation. Thus, if the problem of interest is directly related to variation, it suffices to consider only the first few PCs. (d) When $p^* = rank(Cov_n(\boldsymbol{X}))$, any linear function of X_js can be written in terms of X_j^*. Thus, when focusing on the linear effects of covariates, using PCs can be equivalent to using original covariates. In PCA, the number of PCs can be determined in a similar manner as with PLS. There are also several other *ad hoc* approaches as described in Johnson and Wichern (2001).

Example (MCL). Consider the MCL data described in Section 6.1. We conduct the same preprocessing as described earlier and conduct analysis using the 2000 selected genes. We first conduct PCA using R and extract the first three PCs. We fit the Cox model with the first three PCs as covariates using the R function `coxph`. The response variables are stored in the file `survival.dat` (available at the first author's Web site), with (`Followup`, `Status`)$=(min(T, C), I(T \leq C))$. In this model, the regression coefficient of the first PC (denoted by `exp.PC1`) is highly significant with p-value 2.71×10^{-5}, and the regression coefficients of the other two PCs (denoted by exp.PC2 and exp.PC3) are not significant under the significance level 0.05.

```
coxph(formula = Surv(survival.dat$Followup, survival.dat$Status)
    ~exp.PC)
  n= 92
```

```
              coef exp(coef) se(coef)      z Pr(>|z|)
exp.PC1  0.06159    1.06353  0.01468  4.197 2.71e-05 ***
exp.PC2 -0.02481    0.97550  0.02160 -1.149  0.2507
exp.PC3  0.04606    1.04714  0.02528  1.822  0.0685 .
---
Signif. codes:  0 *** 0.001 ** 0.01 * 0.05 . 0.1   1

Rsquare= 0.191   (max possible= 0.993 )
Likelihood ratio test= 19.48  on 3 df,   p=0.0002178
Wald test            = 20.26  on 3 df,   p=0.0001497
Score (logrank) test = 20.71  on 3 df,   p=0.0001210
```

Motivated by the above results, we choose the Cox model with only the first PC as the final model. The estimation results for the first PC are only slightly different from those in the above model.

```
              coef exp(coef) se(coef)     z Pr(>|z|)
exp.PC.1 0.05863    1.06038  0.01398 4.195 2.73e-05 ***
---
Signif. codes:  0 *** 0.001 ** 0.01 * 0.05 . 0.1   1

Rsquare= 0.154   (max possible= 0.993 )
Likelihood ratio test= 15.39  on 1 df,   p=8.742e-05
Wald test            = 17.6  on 1 df,   p=2.728e-05
Score (logrank) test = 17.1  on 1 df,   p=3.556e-05
```

To further evaluate the Cox regression model with the first PC, we create two hypothetical risk groups with equal sizes based on the first PC. The survival functions of the two groups are plotted in Figure 6.2. It is clear that the two groups are well separated. The log-rank test statistic (χ^2 distributed with degree of freedom one), which measures the difference between survival, is 16.7 (p-value 4.38×10^{-5}). This result suggests that the first PC, which explains the most of the variation in gene expression, can provide an effective way of assigning subjects into different risk groups. Such a result is interesting considering the simplicity and unsupervised nature of PCs.

A remarkable advantage of PCA-based approaches is their low computational cost. When coded using R on a standard desktop computer, the computational time for the analysis of MCL data is almost negligible. In addition, PCA usually does not demand new software packages. It is interesting to note

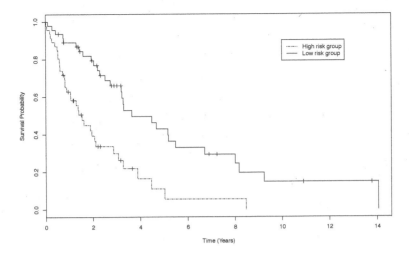

FIGURE 6.2: Analysis of MCL data: survival functions for high and low risk groups computed using the first PC.

that in the analysis of many data sets, the first few PCs, which are constructed independent of survival, can have satisfactory estimation and prediction performance.

On the other hand, PCA also suffers certain drawbacks. First, as the PCs are constructed in an unsupervised manner, the "order" of PCs, which is determined by the magnitude of eigenvalues, may have no direct implication in the strength of association with survival. Thus, in the analysis of MCL data, although the regression coefficients for the second and third PCs are not significant, we cannot deduce that the regression coefficients of higher-order PCs are all insignificant. In addition, as mentioned earlier, PCA-based analysis may suffer a lack of interpretability. PCs are linear combinations of all original covariates. The Cox model constructed using the first PC includes all original covariates with nonzero coefficients. Thus, if we simply look at the regression coefficients, we would conclude that all covariates are associated with the response variable. Such a conclusion may be unreasonable. In pangenomic studies, all genes are surveyed, whereas it is commonly accepted that only a small number of genes are "cancer genes." In Figure 6.3, we show the absolute values of loadings (sorted from smallest to largest) for the first PC. We can see that the loadings are "continuously distributed" across the 2000 genes. Therefore, it is difficult to represent or approximate the first PC using just a small number of covariates. The aforementioned two sparse PCA approaches (Zou et al., 2006; Johnstone and Lu, 2009) can thus be more appealing in

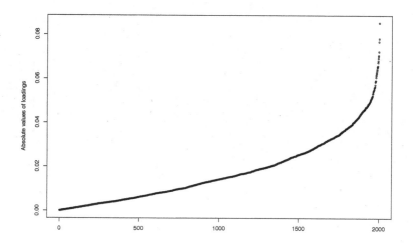

FIGURE 6.3: Analysis of MCL data: absolute values of loadings of the first PC.

these cases. One additional option is to use a supervised PCA approach (Bair and Tibshirani (2004), Bair et al. (2006)) which is closely related to PLS.

Example (Follicular Lymphoma). We revisit the follicular lymphoma study. We first use the ordinary PCA and construct the first PC which is a linear combination of all $p = 6506$ gene expressions. A binary indicator is formed by using the mean of the first PC as a threshold. Comparing survival probabilities between the two groups (high and low levels of the first PC) with a log-rank test, we obtain a p-value of 0.35 (see the left panel of Figure 6.4). Next we consider using a sparse PCA approach (following the elastic net approach in Zou et al., 2006) and produce the first PC with 30 nonzero loadings. This can be achieved in R using function spca in library elasticnet. After dichotomizing in the same manner, we obtain a p-value of 0.16 in a log-rank test. Although the difference may not be significant, the sparse PCA approach may be more helpful to build a latent characteristic variable that distinguishes the survival experiences. In this example, we fix the number of nonzero loadings to be 30. In practice this number or, alternatively, a regulating parameter needs to be selected by a cross-validation procedure.

Remark. Many simulation studies confirm the superiority of the sparse PCA over ordinary PCA where data are indeed generated according to a sparse design (Bair et al., 2006; Zou et al., 2006; Nygard et al., 2008). However, comparisons made on practical data sets suggest the other way around (Bovelstad et al., 2007, 2009). For a real data set involving patients with cancer or other fatal diseases, accurate survival predictions are critical for better treatment options and prolonged survival. The prediction performance (in terms of log-rank test, prognostic index, and model deviance) is better when ordinary PCA is used. One plausible explanation behind this empirical discrepancy is that real gene expressions may come with continuous effects ranging from small to large. Sparse PCA preselects a subgroup of genes with large effects and completely excludes others with small, yet nonzero, effects from consideration, while all genes are represented in an ordinary PCA with small to large loadings in a more "continuous manner" (see for example, Figure 6.3).

The above applications of PCA are "standard" in the sense that PCA is conducted straightforwardly with covariates. In the following, we introduce two "nonstandard" PCA-based methods.

Conducting PCA with Estimating Equations. The additive risk and AFT models have been introduced in Chapter 2. Under those two models, it is possible to directly apply the PCA in a similar manner as with the Cox model. However, it is also possible to take advantage of the estimating equations and conduct PCA.

Additive Risk Model Consider the additive risk model where the conditional hazard function $h(t|\boldsymbol{X}) = h_0(t) + \boldsymbol{X}^T\boldsymbol{\beta}$. Assume n i.i.d. observations $\{(Y_i = \min(T_i, C_i), \delta_i = I(T_i \leq C_i), \boldsymbol{X}_i), i = 1, \ldots, n\}$. For the ith subject, denote $X_{i,j}$ as the jth component of $\boldsymbol{X}$. For the ith subject, denote $\{N_i(t) = I(Y_i \leq$

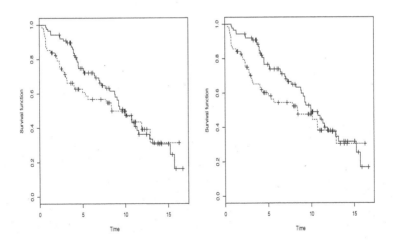

FIGURE 6.4: Kaplan-Meier curves for two groups corresponding to high and low values of the first PC. The left panel is based on the ordinary PCA, while the right panel is based on the sparse PCA.

$t, \delta_i = 1); t \geq 0\}$ and $\{A_i(t) = I(Y_i \geq t); t \geq 0\}$ as the observed event process and at-risk process, respectively.

Lin and Ying (1994a) show that $\boldsymbol{\beta}$ can be estimated by solving

$$U(\boldsymbol{\beta}) = \sum_{i=1}^{n} \int_0^{\infty} \boldsymbol{X}_i \{dN_i(t) - A_i(t)d\hat{H}(\boldsymbol{\beta}, t) - A_i(t)\boldsymbol{X}_i^T \boldsymbol{\beta} dt\} = 0.$$

Here, $\hat{H}(\boldsymbol{\beta}, t)$ is the estimate of $H_0(t)$ (cumulative baseline hazard function) satisfying

$$\hat{H}(\boldsymbol{\beta}, t) = \sum_i \int_0^t \frac{dN_i(u) - A_i(u)\boldsymbol{X}_i^T \boldsymbol{\beta} du}{\sum_i A_i(u)}.$$

The resulting estimate of $\boldsymbol{\beta}$ satisfies the estimating equation

$$\left[\sum_{i=1}^{n} \int_0^{\infty} A_i(t)(\boldsymbol{X}_i - \bar{\boldsymbol{X}}(t))^{\otimes 2} dt \right] \hat{\boldsymbol{\beta}} = \left[\sum_{i=1}^{n} \int_0^{\infty} (\boldsymbol{X}_i - \bar{\boldsymbol{X}}(t)) dN_i(t) \right].$$

Here, $\bar{\boldsymbol{X}}(t) = \frac{\sum_{i=1}^{n} \boldsymbol{X}_i A_i(t)}{\sum_{i=1}^{n} A_i(t)}$. Note that the term

$$L = \left[\sum_{i=1}^{n} \int_0^{\infty} A_i(t)(\boldsymbol{X}_i - \bar{\boldsymbol{X}}(t))^{\otimes 2} dt \right]$$

is symmetric, semipositive definite, and mimics a variance-covariance matrix. Following a similar strategy as with PCA and conducting SVD, there exist matrices P and $M = diag(m_1, \ldots, m_k, 0, \ldots, 0)$ such that $L = PMP^T$ and $PP^T = I_p$ (the $p \times p$ identity matrix). Denote $R = \left[\sum_{i=1}^{n} \int_0^{\infty} (\boldsymbol{X}_i - \bar{\boldsymbol{X}}(t)) dN_i(t) \right]$. Define the general inverse of M as $M^G = diag(\frac{1}{m_1}, \ldots, \frac{1}{m_k}, 0, \ldots, 0)$. Then from the estimating equation, we have

$$\hat{\boldsymbol{\beta}} = P\hat{\boldsymbol{\gamma}}, \text{ where } \hat{\boldsymbol{\gamma}} = M^G P^T R.$$

In Ma et al. (2006), it is suggested that with high-dimensional covariates, some components of $\hat{\boldsymbol{\gamma}}$ may have estimated variances several orders larger than the other components, which suggest unreliable estimates. With the standard PCA, using only the first few PCs with large eigenvalues may improve the stability of estimates. Motivated by such considerations, PCA-based estimates can be defined by removing PCs with small eigenvalues in the decomposition of L. Specifically, denote $S = diag(I_{p^*}, 0)$, where I_{p^*} is the p^*-dimensional identity matrix with $p^* \leq rank(L)$. Define the PCA-based estimate as

$$\hat{\boldsymbol{\gamma}}_{PC} = S\hat{\boldsymbol{\gamma}} \text{ and } \hat{\boldsymbol{\beta}}_{PC} = P\hat{\boldsymbol{\gamma}}_{PC}.$$

Ma et al. (2006) analyze two data sets. The first is a "classic" multivariate survival data set with a relatively small number of covariates; and the second is a cancer prognosis study with high-dimensional microarray measurements.

Numerical study shows that the estimating equation-based PCA approach uses only a small number of PCs. The estimates are more stable, and the prediction performance is significantly better than that of several alternatives.

AFT Model   With slight abuse of notations, we use T and C to denote the transformed event and censoring times, respectively. Under the AFT model, $T = \alpha + \boldsymbol{X}^T\boldsymbol{\beta} + \epsilon$, where α is the intercept and ϵ is the unknown random error. Unlike the Cox and additive risk models, the AFT model describes the event time directly (as opposed to survival hazard) and may provide a more lucid interpretation.

When $n >> p$, multiple approaches have been proposed for estimation with a right censored AFT model. Popular examples include the Buckley-James estimator (Buckley and James, 1979) and the rank-based estimator (Ying, 1993). However, the computational cost of these approaches can be too high for high-dimensional data. A computationally more feasible alternative is the weighted least squares approach (Stute, 1993), which has been adopted in several gene expression studies.

Let $\hat{F}_n$ be the Kaplan-Meier estimator of the distribution function F of T. Following Stute (1993), $\hat{F}_n$ can be written as $\hat{F}_n(y) = \sum_{i=1}^{n} w_i I(Y_{(i)} \leq y)$, where w_is are the Kaplan-Meier weights defined as

$$w_1 = \frac{\delta_{(1)}}{n} \quad \text{and} \quad w_i = \frac{\delta_{(i)}}{n-i+1} \prod_{j=1}^{i-1} \left(\frac{n-j}{n-j+1} \right)^{\delta_{(j)}}, j = 2, \ldots, n.$$

Here, $Y_{(1)} \leq \cdots \leq Y_{(n)}$ are the order statistics of Y_i's, and $\delta_{(1)}, \ldots, \delta_{(n)}$ are the associated censoring indicators. Similarly, let $\boldsymbol{X}_{(1)}, \ldots, \boldsymbol{X}_{(n)}$ be the associated covariates of the ordered Y_i's.

Stute (1993) proposed the weighted least squares estimator $(\hat{\alpha}, \hat{\boldsymbol{\beta}})$ that minimizes

$$M(\alpha, \boldsymbol{\beta}) = \frac{1}{2} \sum_{i=1}^{n} w_i (Y_{(i)} - \alpha - \boldsymbol{X}_{(i)}^T\boldsymbol{\beta})^2.$$

Let $\bar{\boldsymbol{X}}_{wi} = \frac{\sum_{i=1}^{n} w_i \boldsymbol{X}_{(i)}}{\sum_{i=1}^{n} w_i}$ and $\bar{Y}_{wi} = \frac{\sum_{i=1}^{n} w_i Y_{(i)}}{\sum_{i=1}^{n} w_i}$. To obtain a simplified format of the objective function, we make the transformation $\tilde{\boldsymbol{X}}_{(i)} = w_i^{1/2}(\boldsymbol{X}_{(i)} - \bar{\boldsymbol{X}}_{wi})$ and $\tilde{Y}_{(i)} = w_i^{1/2}(Y_{(i)} - \bar{Y}_{wi})$. Using the weighted centered values, the intercept estimate $\hat{\alpha}$ is zero. So the weighted least squared objective function can be written as

$$M(\boldsymbol{\beta}) = \frac{1}{2} \sum_{i=1}^{n} (\tilde{Y}_{(i)} - \tilde{\boldsymbol{X}}_{(i)}^T\boldsymbol{\beta})^2. \tag{6.1}$$

Consider the following PCA-based approach. Denote $\mathbb{X}^T$ as the $n \times p$ matrix composed of $\tilde{\boldsymbol{X}}_{(i)}^T$s and $\mathbb{Y}$ as the length n vector composed of $\tilde{Y}_{(i)}$s. Note that $\mathbb{X}$ is composed of "weighted covariates," which may differ significantly from the ordinary design matrix. The estimate $\hat{\boldsymbol{\beta}}$ defined as the minimizer of (6.1) satisfies $\{\mathbb{X}\mathbb{X}^T\}\hat{\beta} = \mathbb{X}\mathbb{Y}$. Since $\mathbb{X}\mathbb{X}^T$ is a semipositive-definite matrix, there exists a p dimensional square matrix P satisfying

$$P^T\mathbb{X}\mathbb{X}^T P = M = diag(m_1, m_2, \ldots, m_k, 0, \ldots, 0) \text{ and } PP^T = I_p. \quad (6.2)$$

Here, I_p denotes the p-dimensional identity matrix. k is the rank of $\mathbb{X}\mathbb{X}^T$. For data with high-dimensional covariates, $n << p$ and thus $k << p$. P is composed of eigenvectors of $\mathbb{X}\mathbb{X}^T$, and m_is are the eigenvalues of $\mathbb{X}\mathbb{X}^T$.

So we have $P^T\mathbb{X}\mathbb{X}^T PP^T\hat{\beta} = P^T\mathbb{X}\mathbb{Y}$ and $MP^T\hat{\beta} = P^T\mathbb{X}\mathbb{Y}$. If we denote $P^T\hat{\beta} = \hat{\gamma}$ and $M^G = diag(1/m_1, \ldots, 1/m_k, 0, \ldots, 0)$, it can be seen that one solution to the Stute estimating equation when $n << p$ is $\hat{\gamma} = M^G P^T\mathbb{X}\mathbb{Y}$ and $\hat{\beta} = P\hat{\gamma}$.

Empirical studies have shown that for data with small to moderate sample sizes, when p is comparable to or larger than n, some components of $\hat{\gamma}$ can have estimated variances several orders larger than the other components, which indicates unstable estimates. For ultrahigh-dimensional data such as the MCL data, this poses especially serious concerns. This phenomenon motivates using the PCA to yield more reliable estimators by excluding certain principal components from the regression.

Denote S as the component-selection matrix with certain diagonal elements equal to 1 and all other elements equal to 0. For example, if only the principal components corresponding to the first p^* elements of $\hat{\gamma}$ are selected, then $S = diag(I_{p^*}, 0)$, where I_{p^*} denotes the p^*-dimensional identity matrix. The PCA estimator can then be defined as

$$\hat{\gamma}_{pc} = S\hat{\gamma} \text{ and } \hat{\beta}_{pc} = P\hat{\gamma}_{pc}.$$

Accommodating Interaction Effects Using PCA. In classic survival analysis with $n >> p$, it is straightforward to deal with higher-order terms such as quadratics of covariate effects and second-order interactions. With high-dimensional covariates, analyzing higher-order terms can be nontrivial. Consider, for example, the MCL data. Incorporating all second-order terms leads to a model with $\sim 10^6$ covariates, which is both theoretically and computationally prohibitive. PCA may provide an effective way of studying higher-order terms.

Ma and Kosorok (2010) and Ma et al. (2011b) propose the following ways of constructing PCA-based covariates, which are then used in survival analysis as opposed to the original covariates:

(A1) Conduct PCA on the original covariates. Denote $\{X_1^*, \ldots, X_{p^*}^*\}$ as the p^* PCs. Note that $p^* << p$. The new set of covariates is $\{X_1^*, \ldots, X_{p^*}^*\} \cup \{X_i^* \times X_j^* : i, j = 1 \ldots, p^*\}$, which has dimensionality $p^* + p^{*2}$.

(A2) Conduct PCA on the set $\{X_1, \ldots X_p\} \cup \{X_i \times X_j : i, j = 1 \ldots p\}$. Denote $\{X_1^{**}, \ldots, X_{p^{**}}^{**}\}$ as the p^{**} PCs. The new covariate set is composed of the p^{**} PCs or its subset.

Using A1 and A2 as opposed to the simple PCs has been proposed in recent studies. With A1, interactions among covariates are accommodated via the interactions among PCs. This approach is feasible as the dimensionality of PCs is usually much smaller than that of original covariates. With A2, interactions among covariates are first computed, and then PCA is conducted. Although the dimensionality of $\{X_1, \ldots X_p\} \cup \{X_i \times X_j\}$ can be very high, many software packages are capable of conducting SVD effectively and carrying out PCA.

Example (MCL). We reanalyze the MCL data. Specifically, we fit a Cox model with the first PC (denoted as `PC.1`) and its quadratic (denoted as `PC.1.squared`) as covariates and present the analysis result below. The first PC remains highly significantly, and its quadratic is borderline significant with p-value 0.07.

```
coxph(formula = Surv(survival.dat$Followup, survival.dat$Status)~
   PC.1 + PC.1.squared)
  n= 92

                 coef exp(coef)  se(coef)       z Pr(>|z|)
PC.1         0.084079  1.087715  0.021507   3.909 9.25e-05 ***
PC.1.squared -0.001932  0.998070  0.001070  -1.805   0.0711 .
---
Signif. codes:  0 '***' 0.001 '**' 0.01 '*' 0.05 '.' 0.1 ' ' 1

Rsquare= 0.19   (max possible= 0.993 )
Likelihood ratio test= 19.4  on 2 df,    p=6.124e-05
Wald test          = 15.69  on 2 df,    p=0.0003912
Score (logrank) test = 17.15  on 2 df,    p=0.0001885
```

Theoretical Validity of PCA-Based Approaches. Except for conducting PCA with estimating equations, the PCA-based approaches described above consist of two main steps. The first step is the construction of PCA-based covariates. The second step is survival model fitting. As the dimensionality of PCA-based covariates is usually low, the validity of survival model fitting follows from standard likelihood- or estimating equation-based theories. The validity of the first step, which involves constructing the PCs, can be highly nontrivial. In analysis with a small to moderate number of covariates (where p can be taken as fixed), the consistency of the PCs may follow Johnson and Wichern (2001). With high-dimensional covariates, it has been proved that under mild assumptions, if $p/n \to 0$ as $p, n \to \infty$, the variance-covariance matrix and the first fixed number of PCs can be consistently estimated. Such a consistency result, although insightful, may not be quite useful in, for example, cancer microarray studies. In recent studies, it has been shown that if certain assumptions on the variance matrix hold, then less strict requirements on p can be assumed. For example, in Bickel and Levina (2008), the "bandable" assumption is made, and it is proved that if $\log(p)/n \to 0$, the estimates of the first fixed number of PCs are consistent. Note that this result allows the number of covariates to be much larger than the sample size. The bandable assumption postulates that a covariate is only correlated with a small number of covariates, and not or only weakly correlated with others. The validity of such an assumption needs to be determined on a case-by-case basis depending on the scientific context and properties of the estimated variance-covariance matrix. For example, in cancer genomic studies, this assumption is perhaps reasonable considering that the expression of a gene tends to be correlated only with those of genes with similar biological functions (for example, belonging to the same pathways or network modules). In practical data analysis, an *ad hoc* approach is to compute the variance-covariance matrix first and then count how many elements have absolute values above a certain cut-off, say 0.3. If there are only a small percentage of the elements above the cut-off, then the bandable assumption is perhaps reasonable. The theoretical validity of conducting PCA with estimating equations has been investigated under the assumption of fixed p. Properties under the more challenging setting with $n << p$ are still unclear and warrant further investigations.

Remark. In this section, we introduce the partial least squares approach and provide in-depth discussion of the principal component analysis approach. Between PLS and PCA, one may prefer PLS since more information from the response is used. Indeed the linear combination from PLS is based on the consideration of both high variance and significant correlation with the response. However, for most real data sets these two approaches do not differ significantly in terms of how accurately their generated models predict the survival outcomes (Bovelstad et al., 2007; Nygard et al., 2008; Bovelstad et al., 2009). This is perhaps because the sample variance of **X** dominates over the covariance, a phenomenon witnessed in many empirical studies.

Beyond PLS and PCA, there are several other dimension reduction approaches, including, for example, ICA (independent component analysis) and SIR (slice inverse regression). They share a similar spirit with PLS and PCA, but differ in the way the loadings are constructed. Dimension reduction approaches usually enjoy low computational cost. On the other hand, they may suffer a lack of interpretability. In gene expression studies, the PCs have been referred to as "latent genes" and "super genes." However, there has been little success linking the PCs to certain biological processes. Despite praiseworthy progress, there is still a lag between the development of dimension reduction methodologies and establishment of their statistical properties.

6.3.2 Variable Selection Methods

Variable selection approaches differ from dimension reduction approaches in that they search for a subset of covariates to represent the effects of all original covariates. Simple variable selection approaches such as the forward selection, backward elimination, stepwise regression, best subset, bootstrap (Sauerbrei and Schumacher, 1992) and Bayesian methods (Faraggi and Simon, 1998; Ibrahim et al., 1999) have been extensively used in "classic" survival analysis with a small number of covariates. However, with high-dimensional covariates, those approaches are usually not useful since they lack stability and may invoke high computational cost. In this section, we provide discussions of penalized variable selection approaches, which have attracted tremendous attention in recent statistical and biomedical literature.

Early development in penalized variable (model) selection includes Akaike information criterion (AIC) and Bayesian information criterion (BIC), both introduced in Chapter 2 as common model selection tools. With "standard" likelihood or estimating equation-based approaches, the objective functions for estimation (such as the likelihood function or least square criterion) is usually a measure of goodness-of-fit. With AIC and BIC, the objective functions have two terms. The first term is adopted from the likelihood function or estimating equation and measures goodness-of-fit. The second term is a measure of model complexity. The model obtained by optimizing the so-constructed objective function is expected to provide a balance between model fitting and complexity. Despite their reasonable performance, the AIC and BIC approaches usually demand an intensive evaluation of a large number of models, as the model complexity measures are not continuous (thus prohibit gradient-based algorithms) and cannot be easily optimized. In what follows, we describe a family of penalized variable selection approaches, which may have better statistical properties and/or lower computational cost than stepwise regression, best subset selection, and selection based on AIC and BIC.

Assume that the event time is associated with covariates via the model $T \sim f(\mathbf{X}^T \boldsymbol{\beta})$, where $\boldsymbol{\beta} = (\beta_1, \ldots, \beta_p)^T$ is the length-p unknown regression coefficient. Denote $D = \{(Y_i = \min(T_i, C_i), \delta_i = I(T_i \leq C_i), \mathbf{X}_i) : i = 1 \ldots n\}$ as n i.i.d. observations. In penalized variable selection, the penalized estimate is defined as

$$\hat{\boldsymbol{\beta}} = \operatorname{argmin}_{\boldsymbol{\beta}} \left\{ R(D; \boldsymbol{\beta}) + pen(\lambda, \boldsymbol{\beta}) \right\}. \tag{6.3}$$

In (6.3), $R(D; \boldsymbol{\beta})$ can be regarded as a function of both the observed data and regression coefficient. It measures the (lack of) goodness-of-fit. $pen(\lambda, \boldsymbol{\beta})$ is the penalty function and measures model complexity. It depends on a data-dependent tuning parameter λ and the regression coefficient. With a few penalties, there may be additional tuning/regularization parameters. With a properly chosen tuning parameter and penalty function, $\hat{\boldsymbol{\beta}}$ may have some estimated components *exactly* equal to zero. Only covariates with nonzero estimated regression coefficients are included in the final survival analysis model and identified as being associated with survival. Thus, with penalized variable selection approaches, *model estimation and variable selection are achieved simultaneously.* Variable selection amounts to identifying the nonzero components of penalized estimates. Note that this variable selection strategy may differ significantly from that based on the p-value.

Properties of $\hat{\beta}$ and variable selection are fully determined by the functions $R(\cdot)$, $pen(\cdot)$ and tuning parameter λ. In the following two sections, we provide detailed investigation of R and pen, respectively. The selection of λ is incorporated in the discussion of pen since the meaning of λ usually depends on the specific pen function.

6.3.2.1 Measures of Goodness-of-Fit

To simplify notations, we denote $R(D; \beta)$ as $R(\beta)$. Under parametric survival models, $R(\beta)$ can be simply taken as the negative log-likelihood functions. In the following we briefly describe the $R(\beta)$ functions for three commonly adopted semiparametric models and a nonparametric model. Other semi- and nonparametric models can be studied in a similar manner.

Cox Model. Under the Cox PH model, the conditional hazard function given the covariates is $h(t|\boldsymbol{X}) = h_0(t) \exp(\boldsymbol{X}^T \boldsymbol{\beta})$. Under right censoring, the model fitting is via maximizing the partial likelihood function. We choose $R(\beta)$ as the negative log-partial likelihood function, where

$$R(\beta) = -\sum_{i=1}^{n} \delta_i \left\{ \boldsymbol{X}_i^T \beta - \log \left(\sum_{k \in r_i} \exp(\boldsymbol{X}_k^T \beta) \right) \right\}. \tag{6.4}$$

Here $r_i = \{k : Y_k \geq Y_i\}$ is the at-risk index set at time Y_i.

Additive Risk Model. Under the additive risk model, the conditional hazard function given the covariates is $h(t|\boldsymbol{X}) = h_0(t) + \boldsymbol{X}^T \beta$. When $n > p$, Lin and Ying's approach (Lin and Ying, 1994a) estimates β via the following estimating equation:

$$\left[\sum_{i=1}^{n} \int_0^\infty A_i(t)(\boldsymbol{X}_i - \bar{\boldsymbol{X}}(t))^{\otimes 2} dt \right] \hat{\beta} = \left[\sum_{i=1}^{n} \int_0^\infty (\boldsymbol{X}_i - \bar{\boldsymbol{X}}(t)) dN_i(t) \right].$$

The linear regression format of the above estimating equation motivates the following objective function

$$R(\beta) = \left(\left[\sum_{i=1}^{n} \int_0^\infty A_i(t)(\boldsymbol{X}_i - \bar{\boldsymbol{X}}(t))^{\otimes 2} dt \right] \beta \right.$$
$$\left. - \left[\sum_{i=1}^{n} \int_0^\infty (\boldsymbol{X}_i - \bar{\boldsymbol{X}}(t)) dN_i(t) \right] \right)^2. \tag{6.5}$$

Note that this is just one of the multiple possible choices. Other least-squares type objective functions can be constructed.

AFT Model. With a slight abuse of notation, we use T to denote the transformed event time. Under the AFT model, $T = \alpha + \boldsymbol{X}^T \beta + \epsilon$. When the distribution of the random error ϵ is unknown, Stute (1993) proposes a weighted least squares estimation approach, where the objective function is

$$R(\boldsymbol{\beta}) = \frac{1}{2} \sum_{i=1}^{n} (\tilde{Y}_{(i)} - \tilde{\mathbf{X}}_{(i)}^T \boldsymbol{\beta})^2. \tag{6.6}$$

We refer to Section 6.3.1 for more details on notations.

Concordance. Under the nonparametric model, it is assumed that the link function f is a monotone increasing function. However, its specific form remains unspecified. In Section 6.2, we describe a nonparametric concordance measure, which can assess the *concordance* between the survival time and a covariate (or a linear combination of covariates). The concordance measure can also be used as an objective function in the construction of multivariate predictive model. More specifically,

$$R(\boldsymbol{\beta}) = -\tau = -\frac{\sum_{i=1}^{n} \sum_{k=1}^{n} \frac{\delta_i}{\hat{S}_C^2(Y_i)} I(\mathbf{X}_i^T \boldsymbol{\beta} < \mathbf{X}_k^T \boldsymbol{\beta}, Y_i < Y_k)}{\sum_{i=1}^{n} \sum_{k=1}^{n} \frac{\delta_i}{\hat{S}_C^2(Y_i)} I(Y_i < Y_k)}. \tag{6.7}$$

Unlike with the Cox, additive risk, and AFT models, the above objective function is not continuous. Optimization with respect to discrete functions is computationally challenging even when there are only a small number of covariates. With high-dimensional covariates, the computational challenge can be even more prohibitive. To tackle this problem, we approximate the indicator function $I(x)$ with a continuously differentiable function $g(x)$, which is monotone and satisfies $g(-\infty) = 0$ and $g(+\infty) = 1$. A simple approximation is the sigmoid approximation investigated in Ma and Huang (2007), where

$$g(x) = \frac{1}{1 + \exp(-\frac{x}{\sigma_n})}.$$

Here, $\sigma_n \to 0$ as $n \to \infty$ is a data-dependent tuning parameter and can be chosen via, for example, V-fold cross-validation.

Unlike with the parametric and semiparametric models, under the nonparametric model, the coefficient vector $\boldsymbol{\beta}$ is identifiable only up to a scale constant. Thus, in the penalized estimation, $\hat{\boldsymbol{\beta}}$ needs to be estimated under further constraints. A commonly adopted constraint is the unit norm constraint $||\hat{\boldsymbol{\beta}}||_2 = 1$. A computationally simpler alternative is to reinforce that the estimate of the first component (or another fixed component) satisfies $|\hat{\beta}_1| = 1$.

Remark. The above examples of goodness-of-fit measure include likelihood-based (Cox model), *ad hoc* estimating equation-based (additive risk model), estimating equation-based (AFT model), and a concordance measure which is not based on likelihood or estimating equation. The goodness-of-fit measure needs to be chosen so that the statistical properties of the resulting estimates are warranted. Specifically, first, under the "classic" large-sample condition with $n >> p$, $\tilde{\boldsymbol{\beta}} = \mathrm{argmin}R(\boldsymbol{\beta})$ should be a consistent estimate of the true regression coefficient. Second, due to the consideration on downstream computational cost, it may be more convenient to require that $R(\boldsymbol{\beta})$ is continuously differentiable. Sometimes, other considerations such as efficiency need to be taken into account. For example, under the Cox model, both the negative log-partial likelihood and the concordance measure can be used. However, when the model is properly specified, the likelihood objective function may produce more efficient estimates and hence is preferred.

6.3.2.2 Penalty Functions

Lasso. Lasso is the acronym for least absolute shrinkage and selection operator. Tibshirani (1996) first proposed the Lasso penalty for variable selection with continuous responses and linear regression models. It was later extended to generalized linear models and survival models (Tibshirani, 1997). It is perhaps the most widely used penalty in the research field of variable selection. Under the Lasso penalty, the penalized estimate is defined as

$$\hat{\boldsymbol{\beta}} = \mathrm{argmin}_{\boldsymbol{\beta}} R(\boldsymbol{\beta}) + \lambda \sum_{j=1}^{p} |\beta_j|, \qquad (6.8)$$

which is also equivalent to

$$\hat{\boldsymbol{\beta}} = \mathrm{argmin}_{\boldsymbol{\beta}} R(\boldsymbol{\beta}), \quad \text{subject to} \sum_{j=1}^{p} |\beta_j| \leq u. \qquad (6.9)$$

There is a one-to-one correspondence between the tuning parameters λ and u in the two definitions. However, unless under simple scenarios (for example, linear regression models with orthogonal designs), such correspondence does not have an analytic form.

The penalty function for Lasso may be construed as a special case of the Bridge-type penalty function

$$pen = \lambda \sum_{j=1}^{p} |\beta_j|^{\gamma}, \qquad \gamma > 0 \qquad (6.10)$$

introduced in Frank and Friedman (1993) by setting $\gamma = 1$. Another special case of the Bridge penalty is when $\gamma = 2$ and corresponds to the Ridge penalty (Hoerl and Kennard, 1970). Ridge regression may serve as an appealing alternative to ordinary regression by reducing the estimation variance at the price of invoking a small amount of bias (van Houwelingen et al., 2006). However, for the purpose of variable selection, Ridge regression or more generally, Bridge regression with $\gamma > 1$ may not be a qualified candidate since these penalties do not shrink the estimated coefficients to zero. When $\gamma < 1$, some authors have found satisfactory performance of Lasso-type penalties for variable selection with both small (Knight and Fu, 2000) and large p (Hunter and Li, 2005).

Statistical Properties   Tibshirani (1996) showed that for linear regression models under orthogonal designs, the Lasso estimates have a simple soft thresholding form. Particularly, it shrinks small estimates to be exactly zeros. In general, the Lasso estimates do not have the simple connection with soft-thresholding. However, the shrinkage property in general holds. Statistical properties of the Lasso estimates depend on the tuning parameter λ. As $\lambda \to \infty$, heavier penalties lead to more estimates shrunken to zero, and so fewer covariates are identified as associated with survival. In statistical literature, the plot of estimates as a function of tuning parameter λ has been referred to as "parameter path." In Figure 6.5, we analyze the MCL data and show the Lasso parameter path. We can see that when $\lambda = 25$, only one covariate has a nonzero regression coefficient. When $\lambda = 20$, three covariates are identified as associated with survival. A large number of covariates have nonzero regression coefficients when $\lambda = 6$. Note that with Lasso, the number of nonzero coefficients is bounded by sample size, which differs significantly from the well-known ridge regression (which has all estimated regression coefficients nonzero). Thus, under the $p >> n$ scenario, the Lasso estimate is sparse in that most components of the estimate are exactly zero.

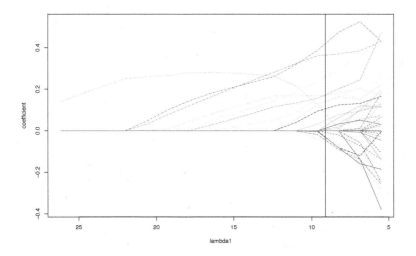

FIGURE 6.5: Analysis of MCL data: Lasso parameter path.

Zhao and Yu (2006) show that under a partial orthogonality condition, the Lasso can consistently discriminate covariates with nonzero effects from those with zero effects. The partial orthogonality condition postulates that "signals" (covariates with nonzero regression coefficients) and "noises" (covariates with zero regression coefficients) are weakly or not correlated. The validity of this assumption is nontrivial and needs to be established on a case-by-case basis. In practical data analysis, this assumption may be very difficult to verify. In Zhang and Huang (2008), it is shown that under conditions weaker than the partial orthogonality condition, the Lasso is capable of identifying all of the true positives. However it tends to overselect, that is, the Lasso may also select a small number of false positives. In Leng et al. (2006), it is shown that in general regression settings, the Lasso is not a consistent approach for variable selection. Other theoretical studies of Lasso include Meinshausen and Yu (2009) and Wang et al. (2006), among others.

Property of the Lasso estimate under the AFT model is rigorously established in Huang and Ma (2010). We outline the main results in the Appendix (Section 6.7) of this chapter.

Computational Algorithms   The Lasso objective function is not differentiable at $\beta_j = 0$. Thus, simple optimization algorithms such as the Newton-Raphson are not directly applicable. Multiple algorithms have been proposed for Lasso optimization. Early algorithms are based on quadratic programming and can be computationally intensive. In Efron et al. (2004), the computationally efficient LARS (least angle regression) algorithm is proposed for linear regression models under the Lasso penalization. For generalized linear models and many survival models, the optimization can be rewritten as an iterative procedure, where at each iteration, a weighted least squares criterion is optimized. Motivated by such a result, researchers have extended the LARS algorithm to censored survival analysis (Gui and Li, 2005). Below we describe two alternative computational algorithms, namely the gradient boosting algorithm and the coordinate descent algorithm. They can be more efficient and more broadly applicable than the LARS algorithm.

Gradient Boosting Algorithm

Kim and Kim (2004) relate the optimization of Lasso estimate to the L_1 boosting algorithm. The gradient boosting algorithm is designed for the constraint optimization problem in (6.9). With a fixed u, this algorithm can be implemented as follows:

1. Initialize $\beta_s = 0$ for $s = 1, \ldots, p$ and $m = 0$.

2. With the current estimate of $\boldsymbol{\beta} = (\beta_1, \ldots, \beta_p)$, compute $\phi(\boldsymbol{\beta}) = \frac{\partial R(\boldsymbol{\beta})}{\partial \boldsymbol{\beta}}$. Denote ϕ_k as the kth component of ϕ.

3. Find k^* that minimizes $min(\phi_k(\boldsymbol{\beta}), -\phi_k(\boldsymbol{\beta}))$. If $\phi_{k^*}(\boldsymbol{\beta}) = 0$, then stop the iteration.

4. Otherwise denote $\gamma = -sign(\phi_{k^*}(\boldsymbol{\beta}))$. Find $\hat{\alpha}$ such that

$$\hat{\alpha} = argmin_{\alpha \in [0,1]} R((1 - \alpha)(\beta_1, \ldots, \beta_p) + \alpha \times u \times \gamma \eta_{k^*}),$$

where η_{k^*} has the k^{*th} element equals to 1 and the rest equal to 0.

5. Let $\beta_k = (1-\hat{\alpha})\beta_k$ for $k \neq k^*$ and $\beta_{k^*} = (1-\hat{\alpha})\beta_{k^*} + \gamma u \hat{\alpha}$. Let $m = m+1$.

6. Repeat Steps 2–5 until convergence or a fixed number of iterations N has been reached.

The β at convergence is the Lasso estimate. The convergence criterion can be defined based on multiple quantities, including, for example, the absolute value of $\phi_{k^*}(\beta)$ computed in Step 3, $R(\beta)$, and the difference between two consecutive estimates. Compared to traditional algorithms, the gradient boosting only involves evaluations of simple functions. Data analysis experiences show that the computational burden for the gradient boosting is minimal. As pointed out in Kim and Kim (2004), one attractive feature of the gradient boosting algorithm is that the convergence rate is relatively independent of the dimension of input. This property of convergence rate is especially valuable for data like the MCL. On the other hand, it has been known that for general boosting methods, overfitting usually does not pose a serious problem (Friedman et al., 2000). So, the overall iteration N can be taken to be a large number to ensure convergence, which may result in a considerable amount of computational cost.

Coordinate Descent Algorithm

In simple linear regressions with orthogonal designs, the estimates obtained by maximizing the likelihood functions are exactly equal to those obtained by maximizing the marginal likelihood functions, which contain only one covariate. Thus, the estimates can be obtained by regressing the response on each covariate separately. Under the Lasso penalization, the estimates have a simple soft-thresholding format. Thus, it is also feasible to obtain the estimates covariate-by-covariate (by first regressing on a single covariate and then thresholding). Although for nonorthogonal designs or more complicated models such a simple approach is not applicable, it motivates the following coordinate descent algorithm.

1. Initialize $\beta = 0$.

2. With slight abuse of notation, use β to denote the current estimate of the regression coefficient. For $j = 1, \ldots, p$, denote β^j as a length p vector with all except its jth components fixed and equal to those of β. Compute the estimate of the jth component of β, β_j, as

$$\hat{\beta}_j = argminR(\beta^j) + \lambda|\beta_j|.$$

3. Iterate Step 2 until convergence.

The above algorithm is iterative. At each iteration, p optimization procedures with one covariate in each optimization procedure are conducted. When computing the marginal Lasso estimate, usually the function $R(\cdot)$ can be written as a weighted least squares criterion, whose Lasso estimate has a closed form. Empirical studies show that usually a small number of overall iterations are needed.

Tuning Parameter Selection   The tuning parameter λ balances sparsity and goodness-of-fit. When $R(\cdot)$ is a negative likelihood function, there are multiple choices for tuning parameter selection methods, including for example AIC, BIC, and GCV (generalized cross-validation). With other goodness-of-fit measures, the choices may be limited. A generically applicable tuning parameter selection approach, which depends less on the format of R, is V-fold cross-validation and proceeds as follows. With a fixed λ,

1. Randomly partition data into V subsets with equal sizes.

2. For $v = 1, \ldots, V$,

   (a) Remove subset v from data;

   (b) Compute $\hat{\beta}^{-v} = \mathrm{argmin}_{\beta} R^{-v}(\beta) + \lambda \sum_{j=1}^{p} |\beta_j|$, where R^{-v} is R evaluated on the reduced data;

   (c) Compute $R^v(\hat{\beta}^{-v})$, which is R evaluated on subset v of the data using the estimate $\hat{\beta}^{-v}$.

3. Compute the cross-validation score $CV(\lambda) = \sum_{v=1}^{V} R^v(\hat{\beta}^{-v})$.

The optimal tuning parameter is defined as $\lambda_{opt} = \mathrm{argmin}_{\lambda} CV(\lambda)$. In general, $CV(\lambda)$ does not have an analytic form. Numerical calculation is needed to search for the minimizer. In practical data analysis, we sometimes search over the discrete grid of $2^{\cdots -3, -2, \ldots, 2, 3, \cdots}$.

   Example (MCL). We use this example to illustrate the Cox PH model with a Lasso penalty. The MCL data has been stored in two files. The first is `clinical` and contains the observed time (variable name `Followup`) and event indicator (variable name `Status`). The second is `exp.dat` and contains the normalized expression levels of 2000 genes.

   We use fivefold cross-validation for tuning parameter selection. Both cross-validation and estimation are carried out using R package `penalized` (*http* : *//cran.r − project.org/web/packages/penalized/*). The same package is also used to obtain the parameter path. We refer to the package Web site for detailed information on the software and functions. The R code proceeds as follows:

```
opt<-optL1(Surv(clinical$Followup, clinical$Status),
           penalized=t(exp.dat), fold=5)
coefficients(opt$fullfit)

prof <- profL1(Surv(clinical$Followup, clinical$Status),
           penalized=t(exp.dat), fold = opt$fold, steps=20)
```

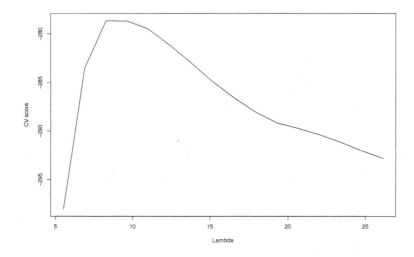

FIGURE 6.6: Analysis of MCL data: CV score as a function of tuning parameter λ (fivefold cross-validation).

```
## Plot of CV score as a function of tuning parameter
## Results shown in Figure 6.6
plot(prof$lambda, prof$cvl, type="l", xlab="Lambda",
        ylab="CV score")

## Parameter path
## Results shown in Figure 6.5
plotpath(prof$fullfit)
abline(v=opt$lambda)
```

Of all the penalized approaches introduced in this section, Lasso has perhaps the most full-fledged implementation support, as demonstrated in the above example. See Johnson (2009) for more practical computing tips for the AFT model.

Elastic Net. Consider genomic studies with gene expression measurements. There exist "groups" of genes, where genes within the same groups tend to have highly correlated expressions. We refer to Section 6.4 for more detailed discussions. When a group of highly correlated covariates are all associated with the response variable, empirical studies in Zou and Hastie (2005) suggest that the Lasso tends to underselect—Lasso sometimes selects only one or a few from this group. That is, the Lasso estimate is "overly sparse." Zou and Hastie (2005) propose solving this problem by "balancing" the sparsity of

Lasso with a "dense penalty." The elastic net penalized estimate is defined as

$$\hat{\boldsymbol{\beta}} = \operatorname{argmin}_{\beta} \left\{ R(\boldsymbol{\beta}) + \lambda_1 \sum_{j=1}^{p} |\beta_j| + \lambda_2 \sum_{j=1}^{p} \beta_j^2 \right\}, \tag{6.11}$$

where λ_1 and λ_2 are data-dependent tuning parameters.

The elastic net penalty is a linear combination of the Lasso and ridge penalties. When $\lambda_2/\lambda_1 \to 0$, the penalty behaves similarly to Lasso, which leads to sparse estimates. When $\lambda_2/\lambda_1 \to \infty$, the penalty behaves similarly to ridge, under which all covariates have nonzero regression coefficients. The elastic net bridges between Lasso and ridge regression, and can be more flexible than both of them. On the negative side, it involves two tuning parameters and can be computationally more expensive.

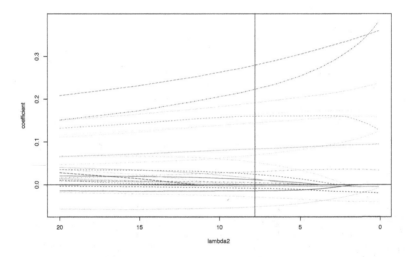

FIGURE 6.7: Analysis of MCL data: parameter path of the elastic net estimate with a fixed λ_1. The vertical line corresponds to the optimal λ_2 selected using fivefold cross-validation.

We analyze the MCL data using the elastic net. As elaborated above, the difference between the Lasso and elastic net is the addition of the ridge penalty. Here, we fix $\lambda_1 = 9.609$, which is the optimal tuning parameter selected using fivefold cross-validation under the Lasso. In Figure 6.7, we display the elastic net parameter path as a function of λ_2. As expected, the estimate becomes denser with more nonzero components as λ_2 increases. In addition, we see a higher level of shrinkage toward zero with a larger λ_2, which is similar to the behavior of ridge regression estimates.

Adaptive Lasso. The Lasso is perhaps the simplest and most extensively used penalty. However, as shown in Leng et al. (2006), Zhang and Huang (2008), and others, it is not selection consistent even with simple linear regression models unless strong conditions on orthogonality are satisfied. For the jth covariate, the Lasso penalty is proportional to $|\beta_j|$. Intuitively, for large coefficients, the level of penalty is "too high," whereas for small coefficients, the penalty is not large enough. In order to fix the selection inconsistency problem, "more (less) penalty" is needed for small (large) coefficients. The *adaptive Lasso* has been proposed by Zou et al. (2006) following such an intuition. The adaptive Lasso estimate is defined as

$$\hat{\boldsymbol{\beta}} = \operatorname{argmin}_{\boldsymbol{\beta}} \left\{ R(\boldsymbol{\beta}) + \lambda \sum_{j=1}^{p} \omega_j |\beta_j| \right\}. \tag{6.12}$$

Here, ω_j is the (possibly data-dependent) weight for covariate j.

From the definition, the adaptive Lasso is simply a "weighted" Lasso. Thus, the computational algorithms for Lasso are applicable to adaptive Lasso with very minor modifications. Some software packages, such as the R package `penalized`, allow for weights in the Lasso estimation. With such software, the adaptive Lasso estimate can be straightforwardly obtained. The application of this penalty in the Cox PH model may be found in Zhang and Lu (2007).

In Zou et al. (2006) and follow-up studies, it is shown that if $\omega_j = |\tilde{\beta}_j|^{-\gamma}$, where $\gamma > 0$ is a fixed constant and $(\tilde{\beta}_1, \ldots, \tilde{\beta}_p)$ is a consistent estimate of the unknown regression coefficient, then the adaptive Lasso is selection consistent. Thus, with the adaptive Lasso, consistent selection amounts to obtaining a consistent initial estimate. When $n >> p$, the consistent estimate can be easily obtained using standard likelihood- or estimating equation-based approaches. With high-dimensional covariates, it can be highly nontrivial to obtain the initial estimate. In Huang and Ma (2010) and several other studies, it is shown that the Lasso estimate

$$\tilde{\boldsymbol{\beta}} = \operatorname{argmin}_{\boldsymbol{\beta}} \left\{ R(\boldsymbol{\beta}) + \lambda \sum_{j=1}^{p} |\beta_j| \right\}$$

is estimation consistent under mild data/model assumptions. More details on the theoretical development are provided in the Appendix (Section 6.7) of this chapter.

Bridge. The second penalty that can enjoy an improved selection consistency property (over the Lasso) is the bridge penalty, under which the estimate is defined as

$$\hat{\boldsymbol{\beta}} = \operatorname{argmin}_{\boldsymbol{\beta}} \left\{ R(\boldsymbol{\beta}) + \lambda \sum_{j=1}^{p} |\beta_j|^{\gamma} \right\}, \tag{6.13}$$

where $0 < \gamma < 1$ is the fixed bridge parameter. From the definition, it can be seen that as $\hat{\gamma} \to 0$, the bridge penalty converges to an AIC/BIC-type penalty; as $\gamma \to 1$, the bridge penalty converges to the Lasso penalty.

Unlike Lasso, the bridge penalty is not convex. Direct minimization of a nonconvex objective function can be difficult. One possible computational solution is proposed in Huang et al. (2008a), where an approximated penalized estimate is proposed as

$$\hat{\beta}_{approx} = \text{argmin}_{\beta} \left\{ R(\beta) + \lambda \sum_{j=1}^{p} (\beta_j^2 + \epsilon)^{\gamma/2} \right\}. \tag{6.14}$$

Here, ϵ is a small positive constant (for example, 0.001). With this approximation, the overall objective function is differentiable everywhere and can be optimized using, for example, gradient searching approaches. We note that the approximation will render all estimates nonzero. Thus, with the obtained estimate, an additional thresholding step needs to be applied. For example, all estimates with absolute values less than $\sqrt{\epsilon}$ are set to be zero. This approximation approach is computationally simple. However, usually it is not perfectly clear how the thresholding should be conducted.

Another relatively less *ad hoc* approach, which is built on the connection between bridge penalization and Lasso, proceeds as follows. Denote $\eta = (1 - \gamma)/\gamma$. Define

$$S(\beta) = R(\beta) + \sum_{j=1}^{p} \theta_j^{-\eta} |\beta_j| + \tau \sum_{j=1}^{p} \theta_j,$$

where $\tau = \eta^{-1/\eta}(1 + \eta)^{-(1+\eta)/\eta} \lambda^{(1+\eta)/\eta}$. $\hat{\beta}$ minimizes the bridge objective function if and only if

$$\hat{\beta} = \text{argmin}_{\beta} S(\beta) \quad \text{subject to} \quad \theta_j \geq 0 \ (j = 1 \dots p).$$

This result can be proved using Proposition 1 of Huang et al. (2009b). Based on this result, the following algorithm is proposed. For a fixed λ,

1. Initialize $\hat{\beta}$ as the Lasso estimate, i.e., $\hat{\beta} = \text{argmin}_{\beta} R(\beta) + \lambda \sum_{j=1}^{p} |\beta_j|$.

2. Compute $\theta_j = (\eta/\tau)^{1/(1+\eta)} |\hat{\beta}_j|_2^{1/(1+\eta)}$ for $j = 1, \dots, p$.

3. Compute $\hat{\beta} = \text{argmin}\{R(\beta) + \sum_{j=1}^{p} \theta_j^{-\eta} |\beta_j|\}$.

4. Repeat Steps 2–3 until convergence.

In the Appendix (Section 3.7), for the AFT model with the Stute estimate, we describe the theoretical result which states that with a high probability, the Lasso can identify all true positives, that is, the set of all covariates that are truly associated with the survival outcome. In addition, the Lasso can significantly reduce the dimensionality, that is, it can remove the majority of true negatives (the set of all covariates that are truly irrelevant to the survival outcome). Thus, it is an appropriate choice for the initial estimate. However, the Lasso tends to overselect, and thus the downstream iterations are needed. In Step 3, we transform the bridge-type optimization to a weighted Lasso-type optimization. The iteration continues until convergence. In numerical studies, we suggest using the l_2 norm of the difference between two consecutive estimates less than a predefined cut-off as the convergence criterion. We note that the above computational algorithm is in general valid. However, the selection properties for both the initial Lasso estimate and iterative estimates need to be established on a case-by-case basis.

Remark. The penalties described above are "bridge-type" penalties. They have the generic form of $|\beta|^{\gamma_1} + |\beta|^{\gamma_2}$ with $0 \le \gamma_1, \gamma_2 \le 2$. In methodological development, the values of γ_1, γ_2 are usually taken as fixed. It may also be possible to develop data-dependent approaches selecting the "optimal" values of γ_1, γ_2. However, they may lead to high computational cost and are usually not pursued. In Figure 6.8, we display various penalties functions for a one-dimensional β. They share the following common features: The penalty is equal to zero when the regression coefficient is equal to zero; it increases as the absolute value of the regression coefficient increases; and it is differentiable everywhere except at zero. At the same time, the differences among different penalties are also obvious. There are also penalty functions that do not have the simple power forms. Below we describe the SCAD and MCP penalties, which also have the desirable selection consistency properties under certain data and model settings.

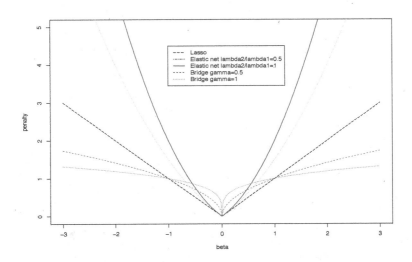

FIGURE 6.8: Different penalty functions.

SCAD. The SCAD (smoothly clipped absolute deviation) penalty is proposed in Fan and Li (2001) for linear regression and Fan and Li (2002) for the Cox PH regression model. Under the SCAD penalty, the estimate of regression coefficients is defined as

$$\hat{\beta} = \mathrm{argmin}_{\beta} \left\{ R(\beta) + \sum_{j=1}^{p} \rho(|\beta_j|; \lambda, a) \right\}, \qquad (6.15)$$

where

$$\rho(t; \lambda, a) = \begin{cases} \lambda|t| & |t| \leq \lambda \\ -(t^2 - 2a\lambda|t| + \lambda^2)/[2(a-1)] & \lambda < |t| \leq a\lambda, \\ (a+1)\lambda^2/2 & |t| > a\lambda \end{cases}$$

where the tuning parameters $\lambda > 0$ and $a > 2$. It is quite evident that the SCAD penalty takes off at the origin mimicking the Lasso penalty and then gradually levels off. As with other penalties, the SCAD penalty is continuously differentiable on $(-\infty, 0) \cup (0, \infty)$ but singular at 0. Its derivative vanishes outside $[-a\lambda, a\lambda]$. As a consequence, it applies a very moderate amount of penalty to large regression coefficients and does not excessively penalize. This is different from the Lasso-based approaches. Empirical studies show that SCAD improves over Lasso by reducing estimation bias (Fan and Li, 2002) under the Cox PH model.

There are two tuning parameters in the SCAD penalty. λ can be chosen using data-dependent approaches such as V-fold cross-validation. a can also be chosen in a similar manner. Extensive simulation studies in Fan and Li (2001, 2002) and other follow-up studies suggest fixing $a = 3.7$, which may come with a Bayesian justification. However, there is no proof of the optimality of this value. In practice, it may still be helpful to experiment with other values for a.

The SCAD penalty function places more computational and theoretical challenges than the Lasso-type penalties since it is nonconvex. Its implementation can be achieved by approximating the SCAD penalty locally with a linear function and thus modifying the problem into an adaptive Lasso penalty (Zou and Li, 2008). The R package `glmnet` can be easily adapted to this end as it allows a weight specification in the Cox regression with Lasso (Simon et al., 2011). In addition to the local linear approximation, some authors have proposed the local quadratic approximation (Fan and Li, 2001), the minorization maximization algorithm (Hunter and Li, 2005), and the difference convex algorithm (Wu and Liu, 2009).

There has also been much progress on the theoretical development for SCAD. As established in Fan and Li (2001) and Fan and Peng (2004) for a few familiar regression models, SCAD penalty retains the selection consistency property for small p and large p. It also enjoys the so-called *oracle* property with a proper choice of the regularization parameter. The oracle property for an estimator, say, $\hat{\beta} = (\hat{\beta}_1, \hat{\beta}_2)$ corresponding to a true parameter $\beta =$

$(\boldsymbol{\beta}_1, \boldsymbol{\beta}_2)$ with each component of $\boldsymbol{\beta}_1$ nonzero and $\boldsymbol{\beta}_2 = 0$, is usually expressed in two parts: (1) the estimated coefficients for the true zero coefficients are equal to zero ($\hat{\boldsymbol{\beta}}_2 = 0$) with probability tending to 1 as $n \to \infty$; (2) it attains the same information bound as the oracle estimator $\hat{\boldsymbol{\beta}}_1$ obtained from fitting the same model with only the covariates with true nonzero coefficients. The first part is also conveniently termed *sparsity* in the literature.

MCP. The MCP (minimax concave penalty) is introduced by Zhang (2010a). The MCP penalized estimate is defined as

$$\hat{\boldsymbol{\beta}} = \operatorname{argmin}_{\boldsymbol{\beta}} \left\{ R(\boldsymbol{\beta}) + \sum_{j=1}^{p} \rho(|\beta_j|; \lambda, \gamma) \right\}, \tag{6.16}$$

where

$$\rho(t; \lambda, \gamma) = \lambda \int_0^{|t|} \left(1 - \frac{x}{\lambda \gamma}\right)_+ dx.$$

Here, λ is the tuning parameter and has a similar interpretation as with other penalized estimates. γ is a regularization parameter. $x_+ = max(0, x)$.

The MCP can be more easily understood by considering its derivative

$$\dot{\rho}(t; \lambda, \gamma) = \lambda \left(1 - \frac{|t|}{\gamma \lambda}\right)_+ sign(t), \tag{6.17}$$

where $sign(a) = -1, 0, 1$ if $a < 0, = 0, > 0$. It begins by applying the same rate of penalization as the Lasso, but continuously reduces the penalization until when $|t| > \gamma \lambda$, the rate of penalization, drops to zero. The MCP provides a continuum of penalties with the L_1 penalty at $\gamma = \infty$ and the hard-thresholding penalty as $\gamma \to 1+$.

The MCP has two tuning parameters: λ and γ. λ has a similar interpretation as that in the previously introduced penalties and can be selected in a similar manner. For γ, with simple linear regression models, Zhang (2010a) conducts extensive simulations and shows that the average value of optimal γ is 2.69. The simulation studies in Breheny and Huang (2011) suggest that $\gamma = 3$ is a reasonable choice. In practice researchers may need to experiment with multiple values and choose the optimal value data-dependently.

The MCP penalty function faces a similar computational challenge as SCAD since it is also nonconvex. One practical reason for statisticians to become intrigued with these nonconvex penalty functions is that they may allow relatively stronger correlations among covariates in lieu of insisting the standard "incoherent condition" (for example, see conditions required by Lasso in Meinshausen and Yu, 2009 and van de Geer and Buhlmann, 2009), and also control the tail bias of the resulting penalized estimators. Even though necessary theoretical preparation for the application of such nonconvex penalized optimization problems (including SCAD and MCP) has been ready *de nos jours* (Fan and Lv, 2011), implementation of more efficient and user-friendly algorithms has barely begun in the literature.

There has been considerable effort to unify the various penalty functions and arguments for their statistical properties (Lv and Fan, 2009; Bradic et al., 2011; Fan and Lv, 2011). A common approach is to consider $\rho(t; \lambda) = \lambda^{-1} pen(\lambda, t)$ for a penalty function $pen(\cdot)$ and assume the following conditions:

The function $\rho(t; \lambda)$ is increasing and concave in $t \in [0, \infty)$, and has a continuous derivative $\rho'(t; \lambda)$ with $\rho'(0+; \lambda) > 0$. In addition, $\rho'(t; \lambda)$ is increasing in $\lambda \in (0, \infty)$ and $\rho'(0+; \lambda)$ is independent of λ.

The two nonconvex penalties SCAD and MCP both meet these conditions while Lasso falls at the boundary of the class of penalties satisfying the above conditions. It has been shown in Bradic et al. (2011) that the Cox PH model with penalties satisfying the above conditions for ρ can yield estimates with the oracle property under the $p >> n$ setting. The proof opens the door to a general justification of the penalization approaches introduced in this section.

Remarks. Penalty-based variable selection is still an active area that attracts abundant attention. Since the early 2000s, applications of the methods introduced in this section have quickly evolved from the initial setting $p << n$ through the relatively nontrivial cases $p = o(n^{1/5})$ or $p = o(n^{1/3})$ to the final arrival at the current most advanced stage $\log p = O(n^a)$ for some $a \in (0, 1)$. Many theoretical achievements have been made for the latest stage of nonpolynomial (NP) order dimensionality which may resemble real data in genetics the most. In addition to the penalty functions summarized in this section, there are also other useful penalization methods for censored survival data, including Wang et al. (2009), Tibshirani (2009), Fan et al. (2010), and Bradic et al. (2011), among others.

6.4   Incorporating Hierarchical Structures

In Sections 6.2 and 6.3, all covariates are treated in an equal manner, and all covariate effects are interchangeable. Under certain scientific contexts, the covariates have hierarchical structures. In this section, we continue to use gene expression studies as an example but note that the hierarchical structures and analysis approaches may be applicable under other more general contexts.

6.4.1   Hierarchical Structure

When examining the correlations among gene expressions, we often see that there exist gene clusters, where genes within the same clusters have highly correlated expressions. From a biological perspective, all functions of living cells are not achieved by individual genes, but rather by groups of genes with coregulated functionalities. There are multiple ways of defining the hierarchical structure. Below we present the most extensively used ones.

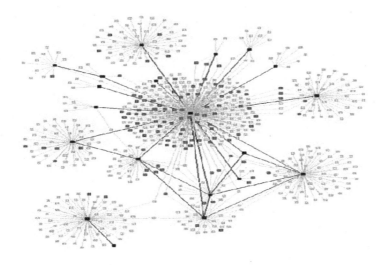

FIGURE 6.9: Gene network.

Statistical Clustering. Clustering gene expressions based on statistical measurements has been implemented in multiple areas of genomics. Examples include data visualization, exploratory analysis, defining new disease subtypes, and others. Commonly used clustering methods may include the hierarchical, K-means, tree-truncated vector quantization, self-organizing map methods, and others. For a comprehensive review, see Gordon (1999). Different clustering methods are built on different data assumptions, and with practical data, may lead to significantly different clustering structures.

Pathway. A "more biological" way of defining the hierarchical structures is via pathways, which are composed of covariates (genes) with coordinated biological functions. Biological functions and hence pathway information of genes have been accumulated from numerous biological experiments and summarized in several large online databases, including KEGG (*www.genome.jp/kegg*), GO (*www.geneontology.org*), BioCarta (*www.biocarta.com*), and others.

With gene expression data, there are moderate correspondences between statistical clusters and biological pathways. That is, covariates (genes) with highly correlated values (expressions) tend to have similar biological functions, and vise versa. However, usually there is no strict one-to-one correspondence. When biological interpretation is of more concern, biological pathways can be preferred over statistical clusters.

Networks. A graphical representation of gene networks is provided in Figure 6.9. In network analysis, nodes represent covariates (genes). Nodes are connected if the corresponding covariates have similar biological functions and/or similar expression patterns across samples. There are subsets of nodes called "modules" that are tightly connected to each other. An example of gene networks is the weighted coexpression network (*www.genetics.ucla.edu/labs/horvath/CoexpressionNetwork*), which is built on the understanding that the coordinated coexpressions of genes encode interacting proteins with closely related biological functions and cellular processes. Other ways of building gene networks include the Boolean network, Bayesian network, use of continuous models, and others. To the best of our knowledge, at this moment, there is a lack of study comprehensively comparing different network construction methods. Performance of different methods has been suggested to be data dependent.

Remark. Different approaches lead to different definitions of hierarchical structures. Among the three described above, some statistical clusters and networks (for example, the weighted coexpression network) are built on the analysis of single data sets and hence can be less reproducible, especially considering the high-dimensional nature of data. In contrast, the biological pathway information has been accumulated over a large number of independent studies and can have better reproducibility. On the negative side, there are still a considerable number of genes not or only partially annotated, thus having no pathway information. Beyond the three introduced above, there are other ways of defining hierarchical structures among covariates. In this chapter, we first focus on a relatively simple structure, where covariates can be separated into *clusters* prior to model fitting. Different clusters may have overlaps. For example, some genes have multiple biological functions and hence belong to multiple pathways. We assume that C^* clusters have been constructed. For covariate $j = 1, \ldots, p$, denote $C(j)$ as its cluster membership. It is noted that in several published studies, approaches that can simultaneously conduct clustering and model fitting have been proposed.

6.4.2   Clusters Marginally Associated with Survival

In Section 6.2, we describe how to identify covariates marginally associated with survival. When the clustering structure is taken into consideration, it is of interest to identify which *clusters* (as opposed to individual covariates) are associated with survival.

One commonly adopted approach is the GSEA (gene set enrichment analysis) (Subramanian et al., 2005). This approach targets finding which clusters contain more, or are "enriched" with, covariates marginally significantly associated with survival. The first step of this approach is to compute an enrichment score for each cluster and proceeds as follows:

1. For covariate $j = 1, \ldots, p$, compute a test statistic r_j, which can measure the strength of association between this covariate and survival. The same test statistics as described in Section 6.2 can be used.

2. Rank the test statistics $r_{(1)} \leq \ldots \leq r_{(p)}$.

3. For $c = 1, \ldots C^*$, denote $|C(c)|$ as the size of cluster c. Compute the statistic

$$N_c = \sum_{C(m)=c} |r_m|^\gamma,$$

where γ is a user-specified parameter. Usually, $\gamma = 1$ is recommended in practical data analysis.

4. For $c = 1, \ldots C^*$, compute

$$P_{hit}(c, i) = \sum_{C(m)=c, m \leq i} \frac{|r_{(m)}|^\gamma}{N_c}, \quad P_{miss}(c, i) = \sum_{C(m) \neq c, j \leq i} \frac{1}{p - |C(m)|},$$

$$M(c) = max\{P_{hit}(c, i) - P_{miss}(c, i) : i = 1 \ldots p\},$$

$$m(c) = min\{P_{hit}(c, i) - P_{miss}(c, i) : i = 1 \ldots p\}.$$

The enrichment score for cluster c is defined as

$$ES(c) = \begin{cases} M(c) & \text{if } M(c) \geq -m(c) \\ m(c) & \text{if } M(c) \leq -m(c) \end{cases}.$$

For cluster $c = 1, \ldots, C^*$, the second step is to compute a permutation p-value and proceeds as follows:

1. Randomly permute the survival outcomes. Note that the observed time and event indicator are treated as a pair and permuted together.

2. With the permuted data, compute $ES(c, 1)$.

3. Repeat Steps 1 and 2 B (e.g., 1000) times and get $ES(c, 1) \ldots, ES(c, B)$.

4. If $ES(c) > 0 (< 0)$, the approximate permutation p-value is computed as the proportion of the positive (negative) $ES(c, b)$ values ($b = 1, \ldots, B$) that are greater (smaller) than or equal to $ES(c)$.

With the p-values so obtained, multiple comparison adjustments can then be conducted and used to identify clusters of covariates significantly associated with survival.

For a cluster, the enrichment score compares the degree of association among covariates in this cluster relative to that of all other clusters/covariates. The permutation testing procedure tests the null hypothesis that there is no association between covariates and survival. A potential limitation of this approach is that it examines only the marginal association of each covariate even though the question of interest involves a cluster of covariates.

In the literature, many cluster-based methods have been proposed. Beyond GSEA, other extensively used methods include the maxmean method (Efron and Tibshirani, 2007) and the global test method (Goeman et al., 2004), among many others. Literature review suggests that different approaches make different model/data assumptions and target different aspects of clusters. In biological studies, our limited experience suggests that the GSEA has been more popular. However, this is still a lack of systematic comparison of different approaches to support the superiority of GSEA or another approach.

6.4.3 Multivariate Prediction Models

6.4.3.1 Dimension Reduction Approaches

Consider PCA as an example of dimension reduction approaches. In Section 6.3.1, PCA is conducted with all covariates. A more sensible cluster-based approach proceeds as follows. First, PCA is conducted within each cluster separately. Second, the first or the first few PCs of each cluster are used as covariates in downstream analysis. Here, the PCs represent the effects of clusters, which are composed of multiple covariates with similar measurement patterns or related biological functions. Thus, they may have more lucid interpretations than those constructed using all covariates. In addition, following a similar strategy as in Section 6.3.1, we can also accommodate the interactions among clusters using the interactions of PCs (from different clusters). With the weighted gene coexpression network, Ma et al. (2011b) have explored this approach and demonstrated its effectiveness.

With other dimension reduction methods such as PLS and ICA, it is also possible to limit the new set of covariates $\{X_j^* : j = 1 \ldots p^*\}$ as the linear combinations of original covariates within the same clusters. In addition, we can follow the strategy in Ma et al. (2011b) and accommodate interactions.

6.4.3.2 Variable Selection

In penalized variable selection, the overall objective function has two components. When the hierarchical structure is taken into consideration, at least one of those components need to be revised. We conjecture that it is possible to revise the goodness-of-fit measure so that it properly reflects the hierarchical structure. However, to the best of our knowledge, there is still little work done in this direction. Most published studies revise the penalty functions. In this section, we take the bridge penalty as a representative example. Other penalty functions can be revised following a similar strategy. The following two penalties have been proposed, extending the bridge penalty to accommodate the clustering structure.

2-Norm Group Bridge Penalty. The *2-norm group bridge* penalized estimate is defined as

$$\hat{\beta} = argmin_\beta \left\{ R(\beta) + \lambda \sum_{c=1}^{C^*} \left\{ \sum_{j:C(j)=c} \beta_j^2 \right\}^{\gamma/2} \right\}, \qquad (6.18)$$

where $0 < \gamma < 1$ is the fixed bridge parameter.

In cluster-based analysis, covariate clusters, as opposed to individual covariates, are the functional units. Thus, the penalty is defined as the summation over C^* individual clusters. With the simple bridge penalty, for each coefficient, the penalty is defined as its l_γ norm. Correspondingly, with a cluster, the penalty is defined as its l_γ *cluster norm*.

When the tuning parameter λ is properly chosen, the 2-norm group bridge penalized estimates may have components exactly equal to zero. However, it is worth noting that this sparsity property is at the cluster level. That is, some clusters may have estimated norms exactly equal to zero. For those clusters with nonzero estimated norms, note that the within-cluster penalty $\sum_{j:C(j)=c} \beta_j^2$ has a ridge form, which has the shrinkage but not selection property. Thus, if a cluster has a nonzero cluster norm, all covariates within this cluster have nonzero estimated regression coefficients. That is, when a cluster is concluded as associated with survival, all covariates within this cluster enter the final model and are concluded as associated with survival.

1-Norm Group Bridge Penalty. The dense property of the 2-norm group bridge estimate within-cluster may not always be reasonable. Consider, for example, when clusters are biological pathways. Here, clusters are constructed using biological information gathered in previous studies and independent of the specific survival outcome being analyzed. Genes within the same clusters have related but never identical functions. Thus, even within clusters associated with survival, there may still exist "noisy" genes. Such a consideration demands variable selection at *both the cluster level and the within-cluster-covariate level*, that is, bi-level selection.

The *1-norm group bridge* penalized estimate has been motivated by such a consideration and defined as

$$\hat{\beta} = argmin_\beta \left\{ R(\beta) + \lambda \sum_{c=1}^{C^*} \left\{ \sum_{j:C(j)=c} |\beta_j| \right\}^\gamma \right\}, \tag{6.19}$$

where $0 < \gamma < 1$ is the fixed bridge parameter.

This penalty has been motivated by similar considerations as with the 2-norm group bridge penalty. The key difference is the within-cluster penalty, $\sum_{j:C(j)=c} |\beta_j|$, which has a Lasso form. As the Lasso has not only shrinkage but also variable selection properties, the 1-norm group bridge penalty can select not only survival-associated clusters but also survival-associated covariates within selected clusters.

Remark. The two penalties introduced above are *composite penalties*. The 2-norm group bridge penalty is the composition of a bridge penalty (at the cluster level) and a ridge penalty (at the within-cluster level). The 1-norm group bridge penalty is the composition of a bridge penalty and a Lasso penalty. At the cluster level, the shrinkage and sparsity properties are similar to those of the simple bridge penalty; At the within-cluster level, the properties are similar to those of the ridge (Lasso) penalty. Following a similar strategy, it is possible to build a large number of composite penalties, for example, adopting the MCP penalties at both the cluster- and within-cluster levels. Compared to simple penalties, the composite penalties may better accommodate the inherent structure of covariates, however, at the price of higher computational cost and more complicated theoretical properties.

6.5   Integrative Analysis

In survival analysis with penalized variable selection, several models/approaches have been shown to have the desired asymptotic variable selection consistency property. As shown in the Appendix (Section 6.7), such a property may hold even when $p >> n$. However, in practical data analysis, multiple studies have shown that variable selection results from the analysis of single survival data sets may suffer a lack of reproducibility. Among multiple contributing factors, the most importance one is perhaps small sample sizes of individual studies. For a few clinically important phenotypes and outcomes, multiple independent studies have been conducted, making it possible to pool data from multiple studies and increase power. Below we describe integrative analysis, which can effectively pool raw data from multiple independent studies. We use the AFT model and bridge-type penalization as an example, and note that other models and penalization methods may be studied in a similar manner. It is noted that integrative analysis differs significantly from classic meta-analysis, which pools summary statistics (lists of identified covariates, effect sizes, p-values) across studies. Below we take gene expression studies as an example and note that other high-dimensional measurements may be studied in a similar manner.

Assume $M(> 1)$ independent studies. For simplicity of notation, assume that the same d covariates (gene expressions) are measured in all studies. We use the superscript "(m)" to denote the mth study. Let $T^{(1)}, \ldots, T^{(M)}$ be the logarithms of failure times, and $X^{(1)}, \ldots, X^{(M)}$ be length-d covariates. For $m = 1, \ldots, M$, assume the AFT model

$$T^{(m)} = \alpha^{(m)} + \beta^{(m)\prime} X^{(m)} + \epsilon^{(m)}. \tag{6.20}$$

Here, $\alpha^{(m)}$ is the unknown intercept, $\beta^{(m)}$ is the regression coefficient, $\beta^{(m)\prime}$ is the transpose of $\beta^{(m)}$, and $\epsilon^{(m)}$ is the random error with an unknown distribution. Denote $C^{(1)}, \ldots, C^{(M)}$ as the logarithms of random censoring times. Under right censoring, we observe $(Y^{(m)}, \delta^{(m)}, X^{(m)})$ for $m = 1 \ldots M$. Here $Y^{(m)} = \min(T^{(m)}, C^{(m)})$ and $\delta^{(m)} = I(T^{(m)} \leq C^{(m)})$.

Consider $\beta = (\beta^{(1)}, \ldots, \beta^{(M)})$, the $d \times M$ regression coefficient matrix. The main characteristics of β are as follows. First, $\beta^{(m)}$s are sparse in that only a subset are nonzero. This feature corresponds to the fact that out of a large number of genes surveyed, only a subset are associated with prognosis, and the rest are noises. Only response-associated genes have nonzero regression coefficients. Second, $\beta^{(1)}, \ldots, \beta^{(M)}$ have the same sparsity structure. That is, elements of β in the same row are either all zero or all nonzero. This feature corresponds to the fact that multiple studies share the same set of markers. Third, for markers with nonzero coefficients, the values of regression coefficients may be different across studies, which can accommodate the heterogeneity across studies.

For estimation, we consider the Stute estimate, which has been described in earlier chapters. In study $m(= 1, \ldots, M)$, assume $n^{(m)}$ i.i.d. observations $(Y_i^{(m)}, \delta_i^{(m)}, X_i^{(m)}), i = 1, \ldots, n^{(m)}$. Let $\hat{F}^{(m)}$ be the Kaplan-Meier estimate of $F^{(m)}$, the distribution function of $T^{(m)}$. It can be computed as $\hat{F}^{(m)}(y) = \sum_{i=1}^{n^{(m)}} w_i^{(m)} I(Y_{(i)}^{(m)} \leq y)$. Here $Y_{(1)}^{(m)} \leq \ldots \leq Y_{(n^{(m)})}^{(m)}$ are the order statistics of $Y_i^{(m)}$s. Denote $\delta_{(1)}^{(m)}, \ldots, \delta_{(n^{(m)})}^{(m)}$ as the associated censoring indicators and $X_{(1)}^{(m)}, \ldots, X_{(n^{(m)})}^{(m)}$ as the associated covariates. $w_i^{(m)}$s are the jumps in the Kaplan-Meier estimate and can be computed as

$$w_1^{(m)} = \frac{\delta_{(1)}^{(m)}}{n^{(m)}}, \quad \text{and} \quad w_{(i)}^{(m)} = \frac{\delta_{(i)}^{(m)}}{n^{(m)} - i + 1} \prod_{j=1}^{i-1} \left(\frac{n^{(m)} - j}{n^{(m)} - j + 1} \right)^{\delta_{(j)}^{(m)}}$$

$$\text{for } i = 2, \ldots, n^{(m)}.$$

For study m, the weighted least squares objective function is defined as

$$R^{(m)} = \frac{1}{2} \sum_{i=1}^{n^{(m)}} w_i^{(m)} \left(Y_{(i)}^{(m)} - \alpha^{(m)} - \beta^{(m)\prime} X_{(i)}^{(m)} \right)^2.$$

We center $X_{(i)}^{(m)}$ and $Y_{(i)}^{(m)}$ as

$$X_{(i)}^{(m)*} = \sqrt{w_i^{(m)}} \left(X_{(i)}^{(m)} - \frac{\sum_j w_j^{(m)} X_{(j)}^{(m)}}{\sum_j w_j^{(m)}} \right) \quad \text{and}$$

$$Y_{(i)}^{(m)*} = \sqrt{w_i^{(m)}} \left(Y_{(i)}^{(m)} - \frac{\sum_j w_j^{(m)} Y_{(j)}^{(m)}}{\sum_j w_j^{(m)}} \right).$$

We define the overall loss function by

$$R(\beta) = \sum_{m=1}^{M} R^{(m)} = \sum_{m=1}^{M} \frac{1}{2} \sum_{i=1}^{n^{(m)}} \left(Y_{(i)}^{(m)*} - \beta^{(m)\prime} X_{(i)}^{(m)*} \right)^2. \tag{6.21}$$

Consider penalized marker selection. Denote $\beta_j^{(m)}$ as the jth component of

$\beta^{(m)}$. $\beta_j = (\beta_j^{(1)}, \dots, \beta_j^{(M)})$ is the jth row of β and represents the coefficients of covariate j across M studies. Define

$$\hat{\beta} = argmin_\beta \left\{ R(\beta) + \lambda_n \sum_{j=1}^d J(\beta_j) \right\}, \tag{6.22}$$

where λ_n is a data-dependent tuning parameter. For the penalty function $J(\cdot)$, consider the 2-norm group bridge penalty proposed by Ma et al. (2011a), where

$$J(\beta_j) = \|\beta_j\|_2^\gamma. \tag{6.23}$$

Here, $\|\beta_j\|_2 = \left[(\beta_j^{(1)})^2 + \dots + (\beta_j^{(M)})^2 \right]^{1/2}$, and $0 < \gamma < 1$ is the fixed bridge index. In numerical studies, we set $\gamma = 1/2$.

The 2-norm group bridge penalty has been motivated by the following considerations. When $M = 1$, it simplifies to the bridge penalty, which has been shown to have the "oracle" properties in the analysis of single data sets. In integrative analysis, for a specific gene, we need to evaluate its overall effects in multiple data sets. To achieve such a goal, we treat its M regression coefficients as a *group* and conduct group-level selection. When $\gamma = 1$, the 2-norm group bridge penalty becomes the group Lasso (GLasso) (Meier et al., 2008). Theoretical investigation in Ma et al. (2012) shows that the 2-norm group bridge penalty has significantly better selection property than the GLasso.

As the 2-norm group bridge penalty is not convex, direct minimization of the objective function can be difficult. Consider the following computational algorithm. Denote $\eta = (1 - \gamma)/\gamma$. Define $S(\beta) = R(\beta) + \sum_{j=1}^d \theta_j^{-\eta} \|\beta_j\|_2 + \tau \sum_{j=1}^d \theta_j$, where $\tau = \eta^{-1/\eta} (1+\eta)^{-(1+\eta)/\eta} \lambda_n^{(1+\eta)/\eta}$. $\hat{\beta}$ minimizes the objective function defined in (6.22) if and only if

$$\hat{\beta} = argmin_\beta S(\beta) \quad \text{subject to} \quad \theta_j \geq 0 \ (j = 1, \dots, d).$$

This result can be proved using Proposition 1 of Huang et al. (2009a). Based on this result, we propose the following algorithm. For a fixed λ_n,

1. Initialize $\hat{\beta}$ as the GLasso estimate, i.e., estimate defined in (6.22) with $\gamma = 1$.

2. Compute $\theta_j = (\eta/\tau)^{1/(1+\eta)} \|\hat{\beta}_j\|_2^{1/(1+\eta)}$ for $j = 1, \dots, d$.

3. Compute $\hat{\beta} = argmin\{R(\beta) + \sum_{j=1}^d \theta_j^{-\eta} \|\beta_j\|_2\}$.

4. Repeat Steps 2 and 3 until convergence.

In Theorem 1 of Ma et al. (2012), it is shown that with a high probability, the GLasso can select all true positives while effectively removing the majority of true negatives. In addition, it is estimation consistent. Thus, it is an appropriate choice for the initial estimate. However, the GLasso tends to overselect, and thus the downstream iterations are needed. In Step 3, the group bridge-type minimization is transformed to a weighted GLasso-type minimization. The iteration continues until convergence. In numerical studies, it is suggested using the ℓ_2 norm of the difference between two consecutive estimates less than 0.01 as the convergence criterion. The GLasso estimate can be computed using the coordinate descent algorithm described in Friedman et al. (2010). An interesting observation is that, for a fixed j and any $k \neq l$, $\partial^2 R(\beta)/\partial\beta_j^{(k)}\partial\beta_j^{(l)} = 0$. Thus, the Hessian for the coefficients in a single group is a diagonal matrix. The unique form of the objective function makes the coordinate descent algorithm computationally less expensive.

6.6 Discussion

There has been abundant development, in terms of upgrading both computational methods and statistical theory, in survival analysis with high-dimensional covariates. Obviously this is well motivated by the advancement in data collection and storage techniques. We note that most methods described here have a history much longer than genetic and genomic data. Their statistical properties have been well established for simpler data/model settings ($p < n$). The nonstandard applications of these methods for high-dimensional data facilitate more sophisticated buildup.

Since early 1990s, there has been dramatic development in genome profiling techniques. From microarray gene expression data to SNP data to deep sequencing data, we are now able to quantify the human genome at a much finer scale. Although it is difficult to precisely predict the development of profiling technique itself, it is "safe" to expect more data of such kind with higher dimension and complexity in the near future.

In this chapter, we focus on the statistical analysis aspect. For many of the discussed methods, multiple published studies have established their satisfactory statistical properties. It should be noted that, ultimately, the analysis results need to be validated by rigorous, independent biomedical studies. As the field of human genetics/genomics is still evolving, validation of biomedical models and corresponding statistical analysis still demands considerable effort.

It has been long recognized that for the same data sets, different models and estimation and inference techniques can lead to different results. For high-dimensional data, the discrepancy (between different analysis methods) is especially obvious. Our limited literature review shows that in most biomedical studies, there is a lack of model checking and/or consideration of alternative analysis methods. In Ma et al. (*http://bib.oxfordjournals.org/content/11/4/385.abstract*), simulation study and data analysis show that marker identification results can heavily depend on the assumed statistical models. However, that study and our own search have not been able to suggest a way of determining proper models. More attention needs to be paid to the impact of analysis strategy on biomedical conclusions in the next step.

For the purpose of demonstrating statistical methods, we focus on a single type of genetic measurement (most of the time, microarray gene expression). It has been recognized that the development and progression of complex human diseases, such as cancer, diabetes, and mental disorders, are a complex process involving a large number of contributing factors. Biomedical studies have shown that prediction of the disease path using a single source of data (e.g., gene expressions only) has only limited success. Thus, it is necessary to adopt a system biology perspective, and integrate multiple sources of data (including, for example, clinical risk factors, environmental exposures, gene expressions, SNPs, methylation, etc.) in disease model building. In principle, the analysis methods discussed in this chapter are still applicable. However, our limited experience suggests that the data analysis performance has been disappointing. For example, when analyzing a model with $\sim 10^4$ gene expressions and ~ 10 clinical covariates, the marker selection results using Lasso are usually dominated by gene expressions. Such a result is not sensible as the clinical covariates have been established to be associated with survival in a large number of epidemiologic studies. At the present stage, there is a need for data integration approaches that can properly accommodate the different dimensionality and distributions of different types of covariates.

6.7 Appendix

6.7.1 Estimation and Inference of the Concordance

In this section, we outline the theoretical properties of the nonparametric concordance measure. Let $\{A_i(t) = I(Y_i \geq t; t \geq 0)\}$ be the at-risk process and $\{B_i(t) = I(Y_i \leq t, \delta_i = 0); t \geq 0\}$ be the counting process for the censoring time. Let $H_C(t) = -\log S_C(t)$ be the cumulative hazard function for the censoring time. Define $M_i(t) = B_i(t) - \int_0^t A_i(s) dH_C(s)$. Let $W_i^j = (X_{i,j}, Y_i, \delta_i)$.

The consistency of $\hat{\tau}_j$ follows from

$$n^{-2}\sum_{i=1}^n\sum_{k=1}^n\frac{\delta_i}{S_C^2(Y_i)}I(X_{i,j}<X_{k,j},Y_i<Y_k) \overset{p}{\to} E\{I(X_{i,j}<X_{k,j},Y_i<Y_k)\},$$

$$n^{-2}\sum_{i=1}^n\sum_{k=1}^n\frac{\delta_i}{S_C^2(Y_i)}I(Y_i<Y_k) \overset{p}{\to} E\{I(Y_i<Y_k)\},$$

$$\sup_u|\hat{S}_C(u)-S_C(u)| \to 0.$$

Define

$$\varphi_{i,j}=h_{i1}^j+h_{i2}^j,$$

$$h_{i1}^j=\int\frac{\left\{q_1^j(u)-q_2(u)\tau_j\right\}}{E\{A_i(u)\}E\{I(T_i<T_k)\}}dM_i(u),$$

$$h_{i2}^j=\frac{q_3^j(W_i^j)}{E\{I(T_i<T_j)\}}-\frac{\tau_jq_4(W_i^j)}{E\{I(T_i<T_j)\}},$$

$$q_1^j(u)=E\left\{\frac{2\delta_iI(X_{i,j}<X_{k,j},Y_i<Y_k)}{S_C^2(Y_i)}I(Y_i\geq u)\right\},$$

$$q_2(u)=E\left\{\sum_{i=1}^n\sum_{k=1}^n\frac{2\delta_iI(Y_i<Y_k)}{S_C^2(Y_i)}I(Y_i\geq u)\right\},$$

$$q_3^j(W_i^j)=E\left\{\frac{\delta_i}{S_C^2(Y_i)}I(X_{i,j}<X_{k,j},Y_i<Y_k)\right.$$
$$+\frac{\delta_k}{S_C^2(Y_k)}I(X_{i,j}<X_{k,j},Y_k<Y_i)$$
$$\left.-2E\{I(X_{i,j}<X_{k,j},T_i<T_k)\}\,|W_i^j,i\neq k\right\},$$

$$q_4(W_i^j)=E\left\{\frac{\delta_i}{S_C^2(Y_i)}I(Y_i<Y_k)+\frac{\delta_j}{S_C^2(Y_k)}I(Y_k<Y_i)\right.$$
$$\left.-2E\{I(T_i<T_k)\}\,|W_i^j,i\neq k\right\}.$$

Note that $n^{1/2}(\hat{\tau}_j-\tau_j)$ can be written as $I_1^j+I_2^j$, where $I_s^j=I_{s1}^j+I_{s2}^j$ for

$s = 1, 2$, and

$$
I_{11}^j = \frac{n^{-3/2} \sum_i \sum_k \frac{\delta_i}{\hat{S}^2(Y_i)} I(X_{i,j} < X_{k,j}, Y_i < Y_k)}{n^{-2} \sum_i \sum_k \frac{\delta_i}{\hat{S}^2(Y_i)} I(Y_i < Y_k)}
$$

$$
- \frac{n^{-3/2} \sum_i \sum_k \frac{\delta_i}{S^2(Y_i)} I(X_{i,j} < X_{k,j}, Y_i < Y_k)}{n^{-2} \sum_i \sum_k \frac{\delta_i}{S^2(Y_i)} I(Y_i < Y_k)},
$$

$$
I_{12}^j = \frac{n^{-3/2} \sum_i \sum_k \frac{\delta_i}{\hat{S}^2(Y_i)} I(X_{i,j} < X_{k,j}, Y_i < Y_j)}{n^{-2} \sum_i \sum_k \frac{\delta_i}{S^2(Y_i)} I(Y_i < Y_j)}
$$

$$
- \frac{n^{-3/2} \sum_i \sum_k \frac{\delta_i}{S^2(Y_i)} I(X_{i,j} < X_{k,j}, Y_i < Y_j)}{n^{-2} \sum_i \sum_k \frac{\delta_i}{S^2(Y_i)} I(Y_i < Y_k)},
$$

$$
I_{21}^j = \frac{n^{-3/2} \sum_i \sum_k \frac{\delta_i}{S^2(Y_i)} I(X_{i,j} < X_{k,j}, Y_i < Y_k)}{n^{-2} \sum_i \sum_k \frac{\delta_i}{S^2(Y_i)} I(Y_i < Y_k)}
$$

$$
- \frac{n^{1/2} P(X_{i,j} < X_{k,j}, T_i < T_k)}{n^{-2} \sum_i \sum_k \frac{\delta_i}{S^2(Y_i)} I(Y_i < Y_k)},
$$

$$
I_{22}^j = \frac{n^{1/2} P(X_{i,j} < X_{k,j}, T_i < T_k)}{n^{-2} \sum_i \sum_k \frac{\delta_i}{S^2(Y_i)} I(Y_i < Y_k)} - \frac{n^{1/2} P(X_{i,j} < X_{k,j}, T_i < T_k)}{P(Z_i^j < Z_k^j)}.
$$

Using a martingale representation of $n^{1/2}(\hat{S} - S)$ (Gill, 1980) and the uniform convergence of

$$
q_{2n}(u) = n^{-2} \sum_i \sum_k \frac{2\delta_i I(Y_i < Y_k)}{S^2(Y_i)} I(Y_i \geq u)
$$

to $q_2(u)$, we can show that

$$
I_{11}^j = -\tau^j n^{-3/2} \sum_i \sum_k \frac{2\delta_i I(Y_i < Y_j)}{S^2(Y_i)} \left\{ \frac{\hat{S}(Y_i) - S(Y_i)}{\hat{S}(Y_i)} \right\} + o_p(1)
$$

$$
= -\tau^j n^{-1} \sum_r \int \frac{q_2(u) dM_r(u)}{E\{A_r(u)\}} + o_p(1),
$$

Similarly, we can show that

$$
I_{12}^j = \frac{1}{P(T_i < T_k)} n^{-1} \sum_r \int \frac{q_1^j(u) dM_r(u)}{E\{A_r(u)\}} + o_p(1).
$$

Using the properties of U-statistics (Shao, 1999), we can show that

$$
I_{21}^j = -\frac{1}{P(T_i < T_k)} n^{-1/2} \sum_i q_4(W_i^j) + o_p(1),
$$

$$
I_{22}^j = \frac{\tau^j}{P(T_i < T_k)} n^{-1/2} \sum_i q_3^j(W_i^j) + o_p(1).
$$

Therefore,

$$n^{1/2} \left(\hat{\tau}_j - \tau_j \right) = n^{-1/2} \sum_{i=1}^{n} \varphi_{i,j} + o_p(1). \qquad (6.24)$$

6.7.2 Asymptotic Properties of Penalized AFT Model

Consider the AFT model where $T = \alpha + \mathbf{X}^T \boldsymbol{\beta} + \epsilon$. Consider the Stute estimate, where the objective function is

$$R(\boldsymbol{\beta}) = \frac{1}{2} \sum_{i=1}^{n} (\tilde{Y}_{(i)} - \tilde{\mathbf{X}}_{(i)}^T \boldsymbol{\beta})^2.$$

We refer to Section 6.3.1 for more details on notations.

In this section, we describe the asymptotic properties of the bridge penalized estimate. In the computational algorithm described in Section 6.3, the bridge optimization problem is transformed into an iterative weighted Lasso optimization, with the Lasso estimate as the initial estimate. Thus, in the process of establishing the properties of bridge estimate, we also establish the properties of Lasso and adaptive Lasso (which generates weights using the Lasso estimate) estimates. The more detailed results and rigorous proofs are presented in Huang and Ma (2010). We note that for a fixed p, the asymptotic results can be obtained relatively easily using standard approaches. Below we focus on the scenario where p is allowed to diverge with n.

In the algorithm for the bridge penalization, covariates not selected by the Lasso initial estimate will not be selected in the final model. Thus, it is crucial to first establish properties of the Lasso estimate. Once the initial estimate is obtained, in each step, an adaptive Lasso estimate is computed. Thus, we are able to use similar methods as in Huang et al. (2008b), which studies properties of the adaptive Lasso in high-dimensional linear regression models, to establish properties of the bridge estimate.

We consider the rescaled $\tilde{\mathbf{X}}_{(i)}$ and $\tilde{Y}_{(i)}$ defined in Section 6.3. For simplicity of notations, we use $\mathbf{X}_{(i)}$ and $Y_{(i)}$ to denote $\tilde{\mathbf{X}}_{(i)}$ and $\tilde{Y}_{(i)}$ hereafter. Let $Y = (Y_{(1)}, \ldots, Y_{(n)})'$. Let X be the $n \times p$ covariate matrix consisting of row vectors $\mathbf{X}'_{(1)}, \ldots, \mathbf{X}'_{(n)}$. Let $X_1, \ldots, X_p$ be the p columns of X. Let $W = diag(nw_1, \ldots, nw_n)$ be the diagonal matrix of the Kaplan-Meier weights. For $A \subseteq \{1, \ldots, p\}$, let $X_A = (X_j, j \in A)$ be the matrix with columns X_j's for $j \in A$. Denote $\Sigma_A = X'_A W X_A / n$. Denote the cardinality of A by $|A|$.

Let $\beta_0 = (\beta_{01}, \ldots, \beta_{0p})'$ be the true value of the regression coefficients. Let $A_1 = \{j : \beta_{0j} \neq 0\}$ be the set of nonzero coefficients and let $q = |A_1|$. We make the following assumptions.

(A1) The number of nonzero coefficients q is finite.

(A2) (a) The observations $(Y_i, X_i, \delta_i), 1 \leq i \leq n$ are independent and identically distributed. (b) The errors $\epsilon_1, \ldots, \epsilon_n$ are independent and identically distributed with mean 0 and finite variance σ^2. Furthermore, they are sub-Gaussian, in the sense that there exist $K_1, K_2 > 0$ such that the tail probabilities of ϵ_i satisfy $P(|\epsilon_i| > x) \leq K_2 \exp(-K_1 x^2)$ for all $x \geq 0$ and all i.

(A3) (a) The errors $(\epsilon_1, \ldots, \epsilon_n)$ are independent of the Kaplan-Meier weights $(w_1, \ldots, w_n)$. (b) The covariates are bounded. That is, there is a constant $M > 0$ such that $|X_{ij}| \leq M, 1 \leq i \leq n, 1 \leq j \leq p$.

(A4) The covariate matrix satisfies the sparse Riesz condition (SRC) with rank q^*: there exist constants $0 < c_* < c^* < \infty$, such that for $q^* = (3+4C)q$ and $C = c^*/c_*$, with probability converging to 1, $c_* \leq \frac{\nu' \Sigma_A \nu}{\|\nu\|^2} \leq c^*$, $\forall A$ with $|A| = q^*$ and $\nu \in \mathbb{R}^{q^*}$, where $\|\cdot\|$ is the l_2 norm.

By (A1), the model is sparse in the sense that although the total number of covariates may be large, the number of covariates with nonzero coefficients is still small. This is a practically reasonable assumption. Consider, for example, cancer genomic studies like the MCL. Although a large number of genes are surveyed, up to date, only < 400 genes have been identified as "cancer genes" (for all types of cancers combined). The tail probability assumption in (A2) has been made with high-dimensional linear regression models. See for example Zhang and Huang (2008). With assumption (A3), it can be shown that the subguassian tail property still holds under censoring. The SRC condition (A4) has been formulated in study of the Lasso with linear regressions without censoring (Zhang and Huang, 2008). This condition implies that all the eigenvalues of any $d \times d$ submatrix of $X'WX/n$ with $d \leq q^*$ lie between c_* and c^*. It ensures that any model with dimension no greater than q^* is identifiable.

We first consider the Lasso estimator defined as

$$\tilde{\beta} = argmin_\beta \left\{ R(\beta) + \lambda \sum_{j=1}^p |\beta_j| \right\}.$$

With $\tilde{\beta} = (\tilde{\beta}_1, \ldots, \tilde{\beta}_p)^T$, let $\tilde{A}_1 = \{j, \tilde{\beta}_j \neq 0\}$ be the set of nonzero Lasso estimated coefficients.

Theorem 1 *Suppose that (A1)–(A4) hold and $\lambda \geq O(1)\sqrt{n \log p}$. Then*

(i) *With probability converging to 1, $|\tilde{A}_1| \leq (2 + 4C)q$.*

(ii) *If $\lambda/n \to 0$ and $(\log p)/n \to 0$, then with probability converging to 1, all the covariates with nonzero coefficients are selected.*

(iii) *$\|\tilde{\beta} - \beta_0\|_2^2 \leq \frac{16\lambda^2 q}{n^2 c_*^2} + O_p\left(\frac{|\tilde{A}_1| \log p}{nc_*^2}\right)$. In particular, if $\lambda = O(\sqrt{n \log p})$, then $\|\tilde{\beta} - \beta_0\|_2^2 = O_p(\log p/n)$.*

This theorem suggests that, with high probability, the number of covariates selected by the Lasso is a finite multiple of the number of covariates with nonzero coefficients. Moreover, all the covariates with nonzero coefficients are selected with probability converging to one. Thus, the first step Lasso estimate can effectively remove the majority of true negatives while keeping all of the true positives. The Lasso estimate is also *estimation* consistent, which makes it a proper choice for weights of the adaptive (weighted) Lasso in downstream iterations. Hence, the Lasso provides an appropriate choice for the initial estimator.

Starting from the initial Lasso estimator $\tilde{\beta}$, we denote $\hat{\beta}$ as the estimate after one iteration (in the algorithm described in Section 6.3). Simple algebra shows that the value of $\theta_j^{(1)}$ computed in Step 2 of the proposed algorithm is $\theta_j^{(1)} = (\lambda/2)|\tilde{\beta}_j|^{-1/2}$. Thus Step 3 of the proposed algorithm is

$$\hat{\beta} = argmin \left\{ R(\beta) + \frac{\lambda}{2} \sum_{j=1}^{p} |\tilde{\beta}_j|^{-1/2} |\beta_j| \right\}.$$

$\hat{\beta}$ computed above takes the form of an adaptive Lasso estimator. Of note, here, the penalty parameter is the same as the λ used in the Lasso estimator.

For any vector $x = (x_1, x_2, \ldots)$, denote its sign vector by $sgn(x) = (sgn(x_1), sgn(x_2), \ldots)$, where $sgn(x_i) = 1, 0, -1$ if $x_i > 0, = 0, < 0$, respectively.

Theorem 2 *Suppose that (A1)–(A4) are satisfied, $(\log p)/n \to 0$, and $\lambda = O(\sqrt{n \log p})$. Then*

$$P(sgn(\hat{\beta}) = sgn(\beta_0)) \to 1.$$

The above theorem shows that the one-step estimator is *sign consistent*. Thus, the one-step estimator is selection consistent, in the sense that it can correctly distinguish covariates with zero and nonzero coefficients with probability converging to 1. Following similar arguments, we can prove that any finite-step estimator (computed from the algorithm described in Section 6.3) is sign consistent and hence selection consistent. We note that, although the one-step estimator is selection consistent, numerical studies suggest that iterating until convergence tends to improve finite sample performance. With the selection consistency property and under the condition that the number of nonzero coefficients q is finite, the estimation consistency can be easily established. We note that in Theorem 2, we allow $\log p = o(n)$ or $p = \exp(o(n))$. Thus the dimension of covariates can be much larger than the sample size.

The above two theorems establish that under the AFT model, penalized estimation (under the Lasso, adaptive Lasso, and bridge) has satisfactory selection properties. However, examining Huang and Ma (2010) also suggests that such properties need to be established on a case-by-case basis. Although we conjecture that similar properties hold for other penalties, there is still a lack of rigorous theoretical study.

6.7.3 Screening

In the identification of covariates marginally associated with survival, computational cost is usually not expensive, as many software packages can carry out marginal estimations very fast and the computation can be conducted in a highly parallel manner. However, in the construction of multivariate predictive models, for both dimension reduction and variable selection approaches, computational cost can be a major concern.

Screening provides a computationally feasible solution. In screening, a computationally simple, rough selection is first conducted. Only covariates that have passed the initial screening are analyzed in downstream analysis. It is worth emphasizing that the goal of screening is *not* to conduct the final covariate selection. Rather, it targets at removing a large number of noises while keeping the majority or all of the true positives. After screening, downstream finer analysis such as Lasso is conducted to achieve a final model. That is, in screening, controlling the false positives is not of major concern.

Screening can be classified as unsupervised and supervised screening. In unsupervised screening, the survival outcome is not used. For example, in microarray gene expression studies, covariates with high missing rates (say, over 30%) are removed from downstream analysis. In addition, usually only genes with a high level of variation are of interest. The variances or interquartile ranges of gene expressions can be first computed, and genes with, for example, variances less than the median variance are removed from analysis. Unsupervised screening is usually *due to technical or scientific considerations that are not relevant to the survival outcome of interest.*

Supervised screening, in contrast, uses the response variable and proceeds as follows:

(S1) For $j = 1, \ldots, p$, compute a ranking statistic for covariate j. One commonly adopted approach is to fit a marginal regression (for example, Cox) model using only covariate j. Then a summary statistic, which may be the magnitude of the regression coefficient or its p-value, is chosen as the ranking statistic.

(S2) Top ranked covariates or covariates with the ranking statistics above a certain cut-off are selected for downstream analysis.

We conduct a small simulation study to demonstrate the properties of supervised screening. We generate the values of 5000 covariates for 100 subjects. Covariates are marginally normally distributed with variance 1 and mean 0. Covariates i and j have correlation coefficient $\rho^{|i-j|}$ with $\rho = 0.3$. There are 30 covariates associated with survival and that have regression coefficients equal to 1. The survival times are generated from the Cox model. The censoring times are independently generated from an exponential distribution and adjusted so that the censoring rate is about 50%. In the supervised screening, we fit a marginal Cox model for each covariate separately. The absolute values of the marginal regression coefficients are chosen as the ranking statistics.

In Figure 6.10, we show the number of true positives versus the number of covariates passed screening. When 1000 (20%) covariates are selected, 20 (out of 30) true positives are included. This observation is generally true, that is, supervised screening is capable of significantly reducing the dimensionality of covariates at the price of losing a small number of true positives.

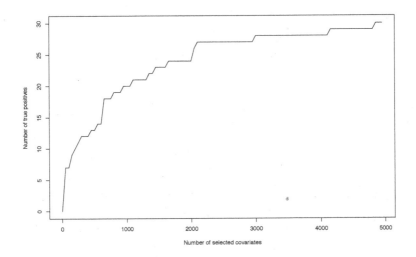

FIGURE 6.10: Simulation study of supervised screening: number of true positives versus number of covariates passed screening.

Example (MCL). As a demonstration of possible effects of supervised screening on analysis, we reanalyze the MCL data as follows. First, we conduct supervised screening and select the top 100 genes. The magnitude of regression coefficients from marginal models is selected as the ranking statistic. Of note, in practice, usually a larger number of genes are selected to pass screening. We intentionally screen a small number of genes for demonstration purpose. Second, we conduct the PCA-based analysis using the first three PCs. The summary of Cox model fitting is as follows:

```
Call:
coxph(formula = Surv(survival.dat$Followup,
    survival.dat$Status)~ exp.PC)

  n= 92


        coef exp(coef) se(coef)       z Pr(>|z|)
exp.PC1 -0.2980    0.7423   0.0474 -6.288 3.22e-10 ***
exp.PC2 -0.4268    0.6526   0.1474 -2.896  0.00378 **
```

```
exp.PC3 -0.3091    0.7341    0.1816 -1.702   0.08871 .
---
Signif. codes:   0 '***' 0.001 '**' 0.01 '*' 0.05 '.' 0.1 ' ' 1

Rsquare= 0.399    (max possible= 0.993 )
Likelihood ratio test= 46.85  on 3 df,   p=3.733e-10
Wald test             = 42.35  on 3 df,   p=3.387e-09
Score (logrank) test = 46.32  on 3 df,   p=4.854e-10
```

The analysis result differs significantly from its counterpart using 2000 unscreened genes. For example, with 2000 genes, the first PC has a positive regression coefficient, whereas with 100 genes, the first PC has a negative regression coefficient. In addition, with 100 genes, the second PC's estimated coefficient is statistically significant. Moreover, there is an increase in the R squared value.

The procedure described above (supervised screening + PCA) is basically the supervised PCA approach proposed in Bair and Tibshirani (2004) and Bair et al. (2006). In PCA-based analysis, the PCs are constructed in an unsupervised manner and target at explaining the data variation, which may have nothing to do with the survival response variable. With the supervised screening, noisy covariates can be effectively removed. Model fitting with the remaining variables, which are "more likely" to be associated with survival, may improve over that without screening.

6.8 Exercises

1. Consider the family-wise Type I error rate (FWER) defined as

$$P(S_1 \cup S_2 \cup \cdots \cup S_p)$$

where $S_j = \{p_j < \alpha | \beta_j = 0\}$ is the event that the jth null hypothesis is incorrectly rejected ($j = 1, \cdots, p$). If the p-values p_j, $j = 1, 2, \cdots, p$, are independent, show that

$$\text{FWER} = 1 - (1 - \alpha/p)^p. \tag{6.25}$$

This suggests that we can use $\alpha = 1 - (0.95)^{1/p}$ as the preplanned cut-off for significance and still retain the overall type I error rate under 0.05. This so-called Sidak adjustment is less conservative for some problems than the Bonferroni adjustment.

2. Using notations in Table 6.1, show that the FDR approach introduced in Section 6.2 preserves that FDR$\leq m_0 q/p$ and therefore FDR$\leq q$ for a prespecified q.

3. Show that the p p-values discussed in Section 6.2 follow a uniform distribution over $[0, 1]$ when all p null hypotheses are true.

4. Given an $n \times p$ matrix $\mathbf{X}$, let

$$
\begin{aligned}
\mathbf{X} &= \mathbf{U}\mathbf{D}\mathbf{V}^T \\
&= \mathbf{R}\mathbf{V}^T
\end{aligned}
$$

be the singular-value decomposition (SVD) of $\mathbf{X}$; that is, $\mathbf{U}$ is $n \times n$ orthogonal, $\mathbf{D}$ is an $n \times n$ diagonal matrix with elements $d_1 \geq d_2 \geq \cdots \geq d_n \geq 0$, and $\mathbf{V}$ is $p \times n$ with orthonormal columns. The matrix $\mathbf{R} = \mathbf{U}\mathbf{D}$ are the principal components (PCs), and the columns of $\mathbf{V}$ are the corresponding loadings of the PCs. The sample variance of the ith PC is d_i^2/n. In gene expression data the standardized PCs $\mathbf{U}$ are called the *eigen-arrays* and $\mathbf{V}$ are the *eigen-genes* (Alter et al., 2000).

(a) Now consider a linear regression problem with a response vector $\mathbf{Y}$ and a design matrix $\mathbf{X}$. Consider estimating the regression coefficients with a least square loss and a quadratic penalty (Section 6.3.2). It is well known that the solution to this Ridge regression is

$$
\hat{\boldsymbol{\beta}} = (\mathbf{X}^T\mathbf{X} + \lambda\mathbf{I})^{-1}\mathbf{X}^T\mathbf{Y}. \tag{6.26}
$$

Show that (6.26) is equivalent to the following

$$
\hat{\boldsymbol{\beta}} = \mathbf{V}(\mathbf{R}^T\mathbf{R} + \lambda\mathbf{I})^{-1}\mathbf{R}^T\mathbf{Y}. \tag{6.27}
$$

We note that the calculation for (6.26) requires the inversion of a matrix of dimension $p \times p$ while that for (6.27) is but $n \times n$. This trick thus reduces the overall computational cost from $O(p^3)$ to $O(pn^2)$ when $p >> n$.

(b) For each i, denote by $\mathbf{R}_i = d_i\mathbf{U}_i$ the ith PC where $\mathbf{U}_i$ is the ith column of $\mathbf{U}$. Consider a ridge regression problem given by

$$
\hat{\boldsymbol{\theta}} = \text{argmax}_{\boldsymbol{\theta}}\|\mathbf{R}_i - \mathbf{X}\boldsymbol{\theta}\|^2 + \lambda\|\boldsymbol{\theta}\|^2.
$$

Show that $\hat{\boldsymbol{\theta}}/\|\hat{\boldsymbol{\theta}}\| = \mathbf{V}_i$.

Bibliography

O. O. Aalen. *Statistical inference for a family of counting processes.* PhD thesis, University of California, Berkeley, 1975.

O. O. Aalen. Nonparametric inference in connection with multiple decrements models. *Scandinavian Journal of Statistics.*, 3:15–27, 1976.

O. O. Aalen. A linear regression model for the analysis of lifetime. *Statistics in Medicine*, 8:907–925, 1989.

O. O. Aalen. Further results on the nonparametric linear regression model in survival analysis. *Statistics in Medicine*, 12:1569–1588, 1993.

O. O. Aalen. Effects of frailty in survival analysis. *Statistical Methods in Medical Research*, 3:227–243, 1994.

J. Abrevaya. Computation of the maximum rank correlation estimator. *Economic Letters.*, 62:279–285, 1999.

A. Agresti. *Categorical Data Analysis.* Hoboken, NJ: Wiley, 1990.

A. Agresti and B. A. Coull. Approximate is better than exact for interval estimation of binomial proportions. *American Statistician*, 52:119–126, 1998.

A. Agresti and C. Franklin. *Statistics: The Art and Science of Learning from Data*, 2nd ed. Upper Saddle River, NJ: Pearson Education, 2009.

M. Aitkin and D. Clayton. The fitting of exponential, Weibull and extreme value distributions to complex censored survival data using GLIM. *Applied Statistics*, 29:156–163, 1980.

M. G. Akritas. Nearest neighbor estimation of a bivariate distribution under random censoring. *Annals of Statistics*, 22:1299–1327, 1994.

A. A. Alizadeh, M. B. Eisen, R. E. Davis, C. Ma, I. S. Lossos, A. Rosenwald, J. C. Broldrick, H. Sabet, R. Tran, X. Yu, et al. Distinct types of diffuse large B-cell lymphoma identified by gene expression profiling. *Nature*, 403:503–511, 2000.

O. Alter, P. Brown, and D. Botstein. Singular value decomposition for genome-wide expression data processing and modeling. *Proceedings of the National Academy of Sciences*, 97:10101–10106, 2000.

D. G. Altman and B. L. de Stavola. Practical problems in fitting a proportional hazards model to data with updated measurements of the covariates. *Statistics in Medicine*, 13:301–344, 1994.

P. K. Andersen and R. D. Gill. Cox's regression model for counting processes: a large sample study. *Annals of Statistics*, 10:1100–1120, 1982.

P. K. Andersen and K. Liestol. Attenuation caused by infrequently updated covariates in survival analysis. *Biostatistics*, 4:633–649, 2003.

J. E. Anderson and T. A. Louis. Survival analysis using a scale change random effects model. *Journal of the American Statistical Association*, 90:669–679, 1995.

P. K. Anderson and R. D. Gill. Cox's regression model for counting processes: a large sample study. *Annals of Statistics*, 10:1100–1120, 1982.

P. K. Anderson and R. B. Ronn. A nonparametric test for comparing two samples where all observations are either left- or right-censored. *Biometrics*, 51:323–329, 1995.

L. Antolini, P. Boracchi, and E. Biganzoli. A time-dependent discrimination index for survival data. *Statistics in Medicine*, 10:1100–1120, 1982.

A. C. Atkinson. A note on the generalized information criterion for choice of model. *Biometrika*, 67:413–418, 1988.

R. R. Bahadur. A note on quantiles in large samples. *Annals of Mathematical Statistics*, 37:577–580, 1966.

E. Bair and R. Tibshirani. Semisupervised methods to predict patient survival from gene expression data. *PLoS Biology*, 2:511–522, 2004.

E. Bair, T. Hastie, D. Paul, and R. Tibshirani. Prediction by supervised principal components. *Journal of the American Statistical Association*, 101:119–137, 2006.

A. I. Bandos, H. E. Rockette, and D. Gur. Incorporating utility weights when comparing two diagnostic systems. *Academic Radiology*, 12:1293–1300, 2005.

M. Banerjee and J. A. Wellner. Confidence intervals for current status data. *Scandinavian Journal of Statistics*, 32:405–424, 2005.

W. E. Barlow and R. L. Prentice. Residuals for relative risk regression. *Biometrika*, 75:54–74, 1988.

J. Bartroff and T. L. Lai. Incorporating individual and collective ethics into phase I cancer trial designs. *Biometrics*, 67:596–603, 2011.

P. Bauer and K. Kohne. Evaluation of experiments with adaptive interim analyses. *Biometrics*, 50:1029–1041, 1994.

Y. Benjamini and Y. Hochberg. Controlling the false discovery rate: a practical and powerful approach to multiple testing. *Journal of the Royal Statistical Society, Series B (Methodological)*, 57:289–300, 1995.

Y. Benjamini and D. Yekutieli. The control of the false discovery rate in multiple testing under dependency. *Annals of Statistics*, 29:1165–1188, 2001.

J. Berger. *Statistical Decision Theory and Bayesian Analysis*. Springer Verlag: New York, 1985.

P. J. Bickel and D. Freedman. Some asymptotic theory for the bootstrap. *Annals of Statistics*, 9:1196–1217, 1981.

P. J. Bickel and E. Levina. Covariance regularization by thresholding. *Annals of Statistics*, 36:2577–2604, 2008.

P. J. Bickel, F. Gotze, and W. R. van Zwet. Resampling fewer than n observations: gains, losses, and remedies for losses. *Statistica Sinica*, 7:1–31, 1987.

P. J. Bickel, C. A. J. Klaassen, Y. Ritov, and J. A. Wellner. *Efficient and Adaptive Estimation for Semiparametric Models*. John Hopkins University Press, 1993.

P. Billingsley. *Convergence of Probability Measure*, 2nd ed. New York: Wiley, 1999.

O. Borgan and K. Leistol. A note on confidence bands for the survival curve based on transformations. *Scandinavian Journal of Statistics*, 17:35–41, 1990.

H. M. Bovelstad, S. Nygard, H. L. Storvold, M. Aldrin, O. Borgan, A. Frigessi, and O. C. Lingjarde. Predicting survival from microarray data—A comparative study. *Bioinformatics*, 23:2080–2087, 2007.

H. M. Bovelstad, S. Nygard, and O. Borgan. Survival prediction from clinico-genomic models—A comparative study. *BMC Bioinformatics*, 10:413, 2009.

J. Bradic, J. Fan, and J. Jiang. Regularization for Cox's proportional hazards model with NP-dimensionality. *Annals of Statistics*, 39:3092–3120, 2011.

J. Braun, T. Duchesne, and J. E. Stafford. Local likelihood density estimation for interval censored data. *Canadian Journal of Statistics*, 33:39–66, 2005.

P. Breheny and J. Huang. Coordinate descent algorithms for nonconvex penalized regression with applications to biologial feature selection. *Annals of Applied Statistics*, 5:232–253, 2011.

N. E. Breslow. A generalized Kruskal-Wallace test for comparing k samples subject to unequal patterns of censorship. *Biometrics*, 57:579–594, 1970.

N. E. Breslow. Binomial confidence interval. *Journal of the American Statistical Association*, 78:108–116, 1983.

N. E. Breslow and J. Crowley. A large sample study of the life table and product limit estimates under random censorship. *Annals of Statistis*, 2: 437–453, 1974.

N. E. Breslow and N. E. Day. *Statistical Methods in Cancer Research. Volume II: The Design and Analysis of Cohort Studies*. Oxford, UK: Oxford University Press, 1987.

L. Brieman, J. H. Friedman, R. Olshen, and C. J. Stone. *Classification and Regression Trees*. Belmont, CA: Wadsworth, 1984.

R. Brookmeyer and J. J. Crowley. A confidence interval for the median survival time. *Biometrics*, 38:29–41, 1982.

J. Buckley and I. James. Linear regression with censored data. *Biometrika*, 66:429–436, 1979.

A. Buja, T. Hastie, and R. Tibshirani. Linear smoothers and additive models (with discussion). *Annals of Statistics*, 17:453–555, 1989.

L. Bullinger, M. Enrich, K. Dohner, R. F. Schlenk, H. Dohner, M. R. Nelson, and D. van den Boom. Quantitative DNA methylation predicts survival in adult acute myoloid leukemia. *Blood*, 115(3):636–642, 2010.

K. P. Burham and D. R. Anderson. Multimodel inference, understanding AIC and BIC in model selection. *Sociological Methods and Research*, 28:49–66, 2004.

J. Cai. Hypothesis testing of hazard ratio parameters in marginal models for multivariate failure time data. *Lifetime Data Analysis*, 5:39–53, 1999.

J. Cai, J. Fan, J. Jiang, and H. Zhou. Partially linear hazard regression for multivariate survival data. *Journal of the American Statistical Association*, 102:538–551, 2007a.

J. Cai, J. Fan, H. Zhou, and Y. Zhou. Hazard models with varying coefficients for multivariate failure time data. *Annals of Statistics*, 35:324–354, 2007b.

T. Cai and Y. Zheng. Nonparametric evaluation of biomarker accuracy under nested case-control studies. *Journal of the American Statistical Association*, 106:569–580, 2011.

G. Campbell. Nonparametric bivariate estimation with randomly censored data. *Biometrika*, 68:417–422, 1981.

C. Cavanagh and R. P. Sherman. Rank estimators for monotonic index models. *Journal Econometrics*, 84:351–381, 1998.

K. M. Chaloner and G. T. Duncan. Assessment of a beta prior distribution: PM elicitation. *Statistician*, 32:174–180, 1983.

L. E. Chambless and G. Diao. Estimation of time-dependent area under the ROC curve for long-term risk prediction. *Statistics in Medicine*, 25:3474–3486, 2006.

S. H. Chang. Estimating marginal effects in accelerated failure time models for serial sojourn times among repeated events. *Lifetime Data Analysis*, 10:175–190, 2004.

M. H. Chen, J. G. Ibrahim, and D. Sinha. A new Bayesian model for survival data with a surviving fraction. *Journal of the American Statistical Association*, 94:909–919, 1999.

M. H. Chen, J. G. Ibrahim, and D. Sinha. Bayesian inference for multivariate survival data with a cure fraction. *Journal of Multivariate Analysis*, 80:101–126, 2002.

M. Y. Cheng, P. Hall, and D. S. Tu. Confidence bands for hazard rate under random censorship. *Biometrika*, 93:357–366, 2006.

Y. Cheng and J. Li. Time-dependent diagnostic accuracy analysis with censored outcome and censored predictor. 2012. Unpublished manuscript.

Y. Cheng, J. P. Fine, and M. R. Kosorok. Nonparametric association analysis of exchangeable clustered competing risks data. *Biometrics*, 65:385–393, 2009.

Y. Cheng, J. P. Fine, and K. Bandeen-Roche. Association analyses of clustered competing risks data via cross hazard ratio. *Biostatistics*, 11:82–92, 2010.

C. T. Chiang and H. Hung. Nonparametric estimation for time-dependent AUC. *Journal of Statistical Planning and Inference*, 140:1162–1174, 2010a.

C. T. Chiang and H. Hung. Nonparametric methodology for the time-dependent partial area under the ROC curve. *Journal of Statistical Planning and Inference*, 141:3829–3838, 2010b.

D. G. Clayton. A model for association in bivariate life tables and its applications in epidemiological studies of familial tendency in chronic disease incidence. *Biometrics*, 65:141–151, 1978.

D. G. Clayton. A Monte Carlo method for Bayesian inference in frailty models. *Biometrics*, 47:467–485, 1991.

D. G. Clayton and J. Cuzick. Multivariate generalizations of the proportional hazards model (with discussion). *Journal of the Royal Statistical Society, Series A*, 148:82–117, 1985.

A. Cnaan and L. Ryan. Survival analysis in natural history studies of disease. *Statistics in Medicine*, 8:1255–1268, 1989.

D. Collet. *Modelling Survival Data in Medical Research*, 2nd ed. New York: Chapman & Hall/CRC Press, 1994.

D. Collet. *Modelling Binary Data*, 2nd ed. New York: Chapman & Hall/CRC Press, 2002.

E. Copelan, J. Biggs, J. Thompson, P. Crilley, J. Szer, K. N. Klein, J. P., B. Avalos, I. Cunningham, K. Atkinson, K. Downs, G. Harmon, M. Daly, I. Brodsky, S. Bulova, and P. Tutschka. Treatment for acute myelocytic-leukemia with allogeneic bone-marrow transplantation following preparation with BUCY2. *Blood*, 78:838–843, 1991.

S. R. Cosslett. Distribution-free maximum likelihood estimator of the binary choice model. *Econometrika*, 51:765–782, 1983.

D. R. Cox. Regression models and life tables (with discussion). *Journal of the Royal Statistical Society, Series B*, 26:103–110, 1972.

D. R. Cox. Partial likelihood. *Biometrika*, 62:269–276, 1975.

D. R. Cox and E. J. Snell. A general definition of residuals (with discussion). *Journal of the Royal Statistical Society, Series B*, 30:248–275, 1968.

J. Crowley and M. Hu. Covariance analysis of heart transplant survival data. *Journal of the American Statistical Association*, 72:27–36, 1977.

J. Cuzick. Rank regression. *Annals of Statistics*, 16:1369–1389, 1988.

D. M. Dabrowska. Kaplan Meier estimate on the plane. *Annals of Statistics*, 16:1475–1489, 1988.

D. M. Dabrowska. Kaplan Meier estimate on the plane: weak convergence, LIL, and the bootstrap. *Journal of Multivariate Analysis*, 29:308–325, 1989.

D. M. Dabrowska and K. A. Doksum. Partial likelihood in transformation models with censored data. *Scandinavian Journal of Statistics*, 15:1–24, 1988.

R. B. D'Agostino and M. A. Stephens. *Goodness-of-Fit Techniques*. New York: Marcel Dekker, 1986.

H. Dai and B. Fu. A polar coordinate transformation for estimating bivariate survival functions with randomly censored and truncated data. *Journal of Statistical Planning and Inference*, 142:248–262, 2012.

S. Datta, J. Le-Rademacher, and S. Datta. Predicting patient survival from microarray data by accelerated failure time modeling using partial least squares and LASSO. *Biometrics*, 63:259–271, 2007.

S. Dave, G. Wright, and B. Tan, et al. Prediction of survival in follicular lymphoma based on molecular features of tumor-infiltrating immune cells. *New England Journal of Medicine*, 351:2159–2169, 2004.

M. H. J. de Bruijne, S. le Cessie, H. C. Kluin-Nelemans, and H. C. van Houwelingen. On the use of Cox regression in the presence of an irregularly observed time-dependent covariate. *Statistics in Medicine*, 20: 3817–3829, 2001.

V. De Gruttola and S. W. Lagokos. Analysis of doubly censored survival data, with application to aids. *Biometrics*, 15:1–24, 1988.

C. deBoor. *A Practical Guide to Splines*. New York: Springer-Verlag, 1978.

D. K. Dey, M. H. Chen, and H. Chang. Bayesian approach for nonlinear random effects models. *Biometrics*, 53:1239–1252, 1997.

I. D. Diamond, J. W. McDonald, and I. H. Shah. Proportional hazards models for current status data: Application to the study of differentials in age at weaning in Pakistan. *Demography*, 23:607–620, 1986.

L. E. Dodd and M. S. Pepe. Partial AUC estimation and regression. *Biometrics*, 59:614–623, 2003.

A. D. Donaldson, L. Razak, J. L. Li, D. A. Fisher, and P. A. Tambyah. Carbapenems and subsequent multiresistant bloodstream infection: Does treatment duration matter? *International Journal of Antimicrobial Agents*, 34:246–251, 2009.

L. Duchateau and P. Janssen. *The Frailty Model*. New York: Springer, 2008.

S. Dudoit, J. Fridlyand, and T. Speed. Comparison of discrimination methods for the classification of tumors using gene expression data. *Journal of the American Statistical Association*, 97:77–87, 2002.

S. Dudoit, M. J. van der Laan, and K. S. Pollard. Multiple testing. Part I. Single-step procedures for control of generaly type I error rates. *Statistical Applications in Genetics and Molecular Biology*, 3:13, 2004.

R. Durrett. *Probability: Theory and Examples*, 3rd ed. Belmont, CA: Thomson Learning. 2005.

R. L. Dykstra and P. W. Laud. A Bayesian nonparametric approach to reliability. *Annals of Statistics*, 9:356–367, 1981.

P. Economou and C. Caroni. Graphical tests for the assumption of gamma and inverse Gaussian frailty distributions. *Lifetime Data Analysis*, 11: 565–582, 2005.

B. Efron. The two sample problem with censored data. In *Proceedings of the Fifth Berkeley Symposium in Mathematical Statistics, IV*. New York: Prentice Hall, 1967.

B. Efron. Efficiency of Cox's likelihood function for censored data. *Journal of the American Statistical Association*, 72:557–565, 1977.

B. Efron. *Large-Scale Inference: Empirical Bayes Methods for Estimation, Testing, and Prediction*. Cambridge, UK: Cambridge University Press, 2010.

B. Efron and R. Tibshirani. On testing the significance of sets of genes. *Annals of Applied Statistics*, 1:107–129, 2007.

B. Efron, R. Tibshirani, J. D. Storey, and V. Tusher. Empirical Bayes analysis of a microarray experiment. *Journal of the American Statistical Association*, 96:1151–1160, 2001.

B. Efron, I. Johnston, T. Hastie, and R. Tibshirani. Least angle regression. *Annals of Statistics*, 32:407–499, 2004.

F. S. Enrique, F. David, and R. Benjamin. Adjusting the generalized ROC curve for covariates. *Statistics in Medicine*, 23:3319–3331, 2004.

R. L. Eubank. *Nonparametric Regression and Spline Smoothing*. New York: Marcel Dekker, 1999.

B. S. Everitt. *The Analysis of Contingency Tables*, 2nd ed. New York: Chapman & Hall/CRC Press, 1992.

G. Faller, R. T. Mikolajczyk, M. K. Akmatov, S. Meier, and A. Kramer. Accidents in the context of study among university students—A multicentre cross-sectional study in North Rhine—Westphalia, Germany. *Accident Analysis and Prevention*, 42:487–491, 2010.

J. Fan and R. Li. Variable selection via noncancave penalized likelihood and its oracle properties. *Journal of the American Statistical Association*, 96: 1348–1360, 2001.

J. Fan and R. Li. variable selection for Cox's proportional hazards model and frailty model. *Annals of Statistics*, 30:74–99, 2002.

J. Fan and J. Lv. Nonconcave penalized likelihood with NP-dimensionality. *IEEE Transactions on Information Theory*, 57:5467–5484, 2011.

J. Fan and H. Peng. Nonconcave penalized likelihood with diverging number of parameters. *Annals of Statistics*, 32:928–961, 2004.

J. Fan, I. Gijbels, and M. King. Local likelihood and local partial likelihood in hazard regression. *Annals of Statistics*, 25:1661–1690, 1997.

J. Fan, H. Lin, and Y. Zhou. Local partial-likelihood estimation for lifetime data. *Annals of Statistics*, 34:290–325, 2006.

J. Fan, Y. Feng, and Y. Wu. High-dimensional variable selection for Cox's proportional hazards model. *IMS Collections*, 6:70–86, 2010.

H. Fang, J. Sun, and M. L. T. Lee. Nonparametric survival comparion for interval-censored continuous data. *Statistica Sinica*, 12:1073–1083, 2002.

H. Fang, G. Li, and J. Sun. Maximum likelihood estimation in a semiparametric logistic/proportional-hazards mixture model. *Scandinavian Journal of Statistics*, 32:59–75, 2005.

D. Faraggi and R. Simon. Bayesian variable selection method for censored survival data. *Biometrics*, 54:1475–1485, 1998.

J. P. Fine and R. J. Bosch. Risk assessment via a robust probit model, with application to toxicology. *Journal of the American Statistical Association*, 95:375–382, 2000.

J. P. Fine and R. J. Gray. A proportional hazards model for the subdistribution of a competing risk. *Journal of the American Statistical Association*, 94:496–509, 1999.

J. P. Fine and H. Jiang. On association in a copula with time transformations. *Biometrika*, 87:559–571, 2000.

J. P. Fine, D. V. Glidden, and K. E. Lee. A simple estimator for a shared frailty regression model. *Journal of Royal Statistical Society Series B*, 65: 317–329, 2003.

D. M. Finkelstein and R. A. Wolfe. A semiparametric model for regression analysis of interval-censored failure time data. *Biometrics*, 41:933–945, 1985.

T. R. Fleming and D. P. Harrington. *Counting Processes and Survival Analysis*. New York: John Wiley & Sons, 1991.

A. M. Foster, L. Tian, and L. J. Wei. Estimation for the Box-Cox transformation model without assuming parametric error distribution. *Journal of the American Statistical Association*, 96:1097–1101, 2001.

I. E. Frank and J. H. Friedman. A statistical view of some chemometrics regression tools. *Technometrics*, 35:109–148, 1993.

J. Friedman, T. Hastie, and R. Tibshirani. Additive logistic regression: a statistical view of boosting. *Annals of Statistics*, 28:337–374, 2000.

J. Friedman, T. Hastie, and R. Tibshirani. A note on the group Lasso and a sparse group Lasso. 20010. Unpublished manuscript. http://arxiv.org/abs/1001.0736.

D. Gamerman. Sampling from the posterior distribution in generalized linear mixed models. *Statistics and Computing*, 7:57–68, 1997.

S. Gao and X. Zhou. An empirical comparison of two semiparametric approaches for the estimation of covariate effects from multivariate failure time data. *Statistics in Medicine*, 16:2049–2062, 1997.

J. J. Gart, D. Krewski, P. N. Lee, R. E. Tarone, and J. Wahrendorf. *Statistical Methods in Cancer Research, Volume III, The Design and Analysis of Longterm Animal Experiments*. IARC Scientific Publications No. 79. Lyon: International Agency for Research on Cancer, 1986.

D. P. Gaver and M. Acar. Analytic hazard representations for use in reliability, mortality and simulation studies. *Communication in Statistics, Simulation and Computation*, 8:91–111, 1979.

E. A. Gehan. A generalized Wilcoxon test for comparing arbitrarily singly-censored samples. *Biometrics*, 21:203–223, 1965.

A. E. Gelfand and A. F. M. Smith. Sampling based approaches to calculating marginal densities. *Journal of the American Statistical Association*, 85: 398–409, 1990.

A. Gelman, J. B. Carlin, H. S. Stern, and D. B. Rubin. *Bayesian Data Analysis*. New York: Chapman & Hall/CRC Press, 2003.

R. Gentleman and J. Crowley. Local full likelihood estimation for the proportional hazards model. *Biometrics*, 47:1283–1296, 1991.

R. Geskus and P. Groeneboom. Asymptotically optimal estimation of smooth functionals for interval censoring, Part 1. *Statistica Neerlandica*, 50:69–88, 1996.

R. Geskus and P. Groeneboom. Asymptotically optimal estimation of smooth functionals for interval censoring, case 2. *Annals of Statistics*, 27:627–674, 1999.

I. Gijbels and U. Gurler. Covariance function of a bivariate distribution function estimator for left truncated and right censored data. *Statistica Sinica*, 8:1219–1232, 1998.

I. Gijbels and J. L. Wang. Strong representations of the survival function estimator for truncated and censored data with applications. *Journal of Multivariate Analysis*, 47:210–229, 1993.

R. D. Gill. *Censoring and Stochastic Integrals.* Mathematical Centre Tract No. 124. Amsterdam: Mathematisch Centrum, 1980.

E. Gine. Lectures on some aspects of the bootstrap. *Lecture Notes in Math,* 1665:37–152, 1997.

D. V. Glidden. Checking the adequacy of the gamma frailty model for multivariate failure time. *Biometrika,* 86:381–393, 1999.

D. V. Glidden. Pairwise dependence diagnostics for clustered failure time data. *Biometrika,* 94:371–385, 2007.

J. J. Goeman, S. van de Geer, F. de Kort, and H. C. van Houwelingen. A global test for groups of genes: Testing association with a clinical outcome. *Bioinformatics,* 20:93–99, 2004.

L. Goldstein and B. Langholz. Asymptotic theory for nested case-control sampling in the Cox regression model. *Annals of Statistics,* 20:1903–1928, 1992.

B. Gompertz. On the nature of the function expressive of the law of human mortality, and on a new mode of determining the value of life contingencies. *Philosophical Transactions of the Royal Society of London,* 115: 513–583, 1825.

M. Gonen and G. Heller. Concordance probability and discriminatory power in proportional hazards regression. *Biometrika,* 92:965–970, 2005.

R. L. Goodall, D. T. Dunn, and A. G. Babiker. Interval-censored survival time data: Confidence intervals for the nonparametric survivor function. *Statistics in Medicine,* 23:1131–1145, 2004.

A. Gordon. *Classification.* New York: Chapman & Hall/CRC Press, 1999.

E. Graf, C. Schmoor, W. Sauerbrei, and M. Schumacher. Assessment and comparison of prognostic classification schemes for survival data. *Statistics in Medicine,* 18:2529–2545, 1999.

P. Grambsch and T. Therneau. Proportional hazards tests and diagnostics based on weighted residuals. *Biometrika,* 81:515–526, 1994.

R. J. Gray. A class of k-sample tests for comparing the cumulative incidence of a competing risk. *Annals of Statistics,* 16:1141–1154, 1988.

P. Groeneboom. Nonparametric maximum likelihood estimators for interval cenosring and deconvolution. Technical Report 378, Department of Statistics, Stanford University, 1991.

P. Groeneboom and J. A. Wellner. *Information Bounds and Nonparametric Maximum Likelihood Estimation.* DMV Seminar, Band 19, Birkhauser, New York, 1992.

L. M. Grummer-Strawn. Regression analysis of current status data: An application to breast feeding. *Journal of the American Statistical Association*, 88:758–765, 1993.

J. Gui and H. Li. Penalized Cox regression analysis in high-dimensional and low-sample size settings, with applications to microarray gene expression data. *Bioinformatics*, 21:3001–3008, 2005.

A. K. Gupta and S. Nadarajah. *Handbook of Beta Distribution and Its Applications*. New York: Marcel Dekker, 2004.

R. C. Gupta and R. D. Gupta. A bivariate random environmental stress model. *Journal of Statistical Planning and Inference*, 139:3277–3287, 2009.

U. Gurler. Bivariate distribution and hazard functions when a component is randomly truncated. *Journal of Multivariate Analysis*, 47:210–229, 1993.

U. Gurler. Bivariate estimation with right truncated data. *Journal of the American Statistical Association*, 91:1152–1165, 1996.

J. E. Haddow, G. E. Palomaki, G. J. Knight, G. C. Cunningham, L. S. Lustig, and P. A. Boyd. Reducing the need for amniocentesis in women 35 years of age or older with serum markers for screening. *New England Journal of Medicine*, 330:1114–1118, 1994.

P. Hall. On Edgeworth expansion and bootstrap confidence bands in nonparametric curve estimation. *Journal of the Royal Statistical Society. Series B (Methodological)*, 55:291–304, 1993.

W. J. Hall and J. A. Wellner. Confidence bands for a survival curve from censored data. *Biometrika*, 67:133–143, 1980.

A. K. Han. A nonparametric analysis of transformations. *Journal of Econometrics*, 35:191–209, 1987.

X. Han, T. Zheng, F. M. Foss, Q. Lan, T. R. Holford, N. Rothman, S. Ma, and Y. Zhang. Genetic polymorphisms in the metabolic pathway and non-Hodgkin lymphoma survival. *American Journal of Hematology*, 85: 51–56, 2010.

F. E. Harrell, R. M. Califf, D. B. Pryor, K. L. Lee, and R. A. Rosati. Evaluating the yield of medical tests. *JAMA*, 247:2543–2546, 1982.

F. E. Harrell, K. L. Lee, R. M. Califf, D. B. Pryor, and R. A. Rosati. Regression modeling strategies for improved prognostic prediction. *Statistics in Medicine*, 3:143–152, 1984.

F. E. Harrell, K. L. Lee, and D. B. Mark. Tutorial in biostatistics: Multi-variable prognostic models: Issues in developing models, evaluating assumptions, and adequacy, and measuring and reducing errors. *Statistics in Medicine*, 15:361–387, 1996.

T. Hastie and R. Tibshirani. Exploring the nature of covariate effects in the proportional hazards model. *Biometrics*, 46:1005–1016, 1990a.

T. Hastie and R. Tibshirani. *Generalized Additive Models*. London: Chapman & Hall, 1990b.

T. Hastie, R. Tibshirani, and J. Friedman. *The Elements of Statistical Learning*. New York: Springer, 2009.

L. Hatcher and E. J. Stepanski. *A Step-by-step Approach to Using the SAS System for Univariate and Multivariate Statistics*. Cary, NC: SRS Institute, 1994.

S. He and G. Yang. Estimation of the truncation probability in the random truncation model. *Annals of Statistics*, 26:1011–1027, 1998.

P. Heagerty and Y. Zheng. Survival model predictive accuracy and ROC curves. *Biometrics*, 61:92–105, 2005.

P. Heagerty, T. Lumley, and M. S. Pepe. Time-dependent ROC curves for censored survival data and a diagnostic marker. *Biometrics*, 56:337–344, 2000.

V. Henschel, J. Engel, D. Holzel, and U. Mansmann. A semiparametric Bayesian proportional hazards model for interval censored data with frailty effects. *BMC Medical Research Methodology*, 9(9), 2009.

U. Hjorth. A reliability distribution with increasing, decreasing and bathtub-shaped failure rate. *Technometrics*, 22:99–107, 1980.

W. Hoeffding. A class of statistics with asymptotically normal distribution. *Annals of Mathematical Statistics*, 19:293–325, 1948.

D. G. Hoel and H. E. Walburg. Statistical analysis of survival experiments. *Journal of National Cancer Institute*, 49:361–372, 1972.

A. E. Hoerl and R. W. Kennard. Ridge regression: Biased estimation for nonorthogonal problems. *Technometrics*, 12:55–67, 1970.

D. W. Hosmer and S. Lemeshow. *Applied Survival Analysis*. New York: Wiley-Interscience, 1999.

D. W. Hosmer and S. Lemeshow. *Applied Logistic Regression*, 2nd ed., New York: Wiley-Interscience, 2000.

P. Hougaard. Modelling multivariate survival. *Scandinavian Journal of Statistics*, 14:291–304, 1987.

F. Hsieh and B. W. Turnbull. Nonparametric and semiparametric estimation of the receiver operating characteristic curve. *Annals of Statistics*, 24: 25–49, 1996.

J. Huang. Efficient estimation for the proportional hazards model with interval censoring. *Annals of Statistics*, 24:540–568, 1996.

J. Huang. Efficient estimation of the partly linear additive Cox model. *Annals of Statistics*, 27:1536–1563, 1999a.

J. Huang. Asymptotic properties of nonparametric estimation based on partly interval-censored data. *Statistica Sinica*, 9:501–519, 1999b.

J. Huang and S. Ma. Variable selection in the accelerated failure time model via the bridge method. *Lifetime Data Analysis*, 16:176–195, 2010.

J. Huang and J. A. Wellner. Asymptotic normality of the NPMLE of linear functionals for interval censored data, case I. *Statistica Neerlandica*, 49: 153–163, 1995.

J. Huang and J. A. Wellner. Interval censored survival data: A review of recent progress. *Proceedings of the First Seattle Symposium in Biostatistics*, pp. 1–47, 1997.

J. Huang, V. J. Vieland, and K. Wang. Nonparametric estimation of marginal distributions under bivariate truncation with application to testing for age-of-onset anticipation. *Statistica Sinica*, 11:1047–1068, 2001.

J. Huang, J. L. Horowitz, and S. Ma. Asymptotic properties of bridge estimators in sparse high-dimensional regression models. *Annals of Statistics*, 36:587–613, 2008a.

J. Huang, S. Ma, and C. Zhang. Adaptive Lasso for high-dimensional regression models. *Statistica Sinica*, 18:1603–1618, 2008b.

J. Huang, S. Ma, H. Xie, and C. Zhang. A group bridge approach for variable selection. *Biometrika*, 96:339–355, 2009a.

J. Huang, S. Ma, H. Xie, and C. Zhang. A group bridge approach for variable selection. *Biometrika*, 96:339–355, 2009b.

M. G. Hudgens. On nonparametric maximum likelihood estimation with interval censoring and left truncation. *Journal of Royal Statistical Society Series B*, 67:573–587, 2005.

H. Hung and C. T. Chiang. Estimation methods for time-dependent AUC models with survival data. *Canadian Journal of Statistics*, 38:8–26, 2010.

D. R. Hunter and R. Li. Variable selection using MM algorithms. *Annals of Statistics*, 33:1617–1642, 2005.

J. G. Ibrahim and M. H. Chen. Prior distributions and Bayesian computation for proportional hazards models. *Sankhya, Series B*, 60:48–64, 1998.

J. G. Ibrahim and M. H. Chen. Power prior distributions for regression models. *Statistical Science*, 15:46–60, 2000.

J. G. Ibrahim, M. H. Chen, and S. N. MacEachern. Bayesian variable selection for proportional hazards models. *Canadian Journal of Statistics*, 27:701–717, 1999.

J. G. Ibrahim, M. H. Chen, and D. Sinha. Bayesian semiparametric models for survival data with a cure fraction. *Biometrics*, 57:383–388, 2001a.

J. G. Ibrahim, M. H. Chen, and D. Sinha. *Bayesian Survival Analysis*. New York: Springer, 2001b.

C. Jennison and B. W. Turnbull. Mid-course sample size modification in clinical trials based on the observed treatment effect. *Statistics in Medicine*, 22:971–993, 2003.

N. P. Jewell and S. C. Shiboski. Statistical analysis of HIV infectivity based on partner studies. *Biometrics*, 46:1133–1150, 1990.

B. A. Johnson. On LASSO for censored data. *Electronic Journal of Statistics*, 3:485–506, 2009.

R. A. Johnson and D. W. Wichern. *Applied Multivariate Statistical Analysis*. Upper Saddle River, NJ: Prentice Hall, 2001.

I. M. Johnstone and A. Y. Lu. On consistency and sparsity for principal components analysis in high dimensions. *Journal of the American Statistical Association*, 486:682–693, 2009.

I. T. Jolliffe. *Principal Component Analysis*. New York: Springer, 1986.

I. T. Jolliffe, N. T. Trendafilov, and M. Uddin. A modified principal component technique based on the Lasso. *JCGS*, 12:531–547, 2003.

G. Jongbloed. The iterative convex minorant algorithm for nonparametric estimation. *Journal of Computational and Graphical Statistics*, 7(3):310–321, 1998.

J. D. Kalbfleisch. Nonparametric Bayesian analysis of survival time data. *Journal of Royal Statistical Society, Series B*, 40:214–221, 1978a.

J. D. Kalbfleisch. Likelihood methods and nonparametric tests. *Journal of the American Statistical Association*, 73:167–170, 1978b.

J. D. Kalbfleisch and J. F. Lawless. The analysis of panel data under a Markov assumption. *Journal of the American Statistical Association*, 80:863–871, 1985.

J. D. Kalbfleisch and J. F. Lawless. Inference based on retrospective ascertainment: An analysis of the data on transfusion related AIDS. *Journal of the American Statistical Association*, 84:360–372, 1989.

J. D. Kalbfleish and Ř. L. Prentice. *The Statistical Analysis of Failure Time Data*. New York: John Wiley & Sons, 1980.

E. L. Kaplan and P. Meier. Nonparametric estimation from incomplete observations. *Journal of the American Statistical Association*, 53:457–481, 1958.

R. E. Kass and L. Wasserman. A reference Bayesian test for nested hypotheses and its relationship to the Schwarz criterion. *Journal of the American Statistical Association*, 90:928–934, 1995.

N. Keiding and R. D. Gill. Random truncation models and Markov processes. *Annals of Statistics*, 18:582–602, 1990.

N. Keiding, P. Andersen, and J. Klein. The role of frailty models and accelerated failure time models in describing heterogeneity due to omitted covariates. *Statistics in Medicine*, 16:215–224, 1997.

M. Kendall. A new measure of rank correlation. *Biometrika*, 30:81–89, 1938.

Y. Kim and J. Kim. Gradient Lasso for feature selection. In *Proceedings of the 21st International Conference on Machine Learning*, 2004.

J. P. Klein. Semiparametric estimation of random effects using the Cox model based on the EM algorithm. *Biometrics*, 48:795–806, 1992.

J. P. Klein and M. L. Moeschberger. *Survival Analysis: Techniques for Censored and Truncated Data*. New York: Springer-Verlag, 2003.

J. P. Klein, C. Pelz, and M. J. Zhang. Random effects for censored data by a multivariate normal regression model. *Biometrics*, 55:497–506, 1999.

K. Knight and W. Fu. Asymptotics for Lasso-type estimators. *Annals of Statistics*, 28:1356–1378, 2000.

L. Knorr-Held and H. Rue. On block updating in Markov random field models for disease mapping. *Scandinavian Journal of Statistics*, 29:597–614, 2002.

S. Knudsen. *Cancer Diagnostics with DNA Microarrays*. New York: Wiley, 2006.

A. Komarek, E. Lesaffre, and C. Legrand. Baseline and treatment effect heterogeneity for survival times between centers using a random effects accelerated failure time model with flexible error distribution. *Statistics in Medicine*, 26:5457–5472, 2007.

C. Kooperberg, C. J. Stone, and Y. Truong. Hazard regression. *Journal of the American Statistical Association*, 90:78–94, 1995a.

C. Kooperberg, C. J. Stone, and Y. Truong. The L2 rate of convergence for hazard regression. *Scandinavian Journal of Statistics*, 22:143–157, 1995b.

M. R. Kosorok. *Introduction to Empirical Processes and Semiparametric Inference*. New York: Springer-Verlag, 2008.

Y. C. A. Kuk, J. Li, and A. J. Rush. Recursive subsetting to identify patients in the STAR*D: A method to enhance the accuracy of early prediction of treatment outcome and to inform personalized care. *Journal of Clinical Psychiatry*, 71:1502–1508, 2010.

J. M. Lachin. Statistical properties of randomization in clinical trials. *Controlled Clinical Trials*, 9:289–311, 1988.

K. F. Lam and Y. W. Lee. Merits of modelling multivariate survival data using random effects proportional odds model. *Biometrical Journal*, 46: 331–342, 2004.

K. F. Lam and H. Xue. A semiparametric regression cure model with current status data. *Biometrika*, 92:573–586, 2005.

K. F. Lam, Y. W. Lee, and T. L. Leung. Modeling multivariate survival data by a semiparametric random effects proportional odds model. *Biometrics*, 58:316–323, 2002.

P. Lambert, D. Collett, A. Kimber, and R. Johnson. Parametric accelerated failure time models with random effects and an application to kidney transplant survival. *Statistics in Medicine*, 23:3177–3192, 2004.

S. Lang and A. Brezger. Bayesian P-splines. *Journal of Computational Graphical Statistics*, 13:183–212, 2004.

J. F. Lawless. *Statistical Models and Methods for Lifetime Data*. New York: John Wiley & Sons, 2003.

M. Lee and J. P. Fine. Inference for cumulative incidence quantiles via parametric and nonparametric approaches. *Statistics in Medicine*, 30:3221–3235, 2011.

M. L. T. Lee and G. A. Whitmore. Threshold regression for survival analysis: modeling event times by a stochastic process reaching a boundary. *Statistical Science*, 21:501–513, 2006.

M. L. T. Lee and G. A. Whitmore. Proportional hazards and threshold regression: Their theoretical and practical connections. *Lifetime Data Analysis*, 16:196–214, 2010.

M. L. T. Lee, G. A. Whitmore, and B. Rosner. Threshold regression for survival data with time-varying covariates. *Statistics in Medicine*, 29:896–905, 2010.

Y. Lee, Y. Lin, and G. Wahba. Multicategory support vector machines, theory, and application to the classification of microarray data and satellite radiance data. *Journal of the American Statistical Association*, 99:67–81, 2004.

C. Leng, Y. Lin, and G. Wahba. A note on the Lasso and related procedures in model selection. *Statistica Sinica*, 16:1273–1284, 2006.

T. Leonard. Density estimation, stochastic processes and prior information. *Journal of Royal Statistical Society Series B*, 40:113–146, 1978.

A. M. Lesk. *Introduction to Bioinformatics*. Oxford: Oxford University Press, 2002.

R. Levine. *Ethics and Regulation of Clinical Research*. New Haven, CT: Yale University Press, 1996.

B. Li, M. G. Genton, and M. Sherman. Testing the covariance structure of multivariate random fields. *Biometrika*, 95:813–829, 2008.

J. Li and J. P. Fine. On sample size for sensitivity and specificity in prospective diagnostic accuracy studies. *Statistics in Medicine*, 23:2537–2550, 2004.

J. Li and J. P. Fine. ROC analysis for multiple classes and multiple categories and its application in microarray study. *Biostatistics*, 9:566–576, 2008.

J. Li and J. P. Fine. Weighted area under the receiver operating characteristic curve and its application to gene selection. *Journal of the Royal Statistical Society Series C (Applied Statistics)*, 59:673–692, 2010.

J. Li and J. P. Fine. Assessing the dependence of sensitivity and specificity on prevalence in meta-analysis. *Biostatistics*, 12:710–722, 2011.

J. Li and M. L. T. Lee. Analysis of failure time using threshold regression with semiparametric varying coefficients. *Statistica Neerlandica*, 65:164–182, 2011.

J. Li and S. Ma. Interval censored data with repeated measurements and a cured subgroup. *Journal of the Royal Statistical Society Series C (Applied Statistics)*, 59:693–705, 2010.

J. Li and W. K. Wong. Selection of covariance patterns for longitudinal data in semiparametric models. *Statistical Methods in Medical Research*, 19: 183–196, 2010.

J. Li and W. K. Wong. Two-dimensional toxic dose and multivariate logistic regression, with application to decompression sickness. *Biostatistics*, 12: 143–155, 2011.

J. Li and W. Zhang. A semiparametric threshold model for censored longitudinal data analysis. *Journal of the American Statistical Association*, 106: 685–696, 2011.

J. Li and X. H. Zhou. Nonparametric and semiparametric estimation of the three way receiver operating characteristic surface. *Journal of Statistical Planning and Inference*, 139:4133–4142, 2009.

J. Li, J. P. Fine, and N. Safdar. Prevalence-dependent diagnostic accuracy measures. *Statistics in Medicine*, 26:3258–3273, 2007.

J. Li, Y. C. A. Kuk, and A. J. Rush. A practical approach to the early identification of antidepressant medication nonresponders. *Psychological Medicine*, 42:309–316, 2012.

A. M. Lilienfeld and D. E. Lilienfeld. *Foundations of Epidemiology*. New York: Oxford University Press, 1980.

E. Lim, S. L. Zhang, J. Li, W. S. Yap, T. C. Howe, B. P. Tan, Y. S. Lee, D. Wong, K. L. Khoo, K. Y. Seto, L. K. A. Tan, T. Agasthian, H. Koong, T. John, C. Tan, M. Caleb, A. Chang, A. Ng, and P. Tan. Using whole genome amplification (WGA) of low-volume biopsies to assess the prognostic role of EGFR, KRAS, p53, and CMET mutations in advanced-stage nonsmall cell lung cancer (NSCLC). *Journal of Thoracic Oncology*, 4:12–21, 2009.

E. Lim, S. L. Ng, J. Li, A. R. Chang, I. Ng, S. Y. J.and Arunachalam, J. H. J. Low, S. Quek, and E. H. Tay. Cervical dysplasia: Assessing methylation status (Methylight) of CCNA1, DAPK1, HS3ST2, PAX1 and TFPI2 to improve diagnostic accuracy. *Gynecologic Oncology*, 119:225–231, 2010.

H. J. Lim and J. Sun. Nonparametric tests for interval-censored failure time data. *Biometrical Journal*, 45:263–276, 2003.

D. Lin and Z. Ying. Semiparametric analysis of the additive risk model. *Biometrika*, 81:61–71, 1994a.

D. Y. Lin and L. J. Wei. The robust inference for the Cox proportional hazards model. *Journal of the American Statistical Association*, 84:1074–1078, 1989.

D. Y. Lin and Z. Ying. Semiparametric analysis of the additive risk model. *Biometrika*, 81:61–71, 1994b.

D. Y. Lin, D. Oakes, and Z. Ying. Additive hazards regression with current status data. *Biometrika*, 85:289–298, 1998.

R. J. A. Little and D. B. Rubin. *Statistical Analysis with Missing Data*. Hoboken, NJ: John Wiley & Sons, 2002.

A. Liu, W. Meiring, and Y. Wang. Testing generalized linear models using smoothing spline methods. *Statistica Sinica*, 15:235–256, 2005.

W. Liu and Y. Yang. Parametric or nonparametric? a parametricness index for model selection. *Annals of Statistics*, 39:2074–2102, 2011.

Y. Liu, B. Mukherjee, T. Suesse, D. Sparrow, and S. Park. Graphical diagnostics to check model misspecification for the proportional odds regression model. *Statistics in Medicine*, 28:412–429, 2009.

T. A. Louis. Confidence intervals for a binomial parameter after observing no success. *American Statistician*, 35:154, 1981.

W. Lu and L. Li. Boosting method for nonlinear transformation models with censored survival data. *Biostatistics*, 9:658–667, 2008.

D. J. Lunn, A. Thomas, N. Best, and D. Spiegelhalter. WinBUGS—A Bayesian modelling framework: Concepts, structure, and extensibility. *Statistics and Computing*, 10:325–337, 2000.

J. Lv and Y. Fan. A unified approach to model selection and sparse recovery using regularized least squares. *Annals of Statistis*, 37:3498–3528, 2009.

S. Ma. Additive risk model for current status data with a cured subgroup. *Annals of the Institute of Statistical Mathematics*, 63:117–134, 2011.

S. Ma and J. Huang. Regularized ROC method for disease classification and biomarker selection with microarray data. *Bioinformatics*, 21:4356–4362, 2005.

S. Ma and J. Huang. Combining multiple markers for classification using ROC. *Biometrics*, 63:751–757, 2007.

S. Ma and J. Huang. Penalized feature selection and classification in bioinformatics. *Briefings in Bioinformatics*, 9(5):392–403, 2008.

S. Ma and M. Kosorok. Adaptive penalized M-estimation with current status data. *Annals of Institute of Statistical Mathematics*, 58:511–526, 2006a.

S. Ma and M. R. Kosorok. Robust semiparametric M-estimation and the weighted bootstrap. *Journal of Multivariate Analysis*, 96:190–217, 2005.

S. Ma and M. R. Kosorok. Adaptive penalized M-estimation for the Cox model with current status data. *Annals of the Institute of Statistical Mathematics*, 58:511–526, 2006b.

S. Ma and M. R. Kosorok. Detection of gene pathways with predictive power for breast cancer prognosis. *BMC Bioinformatics*, 11:1, 2010.

S. Ma and X. Song. Ranking prognosis markers in cancer genomic studies. *Briefings in Bioinformatics*, 12:33–40, 2011.

S. Ma, M. R. Kosorok, and J. P. Fine. Additive risk models for survival data with high-dimensional covariates. *Biometrics*, 62:202–210, 2006.

S. Ma, J. Huang, and X. Song. Integrative analysis and variable selection with multiple high-dimensional datasets. *Biostatistis*, 12:763–775, 2011a.

S. Ma, M. R. Kosorok, J. Huang, and Y. Dai. Incorporating higher-order representative features improves prediction in network-based cancer prognosis analysis. *BMC Medical Genomics*, 4:5, 2011b.

S. Ma, Y. Dai, J. Huang, and X. Y. Identification of breast cancer prognosis markers via integrative analysis. *Computational Statistics and Data Analysis*, 56:2718–2728, 2012.

C. A. MacGilchrist. REML estimation for survival models with frailty. *Biometrics*, 49:221–225, 1993.

R. A. Maller and X. Zhou. Estimating the proportion of immunes in a censored sample. *Biometrika*, 79:731–739, 1992.

R. A. Maller and X. Zhou. *Survival Analysis with Long-Term Survivors*. New York: Wiley, 1996.

A. K. Manatunga and D. Oaks. Parametric analysis of matched pair survival data. *Lifetime Data Analysis*, 5:371–387, 1999.

H. Marten and T. Naes. *Multivariate Calibration*. New York: Wiley, 1989.

T. Martinsussen and T. H. Scheike. Efficient estimation in additive hazards regression with current status data. *Biometrika*, 89:649–658, 2002.

L. Marzec and P. Marzec. On fitting Cox's regression model with time-dependent coefficients. *Biometrika*, 84:901–908, 1997.

M. W. McIntosh and M. S. Pepe. Combining several screening tests: Optimality of the risk score. *Biometrics*, 58:657–664, 2002.

L. Meier, S. van de Geer, and P. Buhlmann. The group Lasso for logistic regression. *Journal of Royal Statistical Society Series B*, 70:53–71, 2008.

N. Meinshausen and B. Yu. Lasso-type recovery of sparse representations of high-dimensional data. *Annals of Statistics*, 37:246–270, 2009.

A. Meucci. A new breed of copulas for risk and portfolio management. *Risk*, 24:122–126, 2011.

D. Mossman. Three-way ROCs. *Medical Decision Making*, 19:78–89, 1999.

H. H. Muller and H. Schafer. Adaptive group sequential designs for clinical trials: Combining the advantages of adaptive and classical group sequential approaches. *Biometrics*, 57:886–891, 2001.

S. A. Murphy. Consistency in a proportional hazards models incorporating a random effect. *Annals of Statistics*, 22:712–731, 1994.

S. A. Murphy. Asymptotic theory for the frailty model. *Annals of Statistics*, 23:182–198, 1995.

S. A. Murphy and P. K. Sen. Time-dependent coefficients in a Cox-type regression model. *Stochastic Processesses and Their Applications*, 39:153–180, 1991.

S. A. Murphy, A. Rossini, and A. W. van der Vaart. Maximum likelihood estimation in the proportional odds model. *Journal of the American Statistical Association*, 92:968–976, 1997.

C. T. Nakas and C. T. Yiannoutsos. Ordered multiple-class ROC analysis with continuous measurements. *Statistics in Medicine*, 23:3437–3449, 2004.

R. B. Nelsen. *An Introduction to Copulas*. New York: Springer, 1999.

W. B. Nelson. Hazard plotting for incomplete failure data. *Journal of Quality and Technology*, 1:27–52, 1969.

W. B. Nelson. Theory and applications of hazard plotting for censored failure data. *Technometrics*, 14:945–965, 1972.

R. G. Newcombe. Two-sided confidence intervals for the single proportion: Comparison of seven methods. *Statistics in Medicine*, 17:857–872, 1998.

W. K. Newey. Efficient instrumental variables estimation of nonlinear models. *Econometrica*, 58:809–837, 1990.

D. V. Nguyen and D. M. Rocke. Tumor classification by partial least squares using gene expression data. *Bioinformatics*, 18:39–50, 2002a.

D. V. Nguyen and D. M. Rocke. Partial least squares proportional hazard regression for application to DNA microarray survival data. *Bioinformatics*, 18:1625–1632, 2002b.

G. G. Nielsen, R. D. Gill, P. K. Andersen, and T. I. Sorensen. A counting process approach to maximum likelihood estimation of frailty model. *Scandinavian Journal of Statistics*, 19:25–43, 1992.

D. J. Nott and J. Li. A sign based loss approach to model selection in nonparametric regression. *Statistics and Computing*, 20:485–498, 2010.

D. Nychka. Bayesian confidence intervals for a smoothing splines. *Journal of the American Statistical Association*, 83:1134–1143, 1988.

S. Nygard, O. Borgan, O. C. Lingjarde, and H. L. Storvold. Partial least squares Cox regression for genome-wide data. *Lifetime Data Anal*, 14: 179–195, 2008.

D. Oaks. Bivariate survival models induced by frailties. *Journal of the American Statistical Association*, 84:487–493, 1989.

N. Obuchowski. Estimating and comparing diagnostic tests accuracy when the gold standard is not binary. *Academic Radiology*, 12:1198–1204, 2005.

F. O'Sullivan. Nonparametric estimation of relative risk using splines and cross-validation. *SIAM Journal on Scientific Statistical Computing*, 9: 531–542, 1988.

W. Pan. A multiple imputation approach to Cox regression with interval-censored data. *Biometrics*, 56:199–203, 2000a.

W. Pan. Smooth estimation of the survival function for interval censored data. *Statistics in Medicine*, 19:2611–2624, 2000b.

W. Pan. Using frailties in the accelerated failure time model. *Lifetime Data Analysis*, 7:55–64, 2001.

E. Parner. Asymptotic theory for the correlated gamma-frailty model. *Annals of Statistics*, 26:183–214, 1998.

M. J. Pencina and R. B. D'Agostino. Overall C as a measure of discrimination in survival analysis: Model specific population value and confidence interval estimation. *Statistics in Medicine*, 23:2109–2123, 2004.

M. J. Pencina, R. B. D'Agostino Sr, R. B. D'Agostino Jr, and R. S. Vasan. Evaluating the added predictive ability of a new marker: From area under the ROC curve to reclassification and beyond. *Statistics in Medicine*, 27: 157–172, 2008.

M. J. Pencina, R. B. D'Agostino Sr., and E. W. Steyerberg. Extensions of net reclassification improvement calculations to measure usefulness of new biomarkers. *Statistics in Medicine*, 30:11–21, 2011.

M. J. Pencina, R. B. D'Agostino, K. M. Pencina, et al. Interpreting incremental value of markers added to risk prediction models. *American Journal of Epidemiology*, 176:473–481, 2012.

M. S. Pepe. Three approaches to regression analysis of receiver operating characteristic curves for continuous test results. *Biometrics*, 56:124–135, 1998.

M. S. Pepe. *The Statistical Evaluation of Medical Tests for Classification and Prediction*. Oxford: Oxford University Press, 2003.

M. S. Pepe and M. L. Thompson. Combining diagnostic test results to increase accuracy. *Biostatistics*, 1:123–140, 2000.

M. S. Pepe, G. Longton, G. L. Anderson, and M. Schummer. Selecting differentially expressed genes from microarray experiments. *Biometrics*, 59: 133–142, 2003.

M. S. Pepe, Z. Feng, and J. W. Gu. Comments on "Evaluating the added predictive ability of a new marker: From area under the ROC curve to reclassification and beyond" by M. J. Pencina et al. *Statistics in Medicine*, 27:173–181, 2008.

R. Peto. Discussion of paper by D. R. Cox. *Journal of the Royal Statistical Society, Series B*, 34:205–207, 1972.

R. Peto and J. Peto. Asymptotically efficient rank invariant procedures (with discussion). *Journal of the Royal Statistical Society, Series A*, 135:185–206, 1972.

A. Ploner, S. Calza, A. Gusnanto, and Y. Pawitan. Multidimensional local false discovery rate for microarray studies. *Bioinformatics*, 22:556–565, 2006.

R. L. Prentice. Linear rank tests with right censored data. *Biometrika*, 65: 167–179, 1978.

R. L. Prentice and J. Cai. Covariance and survivor function estimation using censored multivariate failure time data. *Biometrika*, 79:495–512, 1992.

R. L. Prentice, B. J. Williams, and A. V. Peterson. On the regression analysis of multivariate failure time data. *Biometrika*, 68:373–379, 1981.

R. Rebolledo. Central limit theorems for local martingales. *Z. Washrsch. Verw. Gebiete.*, 51:269–286, 1980.

N. Reid and H. Crepeau. Influence functions for proportional hazards regression. *Biometrika*, 72:1–9, 1985.

H. W. Ressom, R. S. Varghese, S. Drake, G. Hortin, M. Abdel-Hamid, C. A. Loffredo, and R. Goldman. Peak selection from MALDI-TOF mass spectra using ant colony optimization. *Bioinformatics*, 23:619–626, 2007.

H. W. Ressom, R. S. Varghese, L. Goldman, C. A. Loffredo, M. Abdel-Hamid, Z. Kyselova, M. Mechref, Y. Novotny, and R. Goldman. Analysis of MALDI-TOF mass spectrometry data for detection of Glycan biomarkers. *Pacific Symposium on Biocomputing*, 13:216–227, 2008.

C. Robert and G. Casella. *Introducing Monte Carlo Methods with R*. New York: Springer, 2010.

G. O. Roberts and N. G. Polson. On the geometric convergence of the Gibbs sampler. *Journal of Royal Statistical Society Series B*, 56:377–384, 1994.

T. Robertson, F. T. Wright, and R. Dykstra. *Order Restricted Statistical Inference*. New York: John Wiley, 1998.

A. Rosenwald, G. Wright, A. Wiestner, W. C. Chan, J. M. Connors, E. Campo, R. D. Gascoyne, T. M. Grogan, H. K. Muller-Hermelink, E. B. Smeland, M. Chiorazzi, J. M. Giltnane, E. M. Hurt, H. Zhao, L. Averett, S. Henrickson, L. Yang, J. Powell, W. H. Wilson, E. S. Jaffe, R. Simon, R. D. Klausner, E. Montserrat, F. Bosch, T. C. Greiner, D. D. Weisenburger, W. G. Sanger, B. J. Dave, J. C. Lynch, J. Vose, J. O. Armitage, R. I. Fisher, T. P. Miller, M. LeBlanc, G. Ott, S. Kvaloy, H. Holte, J. Delabie, and L. M. Staudt. The proliferation gene expression signature is a quantitative integrator of oncogenic events that predicts survival in mantle cell lymphoma. *Cancer Cell*, 3:185–197, 2003.

R. M. Royall. Model robust confidence intervals using maximum likelihood observations. *International Statistical Review*, 54:221–226, 1986.

D. Rubin. *Multiple Imputation for Nonresponse in Surveys*. New York: Wiley, 1987.

H. Rue. Fast sampling of Gaussian Markov random fields. *JRSSB*, 63:325–338, 2001.

A. J. Rush, M. Fava, S. R. Wisniewski, P. W. Lavori, M. H. Trivedi, H. A. Sackeim, M. E. Thase, A. A. Nierenberg, F. M. Quitkin, T. M. Kashner, D. J. Kupfer, J. F. Rosenbaum, J. Alpert, J. W. Stewart, P. J. McGrath, M. M. Biggs, K. Shores-Wilson, B. D. Lebowitz, L. Ritz, and G. Niederehe. Sequenced treatment alternatives to relieve depression (STAR*D): Rationale and design. *Controlled Clinical Trials*, 25:119–142, 2004.

O. Saarela, S. Kulathinal, E. Arjas, and E. Laara. Nested case-control data utilized for multiple outcomes: A likelihood approach and alternatives. *Statistics in Medicine*, 27:5991–6008, 2008.

Y. Saeys, I. Inza, and P. Larranaga. A review of feature selection techniques in bioinformatics. *Bioinformatics*, 23(19):2507–2517, 2007.

N. Safdar, J. P. Fine, and D. G. Maki. Methods for diagnosis of intravascular device-related bloodstream infection. *Annals of Internal Medicine*, 142: 451–466, 2005.

A. Salim, C. Hultman, P. Sparen, and M. Reilly. Combining data from 2 nested case-control studies of overlapping cohorts to improve efficiency. *Biostatistics*, 10:70–79, 2009.

P. Sasieni. Nonorthogonal projections and their application to calculating the information in a partly linear Cox model. *Scandinavian Journal of Statistics*, 19:215–233, 1992.

A. Satorra and P. M. Bentler. A scaled difference chi-square test statistic for moment structure analysis. *Psychometrika*, 66:507–514, 2001.

W. Sauerbrei and M. Schumacher. A bootstrap resampling procedure for model building: Application to Cox regression model. *Statistics in Medicine*, 11:2093–2109, 1992.

I. R. Savage. Contributions to the theory of rank order statistics—The two sample case. *Annals of Mathematical Statistics*, 27:590–615, 1956.

A. Schafer. The ethics of the randomized clinical trial. *New England Journal of Medicine*, 307:722–723, 1982.

K. F. Schaffner. *Discovery and Explanation in Biology and Medicine*. Chicago: University of Chicago Press, 1993.

M. J. Schervish and B. P. Carlin. On the convergence of successive substitution sampling. *Journal of Computational and Graphical Statistics*, 1:111–127, 1992.

A. Schick and Q. Yu. Consistency of the GMLE with mixed case interval-censored data. *Scandinavian Journal of Statistics*, 27:45–55, 2000.

D. A. Schoenfeld. The asymptotic properties of comparative tests for comparing survival distributions. *Biometrika*, 68:316–319, 1981.

D. A. Schoenfeld. Partial residuals for the proportional hazards regression model. *Biometrika*, 69:239–241, 1982.

D. A. Schoenfeld. Sample size formula for the proportional hazards regression model. *Biometrics*, 39:499–503, 1983.

B. K. Scurfield. Multiple-event forced-choice tasks in the theory of signal detectability. *Journal of Mathematical Psychology*, 40:253–269, 1996.

A. Sen and F. Tan. Cure-rate estimation under case-1 interval censoring. *Statistical Methodology*, 5:106–118, 2008.

V. Seshadri. *Inverse Gaussian Distribution: A Case Study in Exponential Families*. Oxford: Clarendon Press, 1993.

M. Shaked and J. G. Shanthikumar. The multivariate hazard construction. *Stochastic Processes and Their Applications*, 24:85–97, 1997.

J. Shao. An asymptotic theory for linear model selection. *Statistica Sinica*, 7: 221–264, 1997.

J. Shao. *Mathematical Statistics*. New York: Springer-Verlag, 1999.

P. Shen. Estimation of the truncation probability with left-truncated and right-censored data. *Journal of Nonparametric Statistics*, 8:957–969, 2005.

P. Shen and Y. Y. Yan. Nonparametric estimation of the bivariate survival function with left-truncated and right-censored data. *Journal of Statistical Planning and Inference*, 138:4041–4054, 2008.

R. P. Sherman. The limiting distribution of the maximum rank correlation estimator. *Econometrica*, 61:123–137, 1993.

H. Shi, Y. Cheng, and J. Li. Assessing accuracy improvement for competing-risk censored outcomes. 2012a. Unpublished manuscript.

J. Q. Shi, B. Wang, E. J. Will, and R. M. West. Mixed-effects GPFR models with application to dose-response curve prediction. *Statistics in Medicine*, 3165–3177, 2012b.

S. C. Shiboski. Generalized additive models for current status data. *Lifetime Data Analysis*, 4:29–50, 1998.

J. H. Shih and T. A. Louis. Assessing gamma frailty models for clustered failure time data. *Lifetime Data Analysis*, 1:205–220, 1995.

M. A. Shipp, K. Ross, P. Tamayo, A. P. Weng, J. L. Kutok, R. Aguiar, M. Gaasenbeek, M. Angelo, M. Reich, G. S. Pinkus, T. S. Ray, M. Koval, K. W. Last, A. Norton, T. Lister, J. Mesirov, D. S. Neuberg, E. S. Lander, J. Aster, and T. Golub. Diffuse large B-cell lymphoma outcome prediction by gene expression profiling and supervised machine learning. *Nature Medicine*, 8:68–74, 2002.

G. L. Silva and M. A. Amaral-Turkman. Bayesian analysis of an additive survival model with frailty. *Communications in Statistics, Theory and Methods*, 33:2517–2533, 2004.

N. Simon, J. Friedman, T. Hastie, and R. Tibshirani. Regularization paths for Cox's proportional hazards model via coordinate descent. *Journal of Statistical Software*, 39(1):1–13, 2011.

D. Sinha. Semiparametric Bayesian analysis of multiple event time data. *Journal of the American Statistical Association*, 88:979–983, 1993.

D. P. Snustad and M. J. Simmons. *Principles of Genetics*. New York: John Wiley & Sons, 2006.

P. X. K. Song, M. Li, and Y. Yuan. Joint regression analysis of correlated data using Gaussian copula. *Biometrics*, 65:60–68, 2009.

T. Sorlie, C. M. Perou, R. Tibshirani, et al. Gene expression patterns of breast carcinomas distinguish tumor subclasses with clinical implications. *PNAS*, 98:10869–10874, 2001.

C. F. Spiekerman and D. Y. Lin. Marginal regression models for multivariate failure time data. *Journal of the American Statistical Association*, 93: 1164–1175, 1998.

StataCorp. *Survival Analysis and Epidemiological Tables Reference Manual*. College Station, TX: Stata Press, 2011.

M. E. Stokes, C. S. Davis, and G. G. Koch. *Categorical Data Analysis Using the SAS System*, 2nd ed., Cary, NC: SAS Institute, 2000.

C. J. Stone. Consistent nonparametric regression. *Annals of Statistics*, 5: 595–645, 1977.

J. Storey and R. Tibshirani. Statistical significance for genome-wide studies. *PNAS*, 100:9440–9445, 2003.

J. Storey, J. Y. Dai, and J. T. Leek. The optimal discovery procedure for large-scale significance testing with applications to comparative microarray experiments. *Biostatistics*, 8:414–432, 2007.

J. D. Storey. A direct approach to false discovery rates. *Journal of Royal Statistical Society Series B*, 64:479–498, 2002.

D. O. Stram and J. W. Lee. Variance components testing in the longitudinal mixed-effects models. *Biometrics*, 50:1171–1177, 1994.

W. Stute. Consistent estimation under random censorship when covariates are present. *Journal of Multivariate Analysis*, 45:89–103, 1993.

J. Q. Su and J. S. Liu. Linear combinations of multiple diagnostic markers. *Journal of the American Statistical Association*, 88:1350–1355, 1993.

A. Subramanian, P. Tamayo, V. K. Mootha, S. Mukherjee, B. L. Ebert, M. A. Gillette, A. Paulovich, S. L. Pomeroy, T. R. Golub, E. S. Lander, and J. P. Mesirov. Gene set enrichment analysis: A knowledge-based approach for interpreting genome-wide expression profiles. *PNAS*, 102:15545–15550, 2005.

J. Sun. Empirical estimation of a distribution function with truncated and doubly interval-censored data and its application to AIDS studies. *Biometrics*, 51:1096–1104, 1995.

J. Sun. *The Statistical Analysis of Interval-Censored Failure Time Data*. New York: Springer, 2006.

J. Sun and J. D. Kalbfleisch. Nonparametric tests of tumor prevalence data. *Biometrics*, 52:726–731, 1996.

J. Sun and L. J. Wei. Regression analysis of panel count data with covariate-dependent observation and censoring times. *Journal of Royal Statistical Society Series B*, 62:293–302, 2000.

B. C. Tai, J. Lee, and H. P. Lee. Comparing a sample proportion with a specified population proportion based on the mid-P method. *Psychiatry Research*, 71:201–203, 1997.

B. C. Tai, I. R. White, V. Gebski, and D. Machin. On the issue of "multiple" first failures in competing risks analysis. *Statistics in Medicine*, 21:2243–2253, 2002.

R. H. Taplin and A. E. Raftery. Analysis of agricultural field trials in the presence of outliers and fertility jumps. *Biometrics*, 50:764–781, 1994.

R. E. Tarone and J. Ware. On distribution free tests of the equality of survival distributions. *Biometrika*, 64:156–160, 1977.

K. Taylor, R. Margolese, and C. L. Soskolne. Physicians' reasons for not entering eligible patients in a randomized clinical trial of surgery for breast cancer. *NEJM*, 310:1363–1367, 1984.

T. M. Therneau, P. M. Grambsch, and T. R. Fleming. Martingale-based residuals for survival models. *Biometrika*, 77:147–160, 1990.

W. D. Thompson. Effect modification and the limits of biological inference from epidemiologic data. *Journal of Clinical Epidemiology*, 44:221–232, 1991.

L. Tian, D. Zucker, and L. J. Wei. On the Cox model with time-varying regression coefficients. *Journal of the American Statistical Association*, 100:172–183, 2005.

R. Tibshirani. Regression shrinkage and selection via the Lasso. *Journal of Royal Statistical Society Series B*, 58:267–288, 1996.

R. Tibshirani. The Lasso method for variable selection in the Cox model. *Statistics in Medicine*, 16:385–396, 1997.

R. Tibshirani. Univariate shrinkage in the Cox model for high-dimensional data. *Statistical Applications in Genetics and Molecular Biology*, 8:3498–3528, 2009.

R. Tibshirani and T. Hastie. Local likelihood estimation. *Journal of the American Statistical Association*, 82:559–567, 1987.

M. H. Trivedi, M. Fava, S. R. Wisniewski, M. E. Thase, F. M. Quitkin, D. Warden, L. Ritz, A. A. Nierenberg, B. D. Lebowitz, M. M. Biggs, J. F. Luther, K. Shores-Wilson, and A. J. Rush. Medication augmentation after the failure of SSRIs for depression. *New England Journal of Medicine*, 354: 1243–1252, 2006.

C. Tsai, H. Hsueh, and J. J. Chen. Estimation of false discovery rates in multiple testing: Application to gene microarray data. *Biometrics*, 59: 1071–1081, 2003.

S. M. Tse. Strong Gaussian approximations in the left truncated and right censored model. *Statistica Sinica*, 13:275–282, 2003.

S. M. Tse. Lorenz curve for truncated and censored data. *Annals of Institute of Statistical Mathematics*, 58:675–686, 2006.

A. A. Tsiatis. A large sample study of Cox's regression model. *Annals of Statistics*, 9:93–108, 1981.

A. A. Tsiatis and C. R. Mehta. On the inefficiency of the adaptive design for monitoring clinical trials. *Biometrika*, 90:367–378, 2003.

M. C. K. Tweedie. Inverse statistical variates. *Nature*, 155:453, 1945.

S. van de Geer and P. Buhlmann. On conditions used to prove oracle results for the Lasso. *Electronic Journal of Statistics*, 3:1360–1392, 2009.

M. J. van der Laan. Efficient estimation in the bivariate censoring model and repairing NPMLE. *The Annals of Statistics*, 24:596–627, 1996a.

M. J. van der Laan. Nonparametric estimation of the bivariate survival function with truncated data. *Journal of Multivariate Analysis*, 58:107–131, 1996b.

A. W. van der Vaart. *Asymptotic Statistics*. Cambridge: Cambridge University Press, 2000.

A. W. Van der Vaart and J. A. Wellner. *Weak Convergence and Empirical Processes*. New York: Springer, 1996.

H. C. van Houwelingen and H. Putter. *Dynamic Prediction in Clinical Survival Analysis*. New York: Chapman & Hall/CRC Press, 2012.

H. C. van Houwelingen and J. Thorogood. Construction, validation and updating of a prognostic model for kidney graft survival. *Statistics in Medicine*, 14:1999–2008, 1995.

H. C. van Houwelingen, T. Bruinsma, A. Hart, L. van't Veer, and L. Wessels. Cross-validated Cox regression on microarray gene expression data. *Statistics in Medicine*, 25:3201–3216, 2006.

J. M. Vaupel, K. G. Manton, and E. Stallard. The impact of heterogeneity in individual frailty on the dynamics of mortality. *Demography*, 16:439–454, 1979.

C. T. Volinsky and A. E. Raftery. Bayesian information criterion for censored survival models. *Biometrics*, 56:256–262, 2000.

G. Wahba. *Spline Models for Observational Data.* SIAM, Philadelphia. *CBMS-NSF Regional Conference Series in Applied Mathematics*, Vol. 59, 1990.

G. Wahba and J. Wendelberger. Some new mathematical methods for variational objective analysis using splines and cross-validation. *Monthly Weather Review*, 108:1122–1143, 1980.

H. Wang. A note on iterative marginal optimization: A simple algorithm for maximum rank correlation estimation. *Computational Statistics and Data Analysis*, 51:2803–2812, 2006.

H. Wang, G. Li, and G. Jiang. Robust regression shrinkage and consistent variable selection via the LAD-LASSO. *Journal of Business and Economics Statistics*, 25:347–355, 2006.

N. Wang and D. Ruppert. Nonparametric estimation of the transformation in the transform-both-sides regression models. *Journal of the American Statistical Association*, 90:731–738, 1995.

S. Wang, B. Nan, N. Zhou, and J. Zhu. Hierarchically penalized Cox regression with grouped variables. *Biometrika*, 96:307–322, 2009.

Y. Wang and G. Wahba. Bootstrap confidence intervals for smoothing splines and their comparison to Bayesian "confidence intervals." *Journal of Statistical Computation and Simulation*, 51:263–279, 1995.

G. C. G. Wei and M. A. Tanner. Applications of multiple imputation to the analysis of censored regression data. *Biometrics*, 47:1297–1309, 2000.

L. J. Wei, D. Y. Lin, and L. Weissfeld. Regression analysis of multivariate incomplete failure time data by modeling marginal distributions. *Journal of the American Statistical Association*, 84:1065–1073, 1989.

W. Weibull. A statistical distribution function of wide applicability. *Journal of Applied Mechanics ASME*, 18:293–297, 1951.

M. C. Weinstein. Allocation of subjects in medical experiments. *New England Journal of Medicine*, 291:1278–1285, 1974.

J. Whitehead. Fitting Cox's regression model to survival data using GLIM. *Applied Statistics*, 29:268–275, 1980.

G. A. Whitemore. A regression method for censored inverse-Gaussian data. *Canadian Journal of Statistics*, 11:305–315, 1983.

S. Wieand, M. H. Gail, and B. R. James. A family of nonparametric statistics for comparing diagnostic markers with paired or unpaired data. *Biometrika*, 76:585–592, 1989.

A. Wienke. *Frailty Models in Survival Analysis.* New York: Chapman & Hall/CRC Press, 2011.

E. B. Wilson. Probable inference, the law of succession, and statistical inference. *Journal of the American Statistical Association*, 22:209–212, 1927.

R. D. Wolfinger. Heterogeneous variance: Covariance structures for repeated measures. *Journal of Agricultural, Biological and Environmental Statistics*, 1:205–230, 1996.

S. Wong. *The Practical Bioinformatian.* Singapore: World Scientific Publishing, 2004.

C. F. J. Wu and M. Hamada. *Experiments: Planning, Analysis, and Parameter Deisgn Optimization.* New York: John Wiley & Sons, 2000.

Y. Wu and Y. Liu. Variable selection in quantile regression. *Statistica Sinica*, 19:801–817, 2009.

D. Xiang and G. Wahba. Approximate smoothing spline methods for large data sets in the binary case. In *the Proceedings of the 1997 ASA Joint Statistical Meetings, Biometrics Section*, pp. 94–98, 1998.

J. Xu, J. Kabfleisch, and B. C. Tai. Statistical analysis of illness death processes and semicompeting risks data. *Biometrics*, 66:716–725, 2010.

Y. Yang. Prediction/estimation with simple linear models: Is it really that simple? *Econometric Theory*, 23:1–36, 2007.

Z. Ying. A large sample study of rank estimation for censored regression data. *Annals of Statistics*, 21:76–99, 1993.

A. Yu, K. Kwan, D. Chan, and D. Fong. Clinical features of 46 eyes with calcified hydrogel intraocular lenses. *Journal of Cataract and Refractive Surgery*, 27:1596–1606, 2001.

T. Yu, J. Li, and S. Ma. Adjusting confounders in ranking biomarkers: a model-based ROC approach. *Briefings in Bioinformatics*, 13:513–523, 2012.

M. Zelen. Play the winner rule and the controlled clinical trials. *Journal of the American Statistical Association*, 64:131–146, 1969.

M. Zelen. A new design for randomized clinical trials. *New England Journal of Medicine*, 300:1242–1245, 1979.

D. Zeng and D. Y. Lin. Maximum likelihood estimation in semiparametric transformation models for counting processes. *Biometrika*, 93:627–640, 2006.

D. Zeng, Q. Chen, and J. Ibrahim. Gamma-frailty transformation models for multivariate survival times. *Biometrika*, 96:277–291, 2009.

G. O. Zerbe. On Fieller's theorem and the general linear model. *American Statistician*, 32:103–105, 1978.

C. Zhang. Nearly unbiased variable selection under minimax concave penalty. *Annals of Statistics*, 38:894–942, 2010a.

C. Zhang and J. Huang. The sparsity and bias of the lasso selection in high-dimensional linear regression. *Annals of Statistics*, 36:1567–1594, 2008.

C. M. Zhang, J. Li, and J. Meng. On Stein's lemma, dependent covariates and functional monotonicity in multidimensional modeling. *Journal of Multivariate Analysis*, 99:2285–2303, 2008.

H. H. Zhang and W. Lu. Adaptive Lasso for Cox's proportional hazards model. *Biometrika*, 94:691–703, 2007.

J. T. Zhang. Approximate and asymptotic distributions of chi-squared-type mixtures with applications. *Journal of the American Statistical Association*, 100:273–285, 2005.

Y. Zhang. *ROC Analysis in Diagnostic Medicine*. PhD thesis, National University of Singapore, 2010b.

Y. Zhang and J. Li. Combining multiple markers for multicategory classification: An ROC surface approach. *Australian and New Zealand Journal of Statistics*, 53:63–78, 2011.

Y. Zhang, W. Liu, and H. Wu. A simple nonparametric two-sample test for the distribution function of event time with interval censored data. *Journal of Nonparametric Statistics*, 16:643–652, 2003.

P. Zhao and B. Yu. On model selection consistency of Lasso. *Journal of Machine Learning Research*, 7:2541–2563, 2006.

Q. Zhao and J. Sun. Generalized log-rank test for mixed interval-censored failure time data. *Statistics in Medicine*, 23:1621–1629, 2004.

Y. Zheng, T. Cai, M. Pepe, and W. Levy. Time-dependent predictive values of prognostic biomarkers with failure time outcome. *Journal of the American Statistical Association*, 103:362–368, 2008.

X. H. Zhou, N. A. Obuchowski, and D. K. McClish. *Statistical Methods in Diagnostic Medicine*. New York: John Wiley & Sons, 2002.

H. Zou and T. Hastie. Regularization and variable selection via the elastic net. *Journal of Royal Statistical Society Series B*, 67:301–320, 2005.

H. Zou and R. Li. One-step sparse estimates in nonconcave penalized likelihood models (with discussion). *Annals of Statistic*, 36:1509–1566, 2008.

H. Zou, T. Hastie, and R. Tibshirani. Sparse principal component analysis. *Journal of Computational and Graphical Statistics*, 15:265–286, 2006.

K. H. Zou, A. Liu, A. I. Bandos, L. Ohno-Machado, and H. E. Rockette. *Statistical Evaluation of Diagnostic Performance: Topics in ROC Analysis*. New York: Chapman & Hall/CRC Press, 2011.

D. M. Zucker and A. F. Karr. Nonparametric survival analysis with time-dependent covariate effects: A penalized partial likelihood approach. *Annals of Statistic*, 18:329–353, 1990.

Index

accelerated failure time model, 57, 287, 319

additive hazard model, 62, 117, 190, 191, 285

AIC, 48, 298

asymptotic normality, 17, 34, 63, 86

asymptotic relative efficiency, 46

at-risk process, 67, 170

AUC, 212, 274, 316
    partial AUC, 219
    weighted AUC, 219

backfitting algorithm, 143

bandwidth, 142, 253

bathtub hazard, 27

Bayes, 49, 130, 180, 251

Bernoulli distribution, 45

BIC, 48, 298

Bonferroni inequality, 270

bootstrap, 19, 105, 177, 246, 248

bounded cumulative hazard model, 166

Box-Cox transformation, 61

bridge penalty, 301

Brier score, 55

Brownian bridge, 262

case-control study, 5

cause-specific hazard function, 159

censoring, 9, 74, 169
    independent censoring, 11, 23
    interval censoring, 65, 72, 171, 243
    right censoring, 9, 251

central limit theorem, 64

Clayton copula, 150

cluster, 147, 307

cohort study, 5

compensator, 67

competing risks, 159

concordance, 151, 242, 316

confounder, 2, 46

conjugate prior, 193

consistency, 17, 63, 167, 216, 261

convergence rate, 74

copula, 149

correct classification probability, 237

correlation, 147, 305

correlation coefficient, 150

counting process, 18, 66, 170

Cox proportional hazards model, 35, 41, 50, 98, 139, 152, 183, 282, 292

Cox-Snell residual, 50

credible region, 181

cross-sectional study, 5

cross-validation, 56, 293, 298, 304

cure rate model, 165

current status, 74

Dabrowska estimator, 256

deviance, 168

Donsker theorem, 64

elastic net, 300

EM algorithm, 157

empirical process, 17, 64, 100, 262

epidemiology, 1

exponential distribution, 14, 24, 50, 78, 167

false discovery rate, 270

family-wise type I error rate, 270

filtration, 67